Trace Elements
in Soil

Bioavailability, Flux,
and Transfer

Trace Elements in Soil

Bioavailability, Flux, and Transfer

Edited by
I.K. Iskandar *and*
M.B. Kirkham

CRC Press
Taylor & Francis Group
Boca Raton London New York

CRC Press is an imprint of the
Taylor & Francis Group, an **informa** business

CRC Press
Taylor & Francis Group
6000 Broken Sound Parkway NW, Suite 300
Boca Raton, FL 33487-2742

First issued in paperback 2020

ISBN 13: 978-0-367-57883-1 (pbk)
ISBN 13: 978-1-56670-507-3 (hbk)

Library of Congress Cataloging-in-Publication Data

Trace elements in soil: bioavailability, flux, and transfer/editors I.K. Iskandar, M.B. Kirkham
    p. cm.
Includes bibliographical references (p.)
ISBN 1-56670-507-X
1. Soils--Trace element content. I. Iskandar, I.K. (Iskandar Karam), 1938- II. Kirkham, M.B.

S592.6T7 T727 2000
    631.4'1--dc21
                                                            00-048136
                                                            CIP

Library of Congress Card Number 00-048136

Preface

During the past few decades significant progress has been made in several areas of biology, ecology, and the environmental geochemistry of trace elements in soils. In the early 1950s and 1960s research was focused on developing methods to enhance the bioavailability of trace (minor) elements to plants (also called micronutrients) from soils. Research on methods and amounts of trace element application to agricultural soils to correct plant deficiencies received major attention at that time. More recently, due to industrial development and past disposal activities, trace elements are considered to be among the important environmental contaminants that affect all ecosystem components in the atmosphere and in aquatic and terrestrial systems. Concerns about these contaminants are justified in view of an ever-increasing body of information that indicates trace elements are continually accumulating in the food chain. Some pollutants, such as Hg and Cd, have accumulated in the food chain, harming fish, wildlife, and vegetation, and threatening the health of humans. In contrast to the earlier emphasis on mobilizing trace elements in soil to enhance their bioavailability, current emphasis is placed on enhancing their immobilization by developing techniques to predict fate, transport, and bioavailability of trace elements in the soil system. Quantitative information on the factors affecting or controlling trace-element bioavailability in soils is needed to develop cost-effective and innovative techniques for soil remediation, and for enhancing the quality of soil, plants, and the food chain. Methods such as *in situ* immobilization, phytoremediation, and natural attenuation have recently been proposed but are not fully understood.

Bioavailability is a fundamental aspect in organisms for assimilation of nutrients and contaminant compounds, both inorganic and organic. Yet, it has remained a complex process to understand for nutritionists, toxicologists, environmental scientists, policy makers, and regulators. The parameters that measure or predict bioavailability remain diffused, inconsistent, and, at times, unreliable due to variations in organisms (i.e., species, age, gender, etc.), environmental conditions, and soil variability. In addition, the nature and form of the chemicals of concern can have confounding effects on bioavailability. Thus, bioavailability may have different meanings to different disciplines and sectors of society.

Although bioavailability is relatively simple to parameterize under controlled, simulated conditions, this is not true under field conditions. More recently, the scientific community has come to a consensus that, although the "total" content of nutrient and contaminant elements has been infrequently well correlated with elemental uptake by organisms, the "more bioavailable" or "labile" form has more merit since that is the form that can be physically, biologically, and chemically described. The urge to measure the bioavailable forms of chemical compounds in an environmental setting and to be able to predict their uptake by organisms has resulted in numerous and diverse techniques. These techniques are being developed not only to characterize the flow and transfer of chemicals from the substrate to the organism intracellularly or extracellularly, but also to evaluate their potential transfer into the food chain. Whether we are dealing with microorganisms, higher plants and animals, or humans, the endpoint relevance of bioavailability centers on toxicity. This, in turn, requires understanding of the mechanisms of bioavailablity and toxicity at various levels of the biological system: the individual, physiological, and molecular levels. To expand our understanding of the fate and transport of trace elements in soils, the methods of assessing trace element bioavailability, flux, and transfer among the different soil components need to be redefined and developed.

This book has been prepared by a multidisciplinary group of scientists and engineers, and it was written to address the current state of knowledge of trace element bioavailability, flux, and transfer in soils. This volume is the edited proceedings of a workshop held in Vienna, Austria, in July 1999 in conjunction with the 5th International Conference on Biogeochemistry of Trace Elements in the Environment. This series of conferences was dedicated to exploring and discussing emerging issues in biogeochemistry research.

The book consists of 14 chapters and is divided into two major sections. Section I contains five chapters and is focused on trace element bioavailability. Chapter 1 provides a comprehensive review of the bioavailability and fate of trace elements in long-term residual amended soils. The authors are members of a national group evaluating long-term fate and effects of waste treatment (biosolids) on land. The second chapter examines the mobility and bioavailability of trace elements in estimating their risk assessment and pathway. Chapter 3 evaluates the sequential extraction of metals from soil contaminated with organic residue and added inorganic metals. Chapter 4 discusses the results of studies on hyperaccumulation and chemical-induced hyperaccumulation of metals by plant species in relation to chemical forms of metals in soils. Chemicals investigated to induce hyperaccumulation include acetic and citric acids and EDTA. Chapter 5 summarizes a greenhouse study on the bioavailability of Cu, Zn, and Mn to native Australian tree species grown in spiked soil. The authors propose a method to assess the phytotoxicity of metals in contaminated soil.

Section II consists of nine chapters on the fluxes and transfers of trace elements in soils, soil constituents, and soil solution. Chapter 6 examines the partition coefficient (K_d) and development of stochastic models for the prediction of trace element bioavailability by living organisms in soil.

Chapter 7 deals with an isotopic exchange kinetic method to assess the availability of Cd in soils to plants. It is an adaption and improvement of an earlier method and adds to our knowledge of Cd phytoavailability. Chapter 8 is a detailed and comprehensive review of the mobility and the availability of metals in waste-amended soils as influenced by source, time, and soil properties. The sources include swine manure, sewage sludge and municipal wastewater, mine tailings, and metal salts. Sequential extraction techniques are used to follow changes in metal form with time and to study how this may influence metal mobility and bioavailability.

Chapter 9 evaluates the use of poplar trees for phytoremediation of soil contaminated with Cu. It also provides a mathematical model for the description of water, Br, and Cu transport in soils.

Chapter 10 examines the partitioning and reaction kinetics of ^{109}Cd and ^{65}Zn in soils as influenced by organic matter at different temperatures. The increase in organic matter content enhanced Cd and Zn mobility, and the increase in temperature decreased the mobility due to an increase in the chemisorption rate.

Chapter 11 discusses metal speciation in soils and sewage-sludge-amended soils. Sequential extraction was used to speciate Cd, Ni ,and Zn into five groups: water soluble, exchangeable, carbonate oxide, organic, and residual associated groups. Chapter 12 evaluates and discusses the quality of estimated parameters from pedo-transfer functions to predict Cd concentrations of soil solution.

Chapter 13 discusses the effects of sorbed and dissolved organic carbon on Mo retention by iron oxides. The authors conclude that organic C coating on iron oxides has a significant effect on Mo mobility by clogging the pores on iron oxides and slowing down Mo penetration (or fixation). The last chapter (Chapter 14) discusses the speciation of Pb in soils and the mechanisms of Pb binding.

The editors wish to thank Prof. Amos Banin of the Hebrew University, Israel, for chairing a session during the symposium held in Vienna in July 1999, and all the authors for their contributions to this book. Special thanks are due to Drs. A.L. Page, H.M. Selim, D.C. Adriano, and W. Kingery for peer reviews and encouragement. Technical editing, for which we are grateful, was done at CRREL by Maria T. Bergstad and David Cate, and at CRC by William Heyward. Finally, Dr. Iskandar wishes to express his thanks and deep appreciation to his wife Bonnie Iskandar for allowing him to work at home many hours, and for her encouragement. Without the support of the U.S.Army ERDC-CRREL, the Center for Environmental Engineering Science and Technology (CEEST); the University of Massachusetts, Lowell, MA; and Kansas State University, this project could not have been achieved.

Editors

Dr. I.K. Iskandar received his Ph.D. in soil science and water chemistry at the University of Wisconsin, Madison, in 1972. He is currently a Research Physical Scientist at the Cold Regions Research and Engineering Laboratory (CRREL), and a Distinguished Research Professor at the Center for Environmental Engineering Science and Technology (CEEST), University of Massachusetts, Lowell, Massachusetts. During his tenure at CRREL, he developed two major research programs. The first was on land treatment of municipal wastewater, which he successfully coordinated for eight years, and in which he supervised research on the transformation and transport of nitrogen, phosphorus, and heavy metals in soils. The second program dealt with the environmental quality in cold regions. In the early 1980s his research efforts were focused on the fate and transformation of toxic chemicals in soils, the development of nondestructive methods for site assessments and characterization, and the development and evaluation of *in situ* remediation alternatives. He was the first to propose the use of an artificially frozen ground barrier for containment of toxic waste.

Dr. Iskandar has edited or co-edited 13 books on trace element transport and transformations and soil and ground water remediation; written more than 20 chapters of books; published more than 100 technical and reference papers and reports; presented more than 55 invited lectures, seminars, and symposia; and made 45 other presentations. He also has organized and co-organized many national and international conferences, workshops, and symposia. He has received many awards, including the Army Science Conference Award, 1979; the Army R&D Award, 1988; the CRREL Research and Development Award, 1988; several exceptional performance awards; and the 1999 CRREL Technology Transfer Award at the U.S. Army Cold Regions Research and Engineering Laboratory. He is a Fellow of the Soil Science Society of America, a Fellow of the American Society of Agronomy, a member of the International Union of Soil Science, and Vice President of the International Society of Trace Element Biogeochemistry.

Dr. M.B. Kirkham is a graduate of Wellesley College, Wellesley, MA (B.A.), the University of Wisconsin, Madison (M.S. and Ph.D.), and is now a professor in the Department of Agronomy at Kansas State University, Manhattan. Since 1973 Dr. Kirkham has studied the uptake of heavy metals by plants grown on agricultural land spread with sludge from cities. The research was initiated when Dr. Kirkham worked as a plant physiologist in the Ultimate Disposal Research Program, Advanced Water Treatment Research Laboratory of the U.S. Environmental Protection Agency in Cincinnati, Ohio. There, Dr. Kirkham was one of the first to document trace element buildup in soil and plants at a long-term sludge disposal site. Dr. Kirkham's research involves studying the basic processes of water and solute movement in soil and plants, especially under dry conditions. Most recently, Dr. Kirkham and co-workers have been studying chelate-facilitated phytoremediation of trace-element polluted soil. Dr. Kirkham is the author or co-author of more than 180 contributions to scientific journals, and is on the editorial boards of *Plant and Soil, Soil Science, Journal of Crop Production,* and *International Agrophysics.* In addition to conducting research, Professor Kirkham teaches a class on soil-plant water relations, works with graduate students, and participates in national and international meetings. Dr. Kirkham is a Fellow of the American Society of Agronomy, the Soil Science Society of America, the Crop Science Society of America, and the American Association for the Advancement of Science, and is a member of many other scientific societies, including the International Union of Soil Science, the Royal Society of New Zealand, the Society for Environmental Geochemistry and Health, and the International Society of Trace Element Research.

Contributors

Åsegir R. Almås, Ph.D.
Department of Soil and Water Sciences
Agricultural University of Norway
Ås, Norway

Chris Anderson, Ph.D.
Soil and Earth Sciences
Institute of Natural Resources
Massey University
Palmerston North, New Zealand

Jürgen Böttcher, Ph.D.
Institute of Soil Science
University of Hannover
Hannover, Germany

Robert Brooks, Ph.D.
Soil and Earth Sciences
Institute of Natural Resources
Massey University
Palmerston North, New Zealand

Isabel Cardo
Instituto de Recursos Naturales y
  Agrobiologia (CSIC)
Sevilla, Spain

Brent E. Clothier
Environment Group
HortResearch
Palmerston North, New Zealand

Arthur C. de Groot, Ph.D.
National Institute of Public Health and
  the Environment
Laboratory for Ecotoxicology
Bilthoven, The Netherlands

Annabelle Deram
Laboratoire de Génétique et Evolution
  des Populations Végétales
Université de Lille
Villeneuve d'Ascq, France

Encarnación Diaz-Barrientos, Ph.D.
Instituto de Recursos Naturales y
  Agrobiologia (CSIC)
Sevilla, Spain

Guillaume Echevarria
Laboratoire Sols et Environnement
ENSAIA-INRA
Nancy, France

Emilie Gérard
Laboratoire Sols et Environnement
ENSAIA-INRA
Nancy, France

Steven R. Green, Ph.D.
Environment Group
HortResearch
Palmerston North, New Zealand

Feng Xiang Han, Ph.D.
Plant and Soil Science
Mississippi State University
Mississippi State, MS

I.K. Iskandar, Ph.D.
U.S. Army Corps of Engineers
Engineer Research and Development
  Center
Cold Regions Research and
  Engineering Laboratory
Hanover, NH

M. Abul Kashem, Ph.D.
Department of Soil and Water Sciences
Agricultural University of Norway
Ås, Norway

Martin Kaupenjohann, Ph.D.
Institute of Soil Science
University of Hohenheim
Stuttgart, Germany

W.L. Kingery, Ph.D.
Plant and Soil Science
Mississippi State University
Mississippi State, MS

M.B. Kirkham, Ph.D.
Department of Agronomy
Kansas State University
Manhattan, KS

Friederike Lang, Ph.D.
Institute of Soil Science
University of Hohenheim
Stuttgart, Germany

Luis Madrid, Ph.D.
Instituto de Recursos Naturales y
  Agrobiologia (CSIC)
Sevilla, Spain

Neal W. Menzies, Ph.D.
Centre for Mined Land Rehabilitation
University of Queensland
Brisbane, Australia

Eugene V. Mironenko
Institute of Soil Science and
  Photosynthesis
Academy of Sciences of Russia
Pushkino, Moscow Region, Russia

Christian Morel
INRA-Agronomie
Villenave d'Ornon, France

Jean-Louis Morel, Ph.D.
Laboratoire Sols et Environnement
ENSAIA-INRA
Nancy, France

Jan Nemecek, Ph.D.
Department of Soil Science and
  Geology
Czech Agricultural University
Prague, Czech Republic

William J.G.M. Peijnenburg, Ph.D.
National Institute of Public Health and
  the Environment
Laboratory for Ecotoxicology
Bilthoven, The Netherlands

Daniel Petit, Ph.D.
Laboratoire de Génétique et Evolution
  des Populations Végétales
Université de Lille
Villeneuve d'Ascq, France

Gary M. Pierzynski, Ph.D.
Professor of Soil and Environmental
  Chemistry
Department of Agronomy
Kansas State University
Manhattan, KS

Eliska Podlesáková, Ph.D.
Research Institute for Soil and Water
  Conservation
Prague, Czech Republic

Alexander A. Ponizovsky
Institute of Soil Science and
  Photosynthesis
Academy of Sciences of Russia
Pushkino, Moscow Region, Russia

Suzanne M. Reichman
Centre for Mined Land Rehabilitation
University of Queensland
Brisbane, Australia

David L. Rimmer, Ph.D.
Department of Agricultural and
  Environmental Science
University of Newcastle
Newcastle-upon-Tyne, United Kingdom

Brett H. Robinson
Environment Group
HortResearch
Palmerston North, New Zealand

Angela Schön
Institute of Soil Science
University of Hannover
Hannover, Germany

H. Magdi Selim, Ph.D.
Agronomy Department
Louisiana State University
Baton Rouge, LA

Robyn Simcock, Ph.D.
Landcare Research New Zealand,
  Limited
Massey University
Palmerston North, New Zealand

Bal Ram Singh, Ph.D.
Department of Soil and Water Sciences
Agricultural University of Norway
Ås, Norway

Günther Springob, Ph.D.
Institute of Soil Science
University of Hannover
Hannover, Germany

Thibault Sterckeman
Laboratoire Sols et Environnement
ENSAIA-INRA
Nancy, France

Robert (Bob) Stewart, Ph.D.
Soil and Earth Sciences
Institute of Natural Resources
Massey University
Palmerston North, New Zealand

Dörthe Tetzlaff
Institute of Soil Science
University of Hannover
Hannover, Germany

Radim Vácha, Ph.D.
Research Institute for Soil and Water
  Conservation
Prague, Czech Republic

George F. Vance, Ph.D.
Professor of Soil and Environmental
  Chemistry
Department of Natural Resources
University of Wyoming
Laramie, WY

Rens P.M. van Veen, Ph.D.
National Institute of Public Health and
  the Environment
Laboratory for Ecotoxicology
Bilthoven, The Netherlands

Iris Vogeler, Ph.D.
Environment Group
HortResearch
Palmerston North, New Zealand

Table of Contents

Section I

Bioavailability of Trace Elements

1 Bioavailability and Fate of Trace Elements in Long-Term Residual-Amended Soil Studies

George F. Vance and Gary M. Pierzynski

ABSTRACT

The W-170 Multistate Research Committee on the Chemistry and Bioavailability of Waste Constituents in Soils and its predecessors, the W-124 and NC-118 Regional Research Committees, have been involved in conducting, characterizing, and evaluating the bioavailability and fate of trace elements in residual-amended studies for more than 30 years. Results from their research indicate that there are agronomic benefits from land application of various inorganic and organic residuals due to the addition of plant nutrients, improvement in soil chemical and physical properties, and possible use as liming materials. The W-170, W-124, and NC-118 committees have actively participated in research studies that have provided information for regulatory assessments of residual trace elements. For example, the W-170 and W-124 committees were extensively involved in the development of the EPA 503 national sludge rule, with W-170 members continuing their involvement in research to provide information that can be used to evaluate and refine biosolid regulations. As recently as 1998 a subcommittee of W-170 provided a critical peer review of a U.S. EPA risk assessment for the land application of cement kiln dust. W-170 members are currently involved with updating Mo standards in the EPA 503 national sludge rule, although not as part of an official W-170 function. This chapter examines efforts by the W-170 group that involve residual trace element studies, and describes future activities which include a variety of residual materials, biosolids composition, land application of biosolids, and agronomic benefits.

INTRODUCTION

Disposal of residual waste products is a problem that requires practical scientific information to determine if the residual constituents can be safely reused without harming the environment or unfavorably affecting nutrient and trace-element pathways. Land application of a variety of residual materials is known to be an effective means of recycling organic matter and plant nutrients, but these materials must be reused prudently to avoid degradation of the soil as a medium for plant growth. W-170 committee members have been involved in research that has evaluated the biogeochemical cycling of plant nutrients, the movement of trace elements into the food chain, and the long-term bioavailability of trace elements in residuals and residual-amended soils. Research also continues to focus on information related to the EPA 503 rules in order to provide support for risk assessment of land-applied biosolids. Many long-term studies by W-170 members have been and are currently being conducted to address the hypothesis that sequestered metals will be released as the biosolid organic matrix is mineralized. In addition, residual materials will continue to be emphasized in W-170 investigations so that waste is utilized in a manner that protects the sustainability of U.S. agriculture.

Agronomic benefits from the use of various inorganic and organic waste residuals have long been recognized in agriculture, horticulture, and reclamation. These materials can provide nutrients, improve soil physical properties, and potentially have liming value. There is a considerable knowledge base regarding the beneficial reuse of manures and biosolids, but many other residual materials may also be recyclable. Some additional residual materials that have been amended to soils include municipal solid waste (MSW) composts, yard wastes, cement kiln dusts, pharmaceutical biomass, brewery wastes, flue-gas desulfurization by-products, drinking-water treatment residuals, wood ash, and food-processing wastes. With the costs of incineration and disposal in landfills increasing dramatically, the quantity and variety of residuals that are being considered for land application are also increasing. Several key issues have been examined in past W-170 research, such as (1) determining the availability of plant nutrients in the residuals, (2) assessing the bioavailability of trace elements in the residuals and soil-residual mixtures, and (3) evaluating the content and fate of other contaminants, e.g., pathogens, xenobiotics, and salts, in the residual materials.

Efforts are under way to incorporate land application of residuals with assessment of soil quality.[1] The W-170 group and its predecessors (W-124 and NC-118) have been actively involved in research and regulatory aspects. For example, in 1998 a subcommittee of W-170 provided a critical peer review of an EPA risk assessment for the land application of cement kiln dust. The W-124 and W-170 committees were also extensively involved in the development of the EPA 503 national sludge rule[2] and continue to be involved in the refinements of that rule.

Despite the knowledge base that exists, new issues demand further research as more regulations are written and new concerns arise. Several examples will illustrate this point. The EPA 503 national sludge rule originally provided limits for the concentrations of ten trace elements (As, Cd, Cu, Cr, Hg, Mo, Ni, Pb, Se, and Zn) in biosolids and limits on the annual and cumulative loadings of these trace elements to soils. The W-170 group continues to build the database for elements such as As, Cr, Mo, and Se to address critical gaps in our knowledge related to these trace elements. After the publication of the rule and the pathway analysis used in the risk assessment, the protectiveness of certain aspects of the rule was called into question.[3,4] The major issue of concern was the long-term bioavailability of trace elements in biosolids-amended soils. The W-170 group continues to address this issue by utilizing data from several long-term biosolids studies and by working toward the development of new techniques for assessing trace-element bioavailability in soils. Accordingly, the next five-year W-170 project (2000 to 2005) proposes to address several emerging issues, such as

- Utilization of new and novel residual materials
- Expanding the database, when warranted, for trace elements such as As, Cr, Mo, and Se for risk assessment of biosolids, and for all trace elements for other organic and inorganic residuals that may be used in agriculture, horticulture, or reclamation
- Developing methods for estimating trace-element bioavailability in residuals and residual-amended soils
- Assessing the long-term bioavailability of trace elements in residual-amended soils
- Evaluating chemistry and fate of plant nutrients, particularly phosphorus, in residuals and residual-amended soils

New residual materials are continually being considered for land application or for agricultural and reclamation purposes. Commercial blending of a variety of residual materials to produce "synthetic" soils for agricultural and reclamation uses is increasing dramatically. Unprocessed animal manures have been studied extensively, but these materials are being processed or amended more often in an attempt to improve aesthetic issues, reduce volume, or decrease plant nutrient content or availability (e.g., alum amendment). Composting is used on a wider variety of materials that are then considered for land application. These new situations warrant study as the appropriate land application guidelines are developed to fully understand the risk/benefit issues associated with each material.

Development of the U.S. EPA 503 rule relied on an extensive database for trace elements such as Cd, Cu, Cr, Ni, Pb, and Zn in biosolids-amended soils. There is still a shortage of data for elements such as As, Hg, Mo, and Se that needs to be addressed. For example, the original EPA 503 rule provided a cumulative load limit for Mo of only 18 kg ha^{-1}, based on limited data, which would have made Mo a very restricting element for land application programs. Conversely, there is a legitimate concern about Mo-induced Cu deficiency (molybdenosis) in livestock that could develop if the Mo limit is not restrictive enough. To set a limit that is sufficiently protective without being unnecessarily restrictive requires a dataset that encompasses a wide range of soil and climatic conditions. The only study that has significantly added to this database since the EPA 503 rule was written is being conducted by a member of the W-170 group.[5] In addition, there are growing concerns about trace elements in other organic residual materials that have not received much attention in the past. Examples include Cu and Zn in swine manure, and As in poultry manure.

To improve our understanding of the fate and transport of trace elements in residuals and residual-amended soils, the methods for assessing trace element bioavailability need refinement and development. A variety of useful new methods are available that have either not been adequately applied or have been applied only to a limited extent. The bioavailability, fate, and transport of trace elements in residual-amended soils is influenced by the chemical form of the elements: organic vs. inorganic, solid phase vs. adsorbed, solubility of trace element solid phases, coprecipitation with other mineral phases, and so forth. Little is known about how trace elements are actually partitioned into the various chemical forms. Sequential extraction techniques and solubility equilibrium studies have been of some value, but recently developed or improved techniques, such as analytical electron microscopy and synchrotron-based methods like microprobe-extended x-ray fine structure (EXAFS), x-ray absorption near-edge spectroscopy (XANES), and microprobe x-ray fluorescence offer considerable promise and have been used on trace element problems to a limited extent. In addition, procedures more specific to data necessary for risk assessment, such as the physiologically-based extraction technique (PBET[6]), which is an *in vitro* method for assessing the bioaccessibility of As and Pb in soils to humans, have not been adequately utilized for residual-amended soils and may be quite useful.[7]

The degree of protectiveness provided by the EPA 503 national sludge rule has been criticized from several fronts. One is the possibility that the organic carbon added with the biosolids will eventually oxidize, allowing increased trace-element bioavailability over time, a factor that was not considered in the risk assessment for the regulations.[3] This phenomenon has been termed the "time-bomb" hypothesis and has generated considerable discussion in the scientific community and some public opposition to land application of biosolids. This hypothesis is already being considered by the W-170 committee by utilizing long-term biosolids studies,[8,9] and will continue to be addressed in future work. A second concern relates to the possibility of more subtle effects of trace elements on soil microbial populations.[10] Based on these and other issues related to trace-element bioavailability, we are required to improve our methods for assessing the bioavailability of various constituents in residuals and residual-amended soils. For example, trace-element phytoavailability can be predicted fairly well with routine soil extractants for a given soil–residual combination, but we do not have a method that performs satisfactorily across a wide range of soil and climatic conditions.

Much of the earlier work by W-124 and W-170 committee members focused on the availability of nitrogen in organic residuals for crop use. These efforts fulfilled the need mandated by the EPA 503 regulations to determine the agronomic loading rate for biosolids, and the methods that were developed are applicable to many residual materials. Further refinements are still needed as it becomes more important to accurately determine agronomic loading rates.

These emerging issues are currently being investigated by the W-170 committee by addressing the objectives to (1) characterize the chemical and physical properties of residuals and residual-amended soils; (2) evaluate methods for determining the bioavailability of nutrients, trace elements, and organic constituents in residuals; and (3) predict the long-term bioavailability of nutrients, trace elements, and organic constituents in residual-amended soils.

THE W-170 MULTISTATE RESEARCH COMMITTEE

The W-170 Research Committee comprises individuals in academia, federal research and enforcement agencies, and public and private industries who have been involved in short- and long-term residual studies. Their research has emphasized the characterization and understanding of physical, chemical, and biological processes and mechanisms involving a variety of residual materials. One area of extensive research conducted by W-170 committee members is that of the bioavailability and fate of land-applied biosolids. With studies involving both organic and inorganic biosolid constituents, members of the W-170 committee were instrumental in providing scientific data for and assistance in developing the EPA's 503 national sludge rule. Members of the W-170 committee have chaired or have been active participants on several National Research Council, Environmental Protection Agency, and U.S. Department of Agriculture committees to evaluate residual materials and the potential advantages and disadvantages of using them. In the past two years W-170 members have conducted peer review of an EPA risk assessment for land application of cement kiln dust, and an extensive research and coordination project evaluating biosolid Mo risks.

The formal history of the W-170 committee dates back to the early 1970s, when regional research committees were developed in both the western and the north central regions of the U.S. Both committees had as their objective the use of soils for the reception of wastes. In late 1977, the two regional committees formed a cooperative group with the specific objective of studying the land application of sewage sludge (e.g., biosolids). In the mid-1980s W-170 was approved as a regional research committee that continued the work of the W-124 and NC-118 committees. Following is a brief chronology of when the committees were officially approved and their respective titles:

1972	W-124	Soil as a Waste Treatment System
	NC-118	Utilization and Disposal of Municipal, Industrial, and Agricultural Processing Wastes on Land
1977	W-124 and NC-118	Optimum Utilization of Sewage Sludge on Land
1984	W-170	Chemistry and Bioavailability of Waste Constituents in Soils

Earlier work of the W-124 and NC-118 committees, as well as of W-170, focused on agronomic loading rates of biosolid nutrients. Results of committee member research demonstrated that biosolids could be beneficially reused if they were properly managed. In the past 10 years W-170 research has broadened its efforts and now includes modeling programs that evaluate the predictive capabilities for residual nutrients, trace elements and organic constituents, improved management techniques and uses of a variety of residual materials, and uses of residual waste products for land remediation and reclamation of disturbed environments.

The current W-170 Research Committee on the Chemistry and Bioavailability of Waste Constituents in Soils comprises individuals throughout the U.S. There are 48 leaders and cooperators from 18 universities (36 individuals), three USDA–ARS centers (3), two EPA regulatory offices (2), one EPA laboratory (1), one municipal wastewater treatment facility (2), one Army research center (2), and two private industries (2).

PAST W-170 COMMITTEE RESEARCH ACTIVITIES

W-170 research has been used to develop and validate the predictive capabilities of trace element bioavailability, management of soils receiving short- and long-term waste amendments, and remediation of contaminated environments. Results of these studies continue to demonstrate that most organic waste products can be beneficially reused if properly managed. Monitoring plant uptake of trace elements from soils previously treated with high trace-element biosolids for over 20 years indicates that metal uptake by plants on residual sludge-amended plots is minimal.

It has been well established that the bioavailability of metals in soils is higher when the source of the metals is metal salts, compared with metals in biosolids. This "protective" effect of the biosolids was factored into the risk assessment performed for the EPA 503 national sludge rule. The rule provided limits for the concentrations of ten trace elements (As, Cd, Cu, Cr, Hg, Mo, Ni, Se, Pb, and Zn) in biosolids, and on the annual and cumulative loadings of the trace elements to soils. An extensive database was utilized for the trace elements Cd, Cu, Cr, Ni, Pb, and Zn in biosolids-amended soils, but limited data exist for As, Hg, Mo, and Se (for example, Chaney[11]). As noted earlier, the original EPA 503 rule provided a cumulative load limit for Mo of only 18 kg ha^{-1}, which would have made Mo a very restricting element for land application programs, even though there is a legitimate concern about molybdenosis in livestock.

Recent challenges to the appropriateness of the EPA 503 rule were made due to a reduction in the potential protective effect of the biosolids diminishing with time as soil organic carbon oxidizes.[3] Studies by W-170 members, however, suggest there is no increase in plant-available metals over periods of 15 to 20 years.[8,9] For example, the study by Brown and colleagues[9] demonstrated that the carbon losses that occurred with time after biosolids applications ceased did not correspond to an increase in Cd bioavailability. They hypothesized that the protective effect of the biosolids was at least partly due to inorganic constituents in the biosolids and not entirely to organic carbon. Their results suggest that long-term field studies with biosolids are extremely important, and that additional research is needed to address issues that exist for all types of residuals. While both sides have presented compelling arguments, additional research is needed to address that issue, as well as similar questions that exist for residuals other than biosolids.

Earlier work by W-170 committee members on characterizing the chemical properties of residuals and residual-amended soils utilized relatively simple measurements, such as total elemental concentrations or fractions extracted with routine soil-testing procedures. Similar work still needs to be performed for new residual materials where data are lacking. These materials include waste office paper, lake weeds, water treatment residuals, flue-gas desulfurization products, cement kiln dusts, biotechnology residuals, wood ash, and others. More recent work has attempted to refine our characterization methods by making detailed measures of specific soil chemical properties. Candalaria and Chang[12] determined that most of the Cd introduced into soil as biosolids-borne Cd remained in the sludge with only a small portion transferred to solution and to the soil solid phases, yet solution speciation of Cd from sludge and Cd nitrate sources was similar. As the number of soil–residual combinations continues to increase, the need to characterize them will also increase.

Members of the W-170 committee have also investigated methods for determining the bioavailability of nutrients, trace elements, and organic constituents in residuals. Research on the bioavailability of nutrients in residuals and residual-amended soils has focused on nitrogen, in part because its availability is most often used to determine appropriate application rates. Nitrogen is still an important topic as a wider variety of residuals is considered for land application. There has been increased interest in phosphorus because of water quality concerns, and the prospect of phosphorus-based application rates for residuals has been proposed both nationally and in some states. Earlier work by W-170 members characterized inorganic and organic phosphorus in residuals and residual-amended soils using electron microscopy and NMR.[13,14] Little work of this nature has been published since then, and with the increased interest in phosphorus, more research is needed. For example, the use of alum and other residuals to reduce the bioavailability of phosphorus in soils and residual-amended soils has been studied, although little is known about the changes in phosphorus solid phases brought about by the amendments.[15,16] The utility of some recent innovations in assessing phosphorus bioavailability needs to be evaluated for residuals and residual-amended soils. These include iron oxide strip-extractable phosphorus, and assessments of the degree of phosphorus saturation.[17]

Research on assessing the bioavailability of trace elements in residuals and residual-amended soils has not taken full advantage of developments in advanced spectroscopic techniques or procedures for assessing trace element bioavailability in contaminated soils. Examples include the determinative oxidation speciation for S or Mn *in situ* using XANES, a synchrotron radiation-based technique,[18,19] and estimating bioaccessible As and Pb in soils for mammals using a physiologically-based extraction test.[6] Given the interest in long-term bioavailability of trace elements, there is a strong need for research to determine the chemical forms of trace elements in residuals and residual-amended soils, and for methods useful for risk assessment.

Predicting the long-term bioavailability of nutrients, trace elements, and organic constituents in residual-amended soils has been emphasized in W-170 research activities. Bioavailability of metals in soils is higher when the source of the metals is salts rather than biosolids. This "protective" effect of the biosolids was factored into the risk assessment performed for the U.S. EPA 503 national sludge rule.

Cooperative studies have been conducted to examine municipal solid wastes and various types of composts. Phytoremediation of metal-contaminated mine sites was an area of interest in several W-170 member states. For example, sites on the Comprehensive Environmental Response Compensation and Liability Act National

Priority List (CERCLA NPL) were studied to determine ways of reclaiming waste materials. Methods were evaluated for the stabilization of contaminated sites to reduce erosion, leaching, revegetation, and other off-site problems affecting these environments. Collaborative investigations have been helpful in determining ways of reducing trace metal and oxyanion mobility and bioavailability in different soils and under a variety of conditions.

W-170 research has been used to develop and validate the predictive capabilities of nutrient and trace element bioavailability, management of soils receiving short- and long-term waste amendments, and remediation of contaminated environments. W-170 studies continue to demonstrate that most organic waste products can be reused beneficially when properly managed. Research that has monitored plant uptake of trace elements from soils previously treated with high-metal biosolids for up to 20 years indicates metal uptake by vegetables on residual sludge-amended plots has been minimal.

Composts and manures have been shown to increase plant growth, enhance soil properties such as infiltration and permeability, chelate metals, and reduce overland nutrient flow. The W-170 group continues to study the leaching of nutrients and metal-organic complexes, and the determination of nutrient requirements.

Biosolids utilization studies have shown that rangeland vegetation, dryland wheat production, and forest growth are improved with amendments of biosolids. Yard and wastewater composts with high Cu levels had no adverse effects on plant growth. Sludge–Mo bioavailability in pastures was found to be important to Pathway No. 6 (EPA 503 regulations) risk assessment for ruminants foraging sludge-amended lands. Alum-amended manures decreased losses of nitrogen, phosphorus, and organic carbon and improved water quality.

W-170 research has enhanced the understanding of nutrient release mechanisms and developed model parameters for predicting nutrient status for various soils, crop systems and climate conditions, and land loading rates. The use of DTPA-extractable plant nutrients and trace elements was shown to be a useful technique for analyzing soil–plant information. Accurate estimates of the nitrogen-supplying capacity of organic by-products enabled wastes to be used properly and more efficiently, which resulted in greater protection of the environment.

Use of organic and inorganic residuals enhanced the remediation of disturbed environments. Mine sites and smelter-impacted areas were reclaimed by immobilization treatments that utilized biosolids. Selenium in the soil–plant–animal continuum of disturbed environments was found to be regulated by several factors, including environmental conditions. Phytoremediation technology were also found to be useful to decontaminate metal-contaminated soils.

Members of the W-170 committee will continue to work toward obtaining a better understanding of the fundamental properties, reactions, and pathways involving trace elements. The bioavailability, species forms and fractions, and environmental conditions influencing trace-element uptake by plants will remain a major focus of the W-170 group. Additional information on oxyanions is needed to improve state and federal guidelines and rules, e.g., U.S. EPA 503 regulations, for the use of residual materials. In particular, residuals other than biosolids must be evaluated to determine how these materials can be safely and effectively recycled in a sustainable manner.

FUTURE W-170 COMMITTEE RESEARCH ACTIVITIES

Future W-170 Research Committee activities are based on three objectives:

- Characterize the chemical and physical properties of residuals and residual-amended soils.
- Evaluation of methods for determining the bioavailability of nutrients, trace elements, and organic constituents in residuals.
- Prediction of the long-term bioavailability of nutrients, trace elements, and organic constituents in residual-amended soils.

Characterization of Chemical and Physical Properties of Residuals and Residual-Amended Soils

Residual trace-element chemistries and soil quality impacts

W-170 committee members from several states will continue to evaluate the impact of biosolids on oxyanion (e.g., As, Mo, P, Se, and others) retention/release/mobility in soils. Studies will also be conducted to evaluate the forms and bioavailability of several oxyanions. Traditional adsorption isotherms and "single point" isotherms will be developed on greenhouse and field-equilibrated, biosolids-amended soils. Oxyanion solubility and form(s) in residuals and residual-amended soils will be characterized. Common extractants (e.g., water, Mehlich I, etc.) as well as others (e.g., phosphorus sequential extraction schemes) will be evaluated. Existing field studies will be monitored to detect oxyanion movement with depth in both weakly and strongly adsorbing soils.

Committee members will continue to evaluate the chemical properties of residual-amended soils in heavy-metal remediation studies. Initially, the concentration of heavy metals in residual materials (e.g., mining tailings and industrial products such as phosphorus fertilizer, lime, and micronutrient fertilizer) will be determined. The solubility of metals in soils, as affected by repeated additions of soil amendments, and soil pH will also be determined. Sequential extractions will be performed in addition to other methods to characterize the transformation of metals in soils as affected by the quantity and type of residuals and soil amendments added.

Physical properties of residuals and residual-amended soils

Laboratory and field studies will be conducted to evaluate the influence on soil physical properties of residuals alone and in combination with other materials. Physical properties of residual-amended soils will be characterized to evaluate changes over time. Attempts will be made to concentrate trace metals in soils by particle size and density separations to facilitate mineralogical characterization. Compost applications with gypsum-containing flue gas desulfurization residuals are expected to increase Ca migration and exchange in sodic soils and acidic minespoils, thereby enhancing restoration of these degraded systems. Their impact on Ca transport and exchange with other metals, as well as their effects on soil properties and plant growth, will be studied. Infiltration and movement of residual constituents will be examined under different irrigation systems. The mineralogy of phosphorus solid phases will also be determined.

Evaluation of Methods for Determining Bioavailability of Nutrients, Trace Elements, and Organic Constituents in Residuals

Nutrient and trace-element bioavailabilities in residual-amended field and greenhouse studies

Greenhouse, plot, and large-field studies will be conducted to evaluate the effects of residual amendments on plant uptake (leaves and grain) of trace elements (heavy metals, oxyanions) for several growing seasons and with different crops. Plant availability will be correlated with soil metal concentrations to determine the residual bioavailability of the metals in biosolid-amended soils after several years of crop growth, and the ability of the soil-chemical extraction procedures to predict plant uptake. Plant uptake of oxyanions will be related (correlated) to various measures of oxyanion load, including total metal load, extractable metal load, and knowledge of oxyanion "form" or speciation. Lettuce or other types of crops will be planted, and the amount of metals accumulated in the plants will be determined. The relationship between the quantity of metal soluble in dilute salt solution and metal uptake will be used to elucidate the metal availability in various types of soil amendments.

Laboratory studies and soil testing approaches involving evaluation of residual constituent bioavailabilities

Laboratory studies will be used to continue the development of methods, such as the FeO-coated filter paper and *in vitro* gastrointestinal chemical procedures, for determination of the bioavailability of nutrients and trace elements. Physiologically-based extraction techniques for determining bioavailable Pb and As will be applied to residual-amended soils. Soils receiving varying rates of different residual amendments will be analyzed for the quantity of metals soluble in dilute salt solution.

Prediction of Long-Term Bioavailability of Nutrients, Trace Elements, and Organic Constituents in Residual-Amended Soils

Nutrient bioavailability in long-term residual-amended soils

Study of the long-term bioavailability and movement of phosphorus (>5000 kg ha^{-1} yr^{-1} of biosolids-borne phosphorus applied to the area for 20 years) and trace metals within a large watershed where biosolids were applied (1974–1993) will also be continued. A digital elevation model will be constructed to establish water routing, and soil testing will be conducted to evaluate nutrient and metal movement across the watershed landscape. Runoff will be measured and analyzed to quantify losses of nutrients and trace metals from the watershed. Soil and crop samples from fields that have received several applications of biosolids (cumulative applications of up to 50 metric tons per acre) will be analyzed for various trace elements to determine recoveries compared to calculated cumulative loadings, downward transport of trace elements, and plant uptake coefficients.

Effect of time on trace-element chemistry in residual-amended soils

Agricultural fields and trace-element-contaminated sites where residuals have been applied will continue to be monitored. One site to be studied was amended with biosolids having fairly high concentrations of heavy metals (284 Cd, 2040 Ni, 6800 Zn, 1200 Cu, and 1070 Pb mg kg^{-1}) at rates up to 448 metric tons per hectare since 1976, and the plots have been cropped in wheat, oats, soybeans, and corn. Several sites studied by the W-170 group will be intensively sampled and the distribution of metals with depth will be determined. Chemical fractionation will be used to evaluate total metal concentrations as well as water-soluble, exchangeable, and more recalcitrant fractions. An index of availability will be determined by the DTPA extraction procedure. Various methods for estimating bioavailable trace elements, including routine soil extractions and sequential extraction procedures, will be employed.

SPECIFIC REGIONAL EXPERIMENTS

Some prospective regional projects that may be conducted by W-170 committee members include

- Applying a common biosolid to soils at different W-170 member locations
- Developing new analytical procedures for assessing bioavailability and other characteristics of residual materials
- Devising processing recommendations and management practices for the use of waste materials on turf grass and ornamental plants
- Developing and validating computer simulation models to predict the fate of applied nutrients from residual materials.

A strength of the W-170 committee is the participation of soil scientists in locations that represent a wide range in environmental conditions; many of these scientists have developed long-term datasets for residual waste applications. Development of new techniques and the use of simulation models should help integrate these data and develop a capability to predict the long-term bioavailability of residual materials.

Plant uptake slopes of biosolid-borne metals have been the focus of several studies by W-170 investigators, and the group intends to continue these studies to examine long-term processes. Additional cooperative studies to be developed involve testing residual materials such as yardwaste and MSW composts, paper mill wastes, and manufactured soils. Compost research will continue to be conducted as the need for waste residual utilization persists. Pasture and rangeland studies involving applications of biosolids, manures, and waste effluents also require greater collaborative research with W-170 experts.

CONCLUSIONS

Over the past three decades, the W-170 committee has been involved in studies emphasizing the trace-element chemistry of soils and residuals, prediction of plant uptake and movement of trace elements, evaluation of trace-element chemistries in contaminated soils, and examination of the effects of soil remediation on trace-element chemistry, bioavailability, and mobility. W-170 research has enhanced our understanding of trace-element release mechanisms, model parameters for predicting soil trace-element status, conditions related to various crop systems and climate controls, and land loading rates. W-170 long-term studies of high-trace-element biosolids indicate minimal trace element phytoavailability and demonstrate that biosolids are beneficial when properly managed. Scientifically-based findings from the W-170 group are essential for regulatory guidelines to manage the beneficial uses of residual products in a sustainable manner consistent with protecting our environment.

REFERENCES

1. Sims, J.T. and Pierzynski, G.M., Assessing the impacts of agricultural, municipal, and industrial by-products on soil quality, in *Beneficial Uses of Agricultural, Industrial, and Municipal By-Products,* Power, J.F., Ed., Soil Science Society of America, Madison, WI (in press), 2000.
2. U.S. EPA, The standards for use and disposal of sewage sludge, *Title 40 of the Code of Federal Regulations,* Part 503, U.S. Environmental Protection Agency, Washington, D.C., 1993.
3. McBride, M.B., Toxic metal accumulation from agricultural use of sewage sludge: Are USEPA regulations protective? *J. Environ. Qual.,* 24, 5, 1995.
4. Schmidt, J.P., Understanding phytotoxicity thresholds for trace elements in land-applied sewage sludge, *J. Environ. Qual.,* 26, 4, 1997.
5. Nguyen, H.Q. and O'Connor, G.A., Sludge-born molybdenum availability, in *4th Intl. Conf. on Biogeochemistry of Trace Elements Proc.,* Iskandar, I.K., Hardy, S.E., Chang, A.C., and Pierzynski, G.M., Eds., U.S. Army Cold Regions Research and Engineering Laboratory, Hanover, NH, 1997, 695.
6. Ruby, M.V., Davis, A., Schoof, R., Eberle, S., and Sellstone, C.M., Estimation of lead and arsenic bioavailability with a physiologically-based extraction test, *Environ. Sci. Tech.,* 30, 422, 1996.
7. Rodriguez, R. R., Basta, N.T., Casteel, S.W., and Pace, L.W., An *in vitro* gastro-intestinal method to assess bioavailable arsenic in contaminated soils and solid media, *Environ. Sci. Technol.,* 33, 642, 1999.
8. Chang, A.C., Hyun, H., and Page, A.L., Cadmium uptake for Swiss chard grown on composted sewage sludge-treated field plots: plateau or time bomb? *J. Environ. Qual.,* 26, 11, 1997.

9. Brown, S.L., Chaney, R.L, Angle, J.S., and Ryan, J.A., The phytoavailability of cadmium to lettuce in long-term biosolids-amended soils, *J. Environ. Qual.*, 27, 1071, 1998.

10. McGrath, S.P., Chaudri, A.M., and Giller, K.E., Long-term effects of metals in sewage sludge on soils, microorganisms, and plants, *J. Industrial Microbiology*, 14, 94, 1995.

11 Chaney, R.L., Trace metal movement: soil-plant systems and bioavailability of biosolids-applied metals, in *Sewage Sludge: Land Utilization and the Environment*, Clapp, C.E., Larson, W.E., and Dowdy, R.H., Eds., Soil Science Society of America, Inc., Madison, WI, 27, 1994.

12. Candalaria, L.M. and Chang, A.C., Cadmium activities, solution speciation, and solid phase distribution of Cd in cadmium nitrate and sewage sludge-treated soil systems, *Soil Sci.*, 162, 722, 1997.

13. Hinedi, Z.R., Chang, A.C., and Yesinowski, J.P., Phosphorus-31 magic angle spinning nuclear magnetic resonance of wastewater sludges and sludge amended soil, *Soil Sci. Soc. Am. J.*, 53, 1053, 1989.

14. Pierzynski, G.M., Logan, T.J., Traina, S.J., and Bigham, J.M., Phosphorus chemistry and mineralogy in excessively fertilized soils: descriptions of phosphorus-rich particles, *Soil Sci. Soc. Am. J.*, 54, 1583, 1990.

15. Moore, P.A. and Miller, D.M., Decreasing phosphorus solubility in poultry litter with aluminum, calcium, and iron amendments, *J. Environ. Qual.*, 23, 325, 1994.

16. Peters, J.M. and Basta, N.T., Reduction of excessive bioavailable phosphorus in soils using municipal and industrial wastes, *J. Environ. Qual.*, 25, 1236, 1996.

17. Moore, P.A., Joern, B.C., and Provin, T.L., Improvements needed in environmental soil testing for phosphorus, in *Soil Testing for Phosphorus: Environmental Uses and Implications*, Sims, J.T., Ed., Southern Cooperative Series Bulletin No. 389, 1998.

18. Schultze, D.G., McCay-Buis, T., Sutton, S.R., and Huher, D.M., Determination of manganese oxidation states in soils using x-ray absorption near-edge structure (XANES) spectroscopy, *Soil Sci. Soc. Am. J.*, 59, 1540, 1995.

19. Fendorf, S. and Sparks, D., X-ray absorption fine structure spectroscopy, in *Methods of Soil Analysis, Part 3—Chemical Methods*, Sparks, D.L., Ed., SSSA Book Series No. 5, Soil Science Society of America, Madison, WI, 1996, 377.

W-170 COMMITTEE RECENT PUBLICATIONS

Arnold, K., J. Dunn, and D. Sievers, 1995, Biosolids Standards for Metals and Other Trace Substances. *University of Missouri Extension Water Quality Guide Sheets*, pub. WQ 425.

Aynaba, A., M.S. Baram, G.W. Barret, W.G. Boggess, A.C. Chang, R.C. Cooper, R.I. Dick, S.P. Graef, T.E. Long, A.L. Page, C. St. Hilaire, J. Silverstein, S. Clark-Stuart, and P.E. Waggoner (authorship in alphabetical order), 1996, *Use of Reclaimed Water and Sludge in Food Crop Production*, National Academy Press, Washington, D.C.

Barbarick, K.A., J.A. Ippolito, and D.G. Westfall, 1995, Biosolids effect on P, Cu, Zn, Ni, Mo concentrations in dryland wheat, *J. Environ. Qual.*, 24, 608–611.

Barbarick, K.A., J.A. Ippolito, and D.G. Westfall, 1997, Sewage biosolids cumulative effects on extractable-soil and grain elemental concentrations, *J. Environ. Qual.*, 26, 1696–1702.

Barbarick, K.A., J.A. Ippolito, and D.G. Westfall, 1998, Extractable elements in the soil profile after years of biosolids application, *J. Environ. Qual.*, 27, 801–805.

Basta, N.T. and J.J. Sloan, 1998, Application of alkaline biosolids to acid soils: changes in solubility and bioavailability of heavy metals, *J. Environ. Qual.*, 28, 633–638.

Berti, W.R., and L.W. Jacobs, 1996, Chemistry and phytotoxicity of soil trace elements from repeated sewage sludge applications, *J. Environ. Qual.*, 25, 1025–1032.

Brallier, S., R.B. Harrison, and C.L. Henry, 1996, Liming effects on availability of Cd, Cu, Ni, and Zn in a soil amended with sewage sludge 16 years previously, *Water, Air and Soil Pollution*, 86, 195–206.

Brown, S.L., J.S. Angle, and R.L. Chaney, 1997, Correction of limed-biosolid induced Mn deficiency on long-term field plots, *J. Environ. Qual.*, 26, 1375–1384.

Brown, S.L., J.S. Angle, R.L. Chaney, and A.J.M. Baker, 1995, Zinc and cadmium uptake by *Thlaspi caerulescens* and *Silene cucubalis* grown on sludge-amended soils in relation to total soil metals and soil pH, *Environ. Sci. Technol.*, 29, 1581–1585.

Brown, S.L., R.L. Chaney, and J.S. Angle, 1998, Correction of limed-biosolid induced Mn deficiency on a long-term field experiment, *J. Environ. Qual.*, 26, 1375–1384.

Brown, S.L., R.L. Chaney, and J.S. Angle, 1997, Subsurface liming and metal movement in soils amended with lime stabilized biosolids, *J. Environ. Qual.*, 26, 724–732.

Brown, S.L., R.L. Chaney, J.S. Angle, and A.J.M. Baker, 1995, Zinc and cadmium uptake of *Thlaspi caerulescens* grown in nutrient solution, *Soil Sci. Soc. Am. J.*, 59:125–133.

Brown, S.L., R. Chaney, C. Lloyd, J.S. Angle, and J. Ryan, 1996, Relative uptake of cadmium by garden vegetable and fruits grown on long-term biosolid-amended soils, *Environ. Sci. & Technol.*, 30, 3508–3511.

Cai, X.H., R.T. Sayre, S.J. Traina, T.J. Logan, and T. Gustafson, 1995, Applications of eukaryotic algae for the removal of heavy metals from water, *Molecular Marine Biol. and Biochem.*, 4, 338–344.

Camobreco, V.J., B.K. Richards, T.S. Steenhuis, J.H. Peverly, and M.B. McBride, 1996, Movement of heavy metals through undisturbed and homogenized soil columns, *Soil Sci.*, 161, 740–750.

Candalaria, L.M. and A.C. Chang, 1997, Cadmium activities, solution speciation, and solid phase distribution of Cd in cadmium nitrate- and sewage sludge-treated soil systems, *Soil Sci.*, 162, 722–732.

Chaney, R.L., 1994, Trace metal movement: soil-plant system and bioavailability of biosolid-applied metals, in *Sewage Sludge: Land Utilization and the Environment*, C.E. Clapp, W.E. Larson and R.H. Dowdy, Eds., Soil Science Society of America, Madison, WI, 27–31.

Chaney, R.L., et al., 1994, Effect of Fe, Mn, and Zn enriched biosolids compost on uptake of Cd by lettuce from Cd-contaminated soils, in *Sewage Sludge: Land Utilization and the Environment*, C.E. Clapp, W.E. Larson and R.H. Dowdy, Eds., Soil Science Society of America, Madison, WI, 205–207.

Chaney, R.L., S. Brown, Y.M. Li, J.S. Angle, F. Homer, and C. Green, 1995, Potential use of metal hyperaccumulators, *Mining Environment Mag.*, 3, 9–11.

Chaney, R.L. and J.A. Ryan, 1994, *Risk-Based Standards for Arsenic, Lead, and Cadmium in Urban Soils*, DECHEMA, Frankfurt.

Chang, A.C., H. Hyun, and A.L. Page, 1997, Cadmium uptake for Swiss chard grown on composted sewage sludge-treated field plots: plateau or time bomb? *J. Environ. Qual.*, 26, 11–19.

Chang, A.C., A.L. Page, and J.E. Warneke, 1997, An experimental evaluation on the Cd uptake by twelve plant species grown on a sludge-treated soil: is there a plateau? In *Biogeochemistry of Trace Metals*, D.C. Adriano, Z.-S. Chen, S.-S. Yang, and I.K. Iskandar Eds., Science Reviews, Northwood, U.K., 43–60.

Clapp, C.E., R.H. Dowdy, W.E. Larson, D.R. Linden, C.M. Hormann, R.C. Polta, T.R. Halbach, and H.H. Cheng, 1994, Long-term effects of crop, soil, and water quality of sewage sludge applied to an agricultural watershed, in *Proc. 15th Intl. Cong. Soil Sci.,* Acapulco, 3b, 406–407.

Dowdy, R.H. and Sloan, J.J., 1997, Trace metal uptake from biosolids: A 20-year field study at the Rosemount Agricultural Experiment Station, in *Proc. 4th Intl. Conf. Biogeochem. of Trace Metals,* Berkeley, California, 689–690.

Dowdy, R.H., C.E. Clapp, D.R. Linden, W.E. Larson, T.R. Halbach, and R.C. Polta, 1994, Twenty years of trace metal partitioning on the Rosemount sewage sludge watershed, in *Sewage Sludge: Land Utilization and the Environment,* C.E. Clapp, W.E. Larson, and R.H. Dowdy, Eds., Soil Science Society of America, Madison, WI, 149–155.

Freeman, G.B., J.D. Johnson, S.C. Liao, P.I. Feder, A.O. Davis, M.V. Ruby, R.A. Shoof, R.L. Chaney, and P.D. Bergstrom, 1994, Absolute bioavailability of lead acetate and mining waste lead in rats, *Toxicology,* 91, 151–163.

Granato, T.C., R.I. Pietz, J. Gschwind, and C. Lue-Hing, 1994, Mercury in soils and crops from fields receiving high cumulative sewage sludge applications: Validation of U.S. EPA's risk assessment for human ingestion, presented at *Intl. Conf. on Mercury as Global Pollutant,* 10–14 July, 1994, Whistler, British Columbia.

Granato, T.C., L. Kristoff, R.I. Pietz, and C. Lue-Hing, 1995, Changes in concentration of trace metals and radionuclides in Illinois soils since 1935, in *Trace Substances, Environment and Health,* C.E. Cothern, Ed., Science Reviews, Northwood, U.K., 153–164.

Harrison, R.B., X. Dongsen, and C.L. Henry, 1994, Magnesium deficiency in Douglas fir and grand fir growing on a sandy outwash soil amended with sewage sludge, *Water, Air and Soil Poll.,* 75, 37–50.

He, X.T., T.J. Logan, and S.J. Traina, 1995, Physical and chemical characteristics of selected U.S. municipal solid waste composts, *J. Environ. Qual.,* 24, 543–552.

Henry, C.L. and S. Brown, 1997, Restoring a Superfund site with biosolids and fly ash, *Biocycle,* 38, 79–83.

Hoette, G. and J. Brown, 1995, Activity and movement of plant nutrients and other trace substances, *University of Missouri Extension Water Quality Guide Sheets,* pub. WQ 428.

Hoorman, R. and R. Miles, 1995, Benefits and risks of biosolids, *University of Missouri Extension Water Quality Guide Sheets,* pub. WQ 427.

Hue, N.V., 1995, Sewage sludge, in *Soil Amendments and Environmental Quality,* J. Rechcigl, Ed., Lewis Publishers, Boca Raton, 199–247.

Ibekwe, A.M., J.S. Angle, R.L. Chaney, and P. van Berkum, 1995, Sewage sludge-borne heavy metal effects on nodulation and nitrogen fixation, *J. Environ. Qual.,* 24, 1199–1204.

Johnson, C.D. and G.F. Vance, 1994, Accumulation of selenium in thick spick wheatgrass and yellow sweet clover grown on sludge amended alkaline mine spoil, *Communications in Soil Sci. and Plant Analysis,* 25, 2117–2132.

Johnson, C.D. and G.F. Vance, 1998, Accumulation of trace elements in plants and soils: Effects of long-term sludge amendments, *Agricultural Experiment Station Bulletin B-1062,* University of Wyoming, Laramie, WY.

Kuo, S. and J.H. Harsh, 1997, Physiochemical characterization of metal availability in soil, p. 75–120, in *Biogeochemistry of Trace Metals,* D.C. Adriano et al., Eds., Science Reviews, Northwood, U.K.

Lambert, M., G. Pierzynski, L. Erickson, and J. Schnoor, 1997, Remediation of lead, zinc, and cadmium contaminated soils, *Issues in Environ. Science and Technol.,* 7, 91–102.

Laperche, V., T.J. Logan, P. Gaddam, and S.J. Traina, 1997, Effect of apatite amendments on plant uptake of Pb from contaminated soil, *Environ. Sci. & Technol.,* 31, 2745–2753.

Larson, W.E., C.E. Clapp, R.H. Dowdy, and D.R. Linden, 1994, Rosemount watershed study on land application of municipal sewage sludge, in *Sewage Sludge: Land Utilization and the Environment,* C.E. Clapp, W.E. Larson, and R.H. Dowdy, Eds., Soil Science Society of America, Madison, WI, 125–128.

Logan, T.J. 1997, Balancing benefits and risks in biosolids, *Biocycle,* November, 52–57.

Logan, T.J., L.E. Goins, and B.J. Harrison, 1997, Field assessment of trace element uptake by six vegetables from N-Viro soil, *Water Environ. Res.,* 69, 28–33.

Logan, T.J., B.J. Harrison, L.E. Goins, and J.A. Ryan, 1997, Field assessment of sludge metal bioavailability to crops: sludge rate response, *J. Environ. Qual.,* 26, 534–550.

Lue-Hing, C., R.I. Pietz, T.C. Granato, J. Gschwind, and D.R. Zenz, 1994, Overview of the past 25 years: operator's perspective, in *Sewage Sludge: Land Utilization and the Environment,* C.E. Clapp, W.E. Larson, and R.H. Dowdy, Eds., Soil Science Society of America, Madison, WI, 7–14.

Lue-Hing, C., R.I. Pietz, T.C. Granato, and D.R. Zenz, 1997, Thirty Years of Sludge Utilization: The Chicago Contribution, in *Proc. Water Environment Fed. 70th Annual Conf. and Expo., Vol. 2, Residuals and Biosolids Management,* Chicago, IL, 18–22 October, 1997, 55–65.

McGrath, S.P., A.C. Chang, A.L. Page, and E. Witter, 1994, Land application of sewage sludge: Scientific perspectives of heavy metal loading limits in Europe and the United States, *Environ. Review,* 2, 108–118.

McKenna, I.M. and R.L. Chaney, 1995, Characterization of a cadmium-zinc complex in lettuce leaves, *Biol. Trace Element Res.,* 48, 13–29.

Nelson, S.R., and R.I. Pietz, 1995, Land application of sewage sludge: papers and publications by the Research and Development Department and research funded by the District— A bibliography, 1968–1994, Report no. 95–19, Research and Development Department, Metropolitan Water Reclamation District of Greater Chicago, Chicago, IL.

Nguyen, H.Q. and G.A. O'Connor, 1997, Sludge-borne molybdenum availability, *4th Intl. Conf. Biochem. of Trace Metals,* 23–26 June, 1997, Berkeley, CA, 695–696.

O'Connor, G.A., 1997, Molybdenum research: Implications for part 503, *Annual Biosolids Mgt. Conf.,* 14–16 September, 1997, Vancouver, WA, 31.

O'Connor, G.A. and L.R. McDowell, 1997, Understanding fate, transport, bioavailability, and cycling of metals in land-applied biosolids, in *Proc. Workshop No. 104, WEFTEC'97,* 18 Oct., 1997, Chicago, IL, 32–41,

Page, A.L. and A.C. Chang, 1994, Trace elements of environmental concerns, in *15th Intl. Cong. of Soil Science,* Vol. 3a, 568–571.

Page, A.L. and A.C. Chang, 1994, Overview of the past 25 years: technical perspective, in *Sewage Sludge: Land Utilization and the Environment,* C. E. Clapp, W. E. Larson, and R. H. Dowdy, Eds., Soil Science Society of America, Madison, WI, 3–6.

Peverly, J.H., J.M. Surface, and T. Wang, 1995, Growth and trace metal absorption by *Phragmites australis* in wetlands constructed for landfill leachate treatment, *Ecolog. Engin.,* 5, 21–35.

Pierce, B.L., E.F. Redente, K.A. Barbarick, R.B. Brobst, and P. Hegeman, 1998, Plant biomass and elemental changes in shrubland forages following biosolids application, *J. Environ. Qual.,* 27, 789–794.

Pierzynski, G.M. 1997, Strategies for remediating trace element contaminated sites, in *Remediation of Soils Contaminated with Metals,* I.K. Iskander and D.C. Adriano, Eds., Science Reviews, Middlesex, U.K., 67–84.

Pierzynski, G.M., 1997, Past, present and future approaches for testing metals for environmental concerns and regulatory approaches, in *Proc. of 1997 Intl. Soil and Plant Analysis Symp.,* Soil and Plant Analysis Council, Athens, GA, 105–117.

Pierzynski, G.M., G.M. Hettiarachchi, and J.K. Koelliker, 1997, Methods for assessing the inputs of soil degradation on water quality, in *Methods of Assessment of Soil Degradation,*

Advances in Soil Science, R. Lal, W. Blum, C. Valentine, and B.A. Stewart, Eds., CRC Press, Boca Raton, 513–545.

Pierzynski, G.M., J.L. Schnoor, M.K. Banks, J. Tracy, L. Licht, and L.E. Erickson, 1994, Vegetative remediation at Superfund sites, in *Issues in Environmental Science and Technology, Volume 1, Mining and its Environmental Impact,* R.E. Hester and R.M. Harrison, Eds., Royal Society of Chemistry, Cambridge U.K., 49–69.

Pierzynski, G.M., J.T. Sims, and G.F. Vance, 2000, *Soils and Environmental Quality,* 2nd Edition, Lewis Publishers, Inc., Chelsea, MI.

Richards, B.K., J.H. Peverly, T.S. Steenhuis, and B.N. Liebowitz, 1997, Effect of processing mode on trace elements in dewatered sludge products, *J. Environ. Qual.,* 26, 782–788.

Ryan, J.A. and R.L. Chaney, 1995, Development of limits for land application of municipal sewage sludge: risk assessment, in *Proc. Intl. Soil Sci. Soc.,* Acapulco, Mexico, July 1994, 534–553.

Schuman, G.E. and G.F. Vance, Eds., 1995, Decades later: A time for reassessment, in *Proc. 12th Annual Meeting of the Am. Soc. Surface Mining and Reclamation,* ASSMR, Princeton, WV.

Scora, R.W. and A.C. Chang, 1997, Essential oil quality and heavy metal concentrations of peppermint grown on a municipal sludge-amended soil, *J. Environ. Qual.,* 26, 975–979.

Sloan, J.J., R.H. Dowdy, and M.S. Dolan, 1998, Recovery of biosolids-applied heavy metals sixteen years after application, *J. Environ. Qual.,* 27, 1312–1317.

Vance, G.F. and G.M. Pierzynski, 1999, Bioavailability, fluxes and transfer of trace elements in soils and soil components special symposium: bioavailability and fate of trace elements in long-term, residual-amended soil studies, in *5th Intl. Conf. Biogeochem. of Trace Elements,* Vienna, Austria, 116–117.

Vance, G.F. and Schuman (eds.), 1996, Special Symposium – Selenium: Mining, Reclamation, and Environmental Impacts, Office of Surface Mining, Denver, CO.

Vance, G.F., Stevenson, F.J., and Sikora, F.J., 1996, Chapter 5 – Environmental Chemistry of Aluminum-Organic Complexes, In: Sposito, G. (ed.), *The Environmental Chemistry of Aluminum,* 2nd Edition, CRC/Lewis Publishers, New York, 169–220.

Vance, G.F., See, R.B., Reddy, K.J., 1998, Chapter 15 – Selenite Sorption by Coal-Mine Backfill Materials in the Presence of Organic Solutes, In: Frankenberger, W.T., Jr. and Endberg, R.A. (eds.), *Environmental Chemistry of Selenium,* Marcel Dekker, Inc., New York, 259–280.

Wixson, B.G., B.E. Davies, R.L. Bornschein, R.L. Chaney, W.R. Chappell, J.J. Chisolm, Jr., C.R. Cothern, B.T. Kagey, H.W. Mielke, A.L. Page, C.D. Strehlow, I. Thornton, R. Volpe, D.L. Vonberg, and P. Welbourn, 1994, *"Lead in Soil" Task Force Recommended Guidelines* B.G. Wixson and B.E. Davies, Eds., Society for Environmental Geochemistry and Health. Science Reviews Ltd., Northwood, U.K. Summary published in *Env. Sci. Tech.,* 28, 26A–31A.

Yang, L. and C. Henry, 1997, Study on the fixation of heavy metals in tire ash by biosolids, *Agro-Environmental Protection,* 16, 97–101.

Yang, X., V. Römheld, H. Marschner, and R.L. Chaney, 1994, Application of the chelator-buffered nutrient solution technique in studies on zinc nutrition in rice plant (*Oryza sativa* L.), *Plant Soil,* 163, 85–94.

Yang, L., D. Xue, C. Henry, and R. Harrison, 1997, A review: Biosolids applications and the effect on heavy metals, *Agro-Environmental Protection,* 16, 227–231.

2 Mobility and Bioavailability of Trace Elements in Soils

Eliska Podlesáková, Jan Nemecek, and Radim Vácha

Two main transfer pathways are important for the assessment of trace-element (TE) risk in relation to living organisms:

- The soil–plant (fodder and food crop) pathway, which introduces TEs into the food chain and/or phytotoxicity
- Direct (ingest, inhale) uptake by humans and animals

In addition to these ecotoxicological approaches, it is indispensable to consider geogenic and diffuse–anthropogenic background TE concentrations in soil. Knowledge of the background concentration is necessary to evaluate increased inputs of TEs into soils due to airborne emissions, flooding, or sewage sludge disposal. Limits or reference values of critical levels of soil loads have been expressed until now in terms of total TE contents.[1-4]

In the Czech Republic and at geological institutes in Germany, TE background values of the main soil–lithological units have been expressed as the total content in acid extracts (HNO_3 + $HClO_4$ + HF). The soil–lithological groups were classified according to their mineralogical and textural properties. The upper boundaries of the variability of TE values are regarded as soil contamination limits. Extreme geogenic loads are assessed separately.

Expressing soil loads by total content is justified only in the case of statistically derived, geogenically differentiated background values of TEs. Their spatial differentiation can be shown by means of digitized medium-scale maps. In contrast to background limits, the critical values of soil loads in relation to the prevention of food-chain or phytotoxicity exposure must be based on bioavailable forms of TEs in soils, and on the ecotoxicological relevance of plant loads through soils. This chapter focuses primarily on the problem of mobility and bioavailabity of TEs.

The development of TE mobility studies, especially to determine micronutrient status, started with trials that made use of extractions by diluted mineral acids.[5,6] It was later found that some of these extractable concentrations can be used along with total contents to identify the anthropogenic share of TE content.[7] Other methods specific for single TEs were proposed,[8,9] but the mobility studies led to methods that make use of standardized extractants:

- Unbuffered neutral salts for the determination of mobile species ($NaNO_3$, $CaCl_2$, NH_4NO_3)
- Complex-forming extractants (EDTA, DTPA) for the determination of both mobile and potentially mobilizable species

The combination of 1-M NH_4NO_3 or 0.1-M $CaCl_2$[10,11] and 0.025-M Na_2–EDTA is widely used in Germany.[12,13] TE bonding can be approximated by sequential analysis,[14] which is based on experimental studies.[15,16]

TE mobility cannot be systematically correlated with bioavailability; it can be understood only by comparing TE mobility with plant uptake. General rules for TE mobility, as well as the uptake of TEs by plants, can be formulated using both pot experiments and field investigations, the results of which are processed by multivariate statistical methods.[17–19] The rules must predict equations for both TE mobility and transfer into plants. The crucial problem in understanding ecotoxicologically relevant criteria for soil pollution is the evaluation of critical plant loads. It is not so difficult to assess the criteria for soil phytotoxicity,[20] but the more serious and most difficult problem is assessing critical crop loads from the viewpoint of zootoxicity and humanotoxicity. The necessary simplification of this problem uses critical fodder and food-crop pollution standards. Vollmer[21] stresses the fact that these standards are not based on ecotoxicological, but statistical, data. Critical loads are evaluated on the basis of exceeding the standards.

Simulated pollution of soils by TE salts has been used in the past to evaluate critical uptake by crops.[22–24] This method tests the impact of soil loads on soil organisms and their activities. This approach is similar to the common agrochemical investigations of fertilizer effects and allows an exact comparison with nonpolluted soil samples. It eliminates, to some extent, the geogenic load of the sample; however, the disadvantages of the procedure are that

- It introduces conditions of high TE mobilities that do not exist in natural soils
- High mobility decreases only very slowly by diffusion processes; equilibrium is attained only after decades[12]

This procedure, however, cannot be used to set soil pollution criteria. It can assess soil vulnerability, which can be defined as the buffering potential of soil, i.e., a soil's ability to decrease the bioavailability of TEs.[25]

Realistic critical values of soil loads must be assessed on the basis of TE mobilities and TE uptake by crops. The values must be related to soils polluted in the field from different sources (airborne emissions, fluvial loads, waste disposal on soils). In other words, soils must be evaluated based on the stabilized equilibrium of the diffusion processes. This chapter focuses on this evaluation.

STUDIES WITH SPIKED SOIL SAMPLES

Two pot experiments were performed with a simulated pollution of five principal soil units: Regosol (RG), Luvisol (LU), Cambisol (CA), Podzol (PC, cambic), and Chernozem (CH).

The first experiment was with graduated doses of Cd (2, 5, and 10 mg kg^{-1}), Zn (200, 300, and 400 mg kg^{-1}), Ni (50, 100, and 150 mg kg^{-1}), and Cu (50, 100, and 150 mg kg^{-1}); uptake of TEs by barley (*Hordeum vulgare L.*) was studied from Ap horizons of four representative soils adjusted by liming to pH 6.5 over two years.

The second experiment was with graduated doses of Cd (0.4, 1.0, and 3.0 mg kg^{-1}), Zn (150, 200, and 400 mg kg^{-1}), and Ni (60, 100, and 200 mg kg^{-1}). Over a period of four years, transfer to barley from five representative soils and mobilities (1-M NH$_4$NO$_3$, 0.025-M Na$_2$ EDTA) were studied. Sequential analysis of the soils was carried out at the end of the experiment.

The plant-available metals were determined in unbuffered soil extracts: 1-M NH$_4$NO$_3$ (MN), 0.01-M CaCl$_2$ (MC), and the potentially mobile fraction in 0.025 Na$_2$ EDTA (ED). Total metal concentration (TO) in soils was determined using concentrated HNO$_3$ + HF + HClO$_4$, and, in plants, HNO$_3$ and HClO$_4$. All analyses were conducted using atomic absorption spectroscopy (AAS) (Varian AA–300, flame AAS, electrothermal atomizer).

Experiments with metal-spiked soils showed that even after adjusting pH, which is one of the main factors affecting TE mobilities, the soil type had more effect on TE uptake. The results of the first experiment (Figure 2.1) show that the Regosol and albic Luvisol, with low contents of clay and humus, are characterized by higher plant uptakes in comparison with the Chernozem, with higher content of organic matter and clay, and with higher uptake than with the cambic Podzol, which is characterized by

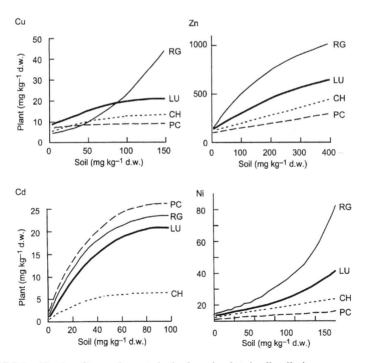

FIGURE 2.1   Uptake of trace elements by barley; simulated soil pollution.

high contents of free Fe and Al. These last two components in cambic Podzols affect the behavior of Cd when compared with the other elements. Phytotoxicity (Table 2.1) was caused by moderate doses of Ni, Cu, and Zn in the sandy Regosol, by medium doses of Zn in the glossalbic Luvisol, and by medium doses of Cd in the cambic Podzol. Transfer factors (also called factor quotients) express the ratio of the TE content in the plant to the TE content in the soil. They generally follow the sequence Cd > Zn > Ni, Cu (Table 2.2) and show the features already described. The transfer-factor values are comparable with data published by Styperek and Sauerbeck.[24]

The main aim of the second experiment was to determine the influence of simulated pollution on Cd, Zn, and Ni speciation, and to compare polluted and unpolluted samples four years after application of TEs into soils. Lower doses of TEs were used in comparison with the first experiment. They better reflect the contamination of soils used for agriculture. The values of effective mobility (Table 2.3), expressed by the relation of TEs determined in 1-M NH_4NO_3 (MN) to their total content (TO), shows (MN/TO) that, even four years after application, a higher value occurred than the original one. The sequence of decreasing capability of soils to immobilize TEs was as follows:

Regosol > glossalbic Luvisol > Cambisol > cambic Podzol > Chernozem

The ratio of the content of the potentially available TE species extracted in 0.025-M Na_2–EDTA (ED) to the total content (TO) is less affected by the peculiarities of soil than MN/TO. Uptake of Ni into barley (Table 2.4) declined in the same sequence. The cambic Podzol is characterized by the highest uptake of Cd and the lowest uptake of Zn. The results of the sequential analysis (Figure 2.2) after simulated pollution also revealed soil-specific features. Sequential analyses showed that the simulated pollution was still observed in mobile fractions, not only immediately after addition of TEs, but also after four years of diffusion of TE salts in soils. In addition to the immobilization sequence from Regosol to Chernozem mentioned above, we also saw increased incorporation of TEs into firmer bonds, especially into occluded or organically bound species in Chernozems, eutric Cambisols, and cambic Podzols. The addition of Ni results in an increase of the mobile and slightly mobilizable bonds in the Regosol, Luvisol, and Podzol, and in an increase in the amorphous sesquioxidic bonds in these soils and in the Chernozem as well.

The data show that the spiked metals cause a distinct increase in the mobile fraction of TEs. The mobility depends upon the soil properties, and it is kept for many years—tens of years, according to Brümmer and his colleagues.[12] The increased mobility is responsible for high plant uptake.

These findings confirm that this procedure (spiked soil) cannot be used to determine critical soil loads for any of the defined pathways (soil–plant, soil–humans). The results of the experiments also indicate that every soil has its own specific response to the addition of TEs in the form of soluble salts, which are incorporated depending upon the specific bonds they make with the soils. These responses are evident even after adjusting the pH, which affects the mobility of most elements in soils. These soil-specific responses determine soil vulnerability. Soil quality depends upon the buffering ability of soils, which immobilizes TEs. Based on our findings, we were able to fix for each element a more (e.g., for Cd, Ni, Zn) or less (e.g., for Cr, As)

TABLE 2.1
Variants with Statistically (test of minimal difference)
Significant Yield (barley) Depressions

Soil	Element	1st year		2nd year	
RG	Cd				
	Ni	100**	150**	100**	150**
	Cu	100**	150**	100**	150**
	Zn		400**	300**	400**
LU	Cd				
	Ni				
	Cu				
	Zn	300*	400*		
CH	Cd				
	Ni				
	Cu				
	Zn				
PC	Cd	5**	10**	5*	10*
	Ni				
	Cu				
	Zn				

References:

* *significant at 90% level*

**significant at 95% level*

5, 10, 100, 150, 300, 400   doses of trace elements in mg.kg^{-1}

TABLE 2.2
Transfer Factor (ratio of trace element contents in plant and soil) of Trace Elements for Representative Soils of the Czech Republic

Element	Arenic Regosol RG	Glossalbic Luvisol LU	Calcic Chernozem CH	Cambic Podzol PC
Cd	2.30[a] (1.1–2.9)[b]	2.26 (1.0–2.6)	1.09 (0.9–1.4)	3.09 (1.8–5.0)
Zn	3.57 (1.7–6.2)	1.76 (0.9–3.1)	1.74 (0.5–1.0)	0.60 (0.3–0.9)
Ni	0.75 (0.1–1.2)	0.22 (0.05–0.6)	0.07 (0.03–0.1)	0.05 (0.02–0.08)
Cu	0.21 (0.16–0.5)	0.18 (0.16–0.4)	0.14 (0.1–0.3)	0.10 (0.08–0.3)

[a]Mean value of all variants

[b]Numbers in parentheses give range of variability (Minimum value: sample without pollution; Maximum value: sample with highest dose of trace elements)

TABLE 2.3
Relative Mobility (in %) of Cd, Zn, and Ni after Simulated Loads of Representative Soils (top soil)

Element	Year	Ratio	Dose (mg kg⁻¹)	RG	LU	Soil CA	PC	CH
Cd	1	ED/TO	0	–	–	–	–	–
			3	93	99	90	76	80
		MN/TO	0	2.6	7.5	3.2	2.0	1.2
			3	40	11	9	4	1
	4	ED/TO	0	–	–	–	–	–
			3	68	60	56	57	49
		MN/TO	0	16.7	15.2	15.7	2.5	3.1
			400	57	20	23	7	0.3
Zn	1	ED/TO	0	24	17	6	12	3
			400	88	99	56	44	48
		MN/TO	0	16	1.5	0.6	0.2	0.2
			400	46	14	13	3	0.2
	4	ED/TO	0	38	29	7	6	4
			400	91	77	44	35	52
		MN/TO	0	12	2.4	0.9	0.2	0.1
			400	33	15	11	6	0.2
Ni	1	ED/TO	0	6.2	6.6	2.5	1.4	3.7
			200	73	75	65	51	35
		MN/TO	0	3.8	0.8	0.5	0.6	0.6
			200	48	18	16	6	1
	4	ED/TO	0	14	13	12	10	15
			200	86	76	48	35	41
		MN/TO	0	2.0	2.1	0.9	0.2	0.4
			200	41	23	10	4	0.7

TO   Total content (HNO_3 + HF + $HClO_4$)

MN   Extractable TEs in 1-M NH_4NO_3

ED   Extractable TEs in 0.025-M Na_2–EDTA

TABLE 2.4
Uptake of Mobile Trace Elements from Soils at pH 6.5 by Barley (mg kg⁻¹)

Soil	Cd	Zn (mg kg⁻¹)	Ni
RG	10.9	495	90
LU	6.1	400	70
CA	5.9	217	42
PC	12.4	195	11
CH	5.0	210	11

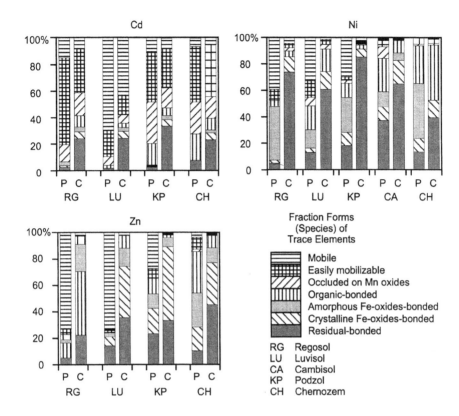

FIGURE 2.2 Sequential analysis of soil samples with simulated pollution after four years (P—Polluted, C—Comparative variant).

detailed scale of responses, and to classify soil units into vulnerability categories. On the basis of this knowledge and digitized soil maps (1:500,000), we displayed the relative vulnerability of soils in the Czech Republic in cartographic form.

STUDIES OF TRACE-ELEMENT AVAILABILITY UNDER FIELD CONDITIONS

Twenty-five samples from great soil groups (arenic Regosols–RG; orthic Luvisols–HM; glossalbic Luvisols–LU; eutrophic (e), modal (m), and dystric (d) Cambisols–CA; calcic Chernozems–CH; and Fluvisols–FL), sampled from typical as well as geogenically extreme parent materials, were used for metal speciation studies using sequential analysis.[14]

For soil mobility and transfer studies, 165 soil samples from Ap horizons of the above-mentioned soil units of the Czech Republic were used, with a wide range of soil properties (pH 3.8–7.4, clay <1 μm 4–30%, oxidizable carbon 0.6–4.2%, free Fe_0 0.2–1.6%); different levels of airborne and fluvial anthropogenic contamination; and high geogenic loads (parent materials from mafic rocks and metallogenic zones of acid rocks).

Pot experiments with four replications were carried out using 54 selected soil samples that reflected the kinds of soil loads, i.e., anthropogenic and geogenic. One half of the samples were adjusted to pH 6.5 by liming (as needed). The test crops were radish (*Raphanus sativus L.*), triticale (*Triticum x Secale*), and spinach (*Spinacea oleracea L.*). Results of transfer studies obtained with radish and triticale are preferred because of their pH tolerance.

There were 111 samples of soils and fodder plants collected under field conditions that were analyzed for total contents of TEs and soluble forms (MC, MN, ED). The results were processed using elementary statistical procedures (geometric means, standard deviation, correlation analysis). Factor analysis was used after logarithmic (ln) transformation of the values. Principal component analysis was applied for the extraction of factors; it takes into account only eigenvalues higher than 1. Orthogonal Varimax rotation was performed. Every table for factor analysis also had a table of communalities that was not published. Multiple regression analysis was used to derive (from the total, and potentially mobilizable TE pool, and pH, as independent variables) relationships with the dependent variables:

- Available species of TEs
- Content of TEs in test plants

Attention was given to the problem of availability of TEs in soils. Our long-term goal was to determine the critical TE loads in plants from the viewpoint of food chain and phytotoxicity threat using principles of soil-available TEs. For this reason, we dealt with soil samples polluted in the field that reflect loads used for agriculture.

Sequential analysis of the samples was carried out using the main soil units of the Czech Republic as well as soils from special parent materials and soil affected by different kinds of anthropogenic pollution. The analysis provided information not only about specific bonds of single trace elements, but also data concerning specific bonds in the soil and parent materials (Figure 2.3).

A high proportion of residual material (the unweatherable portion of the TEs) was evident, especially material containing Cr, Ni, and Co. They were present mainly in soils derived from parent materials of mafic rocks. Mn, Zn, and Cu were also concentrated in residual material. Arsenic was concentrated in Fe–oxide bonds (which, in these cases, were also crystalline), especially those in dystric Cambisols and cambic Podzols from geogenically extreme acid parent materials. Cr, Co, Be, and Zn also had Fe–oxide bonds. Pb and Cu were characterized in general by organic bonds. These bonds predominated, especially in Chernozems and Fluvisols. Increased participation of organic bonds also characterized Cd, Mn, Zn, and Ni in Chernozems; Cd, Ni, and Co in Fluvisols; and Co and Pb in eutrophic Cambisols. The most ecologically important mobile metals are Cd, Mn, Co, Zn, and Ni. For all these trace elements the first fraction extracted depends upon pH. Low content of mobile Cu and As occurred mostly in neutral soils.

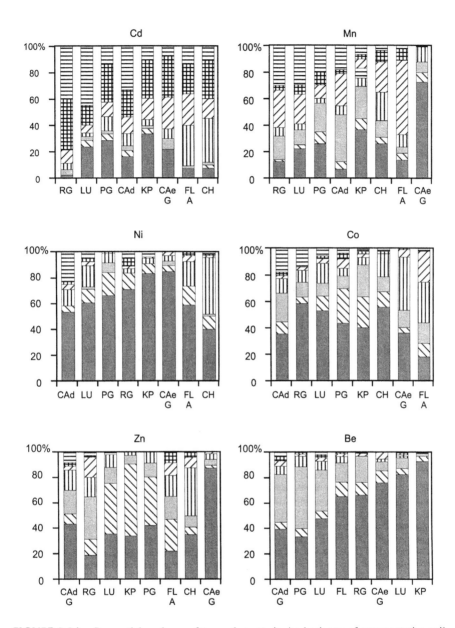

FIGURE 2.3A Sequential analyses of trace elements in Ap horizons of representative soil units.

Statistical analysis (Table 2.5) indicates that the concentrations of metals in these soils generally exceeded not only background concentrations, but also values for food-chain protection. The criteria given in Table 2.5 do not take into account differences in extreme loads from anthropogenic and geogenic sources. When we divided the set into subsets lacking extremes, and into subsets containing geogenic anomalies

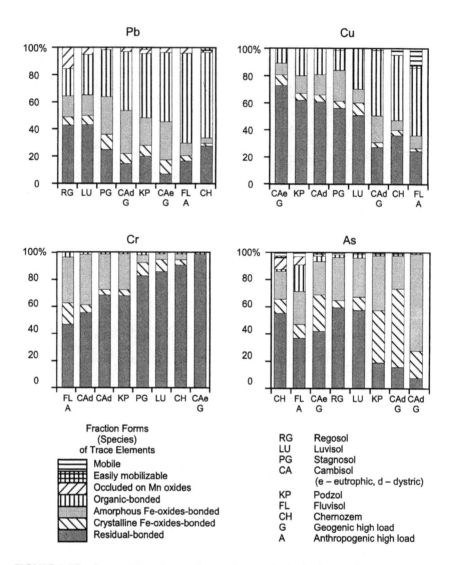

FIGURE 2.3B Sequential analyses of trace elements in Ap horizons of representative soil units.

and a subset of anthropogenic pollution, we concluded that values of potential mobility (ED/TO) differentiated the highest solubility (from anthropogenic loads), even in slightly acid soils such as Fluvisols. Values of effective mobility (MN/TO) did not provide information about differences among the subsets.

Factor analysis of the entire dataset (Table 2.6) showed very clearly that the trace elements can be subdivided into groups according to their different behaviors. The first group involves Mn, Cd, Co, and Zn. In this group of TEs, the first factor is pH-dependent effective mobility (negative relationship), and the second (in the case of Mn, the third)

TABLE 2.5
Elementary Statistics of Trace Elements in Investigated Soils (mg kg^{-1}, ED/TO in %), n = 162

		Mn	Cd	Co	Zn	Ni	Be	Pb	Cu	Cr	V	As
Whole set	GM	752	0.60	14.3	131	26.3	2.26	54	32	73.0	86	30
	St. dev.	540	1.32	22.9	131	244	1.42	228	59	205.0	52	333
TO	Max.	4250	11.10	192.0	736	2955	9.6	2748	433	1846.0	391	3025
	ED/TO × 100	28	41.00	20.0	17	13	10.0	45	22	4.0	5	5
Background reference values[a]	GM	1400	0.50	35.0	160	70	4.50	90	70	155.0	180	30
Maximum permissible value[a]		—	1.00	50.0	200	80	7.00	140	100	200.0	220	30
Intervention values[b]		—	20.00	300.0	2500	250	20.00	300	600	500.0	450	70
Subset lacking	GM	696	0.54	11.7	111	19.5	2.25	50	27	55.0	80	25
all extremes	Max.	1725	3.1	38.0	400	119	1.48	306	244	217.0	290	318
TO	ED/TO × 100	25	40.0	18.0	14	11	9	40	20	4.0	4	4
Subset of	GM	1309	—	34.5	145	107	—	70	56	270.0	133	282
geogenic	Max.	4250	—	192.0	299	2955	—	2748	384	1846.0	391	3025
extremes TO	ED/TO × 100	28	—	21.0	14	11	—	41	16	3.0	5	3
Subset of	GM	719	1.27	13.9	240	41	3.06	74	57	112.0	87	22
significant	Max.	2057	11.10	27.3	736	325	4.12	1898	433	737.0	162	359
anthropogenic loads TO	ED/TO × 100	53	62	37	41	35	11	69	46	14	8	14

[a]Used in Czech Republic

[b]Residential areas

GM Geometric mean

St. dev.   Standard deviation

TO   Total content (mg.kg^{-1})

ED   Extractable TEs in 0.025-M Na$_2$EDTA

Numbers in boldface exceed reference values

TABLE 2.6
Factor Analysis of the Whole Standardized Set of Data concerning the Mobility of Trace Elements (n=288)

variables	Loading of variables for the rotated factors: F1, F2, F3										
	Mn	Cd	Co	Zn	Ni	Be	Pb	Cu	As	V	Cr
pH	**−0.920**	**−0.844**	**−0.845**	**−0.853**	−0.746	**−0.853**	**−0.797**	0.559	0.612	0.647	**0.438**
					0.555						0.471
MC	**0.902**	**0.851**	**0.856**	**0.831**	0.738	0.607	0.664	**0.815**	**0.827**	0.821	**0.514**
MN	**0.958**	**0.823**	**0.898**	**0.903**	**0.893**	**0.848**	0.803	**0.872**	**0.821**	**0.484**	**0.632**
ED	*0.887*	*0.926*	*0.909*	*0.866*	**0.901**	*0.755*	**0.885**	**0.906**	**0.788**	*0.891*	**0.852**
TO	*0.853*	*0.907*	*0.913*	*0.787*	**0.847**	*0.708*	**0.844**	**0.787**	**0.752**	**0.459**	**0.717**
										0.579	
<1μm	*0.925*	*0.928*	*0.897*	*0.946*	0.900	**0.942**	0.913	0.920	0.862	**0.914**	0.933
<10μm	0.911	0.921	0.877	0.920	0.909	**0.921**	0.904	0.883	0.841	**0.909**	0.890
Cox	−0.441	-	−0.517	0.599	−0.499	0.601	**0.484**	**0.462**	–	*0.807*	−0.416
	0.449										
Weight of factor											
F1	**33.1**	**27.4**	**29.4**	**29.1**	**28.2**	**25.6**	**29.6**	**40.6**	**34.9**	**28.5**	**27.6**
F2	24.7	25.1	24.0	24.1	25.1	24.0	23.6	25.8	25.3	24.6	26.9
F3	*22.1*	*24.1*	*24.0*	*23.3*	*24.9*	*19.2*	*17.9*			*15.1*	

the total and potentially mobilizable pool of TEs. The weight of these two factors accounted for 52–58% of the total variance. The third factor (or second in the case of Mn) reflected primarily the content of fine particles and partly that of humus (lowest communality). For Ni and Be, the role of pH-dependent effective mobility diminishes slightly and shifts to the second (Be) or third (Ni) factor. In the case of Ni, the role of the TE pool increased, and in the case of Be it decreased. Pb has some similarity to the mentioned groups, with the increased role of the TE as for Ni. As far as the other elements are concerned, both mobilities and pools are associated in the first factor without any relation to pH.

We also investigated the influence of liming (Figure 2.4) on TE mobility. Liming diminished by 60 to 80% the effective mobility of TEs that were highly pH dependent, such as Mn, Zn, Cd, Ni, Co, and Be. The potential mobility of these TEs or the mobility of the less soluble elements such as Pb were affected to a lesser degree (6–20%). For the other elements, the effect of liming on TE mobility was negligible. In the cases of As and Cu, liming caused an increase in mobility.

To predict TE mobilities, we used multiple regression analyses (Table 2.7). Equations were characterized by significant values of all coefficients (i.e., $p < 0.05$). We found prediction equations for all elements except Cr. We prepared examples of prediction equations for elements with high mobility. We selected equations that included both the total or the potentially mobilizable pool of elements and pH.

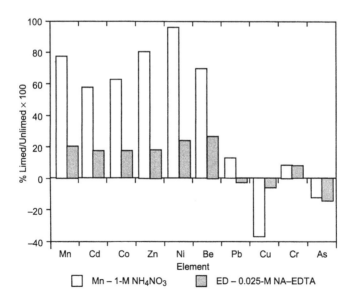

FIGURE 2.4 Influence of liming on trace element mobility.

TABLE 2.7
Prediction Equations of Effective Mobilities in Soils (multiple regression analysis; sig T of all coefficients <0.05, n = 290)

Element	Equation
Mn	$\ln MN_{Mn} = 0.477 \ln ED_{Mn} - 1.413\ pH + 7.366$
Cd	$\ln MN_{Cd} = 0.634 \ln TO_{Cd} - 0.818\ pH + 1.555$
Co	$\ln MN_{Co} = 0.763 \ln TO_{Co} - 0.937\ pH - 0.178$
Zn	$\ln MN_{Zn} = 1.197 \ln TO_{Zn} - 1.354\ pH + 0.782$
Ni	$\ln MN_{Ni} = 1.457 \ln TO_{Ni} - 1.071\ pH - 1.178$

MN Extractable TEs in 1-M NH_4NO_3

ED Extractable TEs in 0.025-M Na–EDTA

TO Total content (HNO_3 + HF + $HClO_4$)

TRANSFER OF TRACE ELEMENTS INTO PLANTS

Trace-element mobility cannot be compared automatically with bioavailability. TE mobility must include uptake by plants. Common features of both TE mobilities and their transfer to plants can be found from pot experiments, with well-defined and uniform conditions, and from field studies where many factors interfere. Under both conditions, we studied critical TE loads of plants from the viewpoint of food-chain and phytotoxicity threat. For this chapter we present results from pot experiments. Under field conditions, the behavior of TEs and their transfer to plants was confounded due to variations in hydrothermic factors, cultivation techniques, and airborne emissions.

In the pot experiments, standards for zootoxicity and phytotoxicity of all elements were exceeded in most cases for radish and triticale (Table 2.8). For evaluation, we used data presented by Vollmer.[21] (Data are missing for Be.) Table 2.8 also shows that radish and triticale took up significantly different amounts of most mobile trace elements compared with spinach. The results were

Cd > Zn > Co, Mn (Ni).

For the other elements, the difference was not significant. This fact is important in assessing hyperaccumulators. Transfer factors (the ratio of plant TE content to soil TE content) indicated the following sequence:

Cd > Zn > Mn > Cu > Co > Pb, Cr > As.

Transfer factors determined with field soils are much lower than transfer factors determined from soils with simulated pollution. This result is due to differences in

TABLE 2.8

Elementary Statistics of Trace-Element Contents in Crops and Their Relation to TE Contents in Soils. Pot Experiment, Unlimed Variants (n = 108)

		Mn	Cd	Co	Zn	Ni	Be	Pb	Cu	As	Cr	V
Radish (mg kg⁻¹ d.w.)	GM	30	0.86	0.27	41.2	0.88	0.023	0.397	3.01	0.149	0.612	1.41
	St. dev.	84	1.08	1.39	40.0	4.72	0.050	3.69	4.61	2.07	1.60	2.01
	Max.	**400**	**6.36**	**12.4**	**299**	**42.3**	0.33	**36.7**	**41.7**	**13.54**	**12.90**	**18.10**
Triticale (mg kg⁻¹ d.w.)	GM	169	0.78	0.15	36.2	0.67	0.006	0.570	8.35	0.322	0.735	1.06
	St. dev.	340	0.83	3.28	16.1	1.25	0.006	0.436	2.73	1.48	0.297	0.56
	Max.	**1913**	**3.15**	**21.0**	115	6.05	0.05	2.13	17.4	**8.76**	1.83	2.72
Spinach (mg.kg⁻¹ d.w.)	GM	207	6.03	0.61	182	1.03	0.011	0.312	7.79	0.092	0.948	1.84
	Max.	**7356**	**48.4**	**66.3**	**880**	**26.1**	0.14	**21.4**	**34.2**	**79.0**	**10.42**	**8.05**
Reference values	Zootoxicity	250	1.1	6	500	—	—	10	30	4	10	—
	Phytotoxicity	—	—	10	250	10	—	10	30	—	—	—
Transfer factors[a]	Radish	0.037	1.30	0.017	0.271	0.030	0.008	0.008	0.088	0.004	0.007	0.016
	Triticale	0.208	1.17	0.009	0.236	0.021		0.011	0.241	0.010	0.008	0.012
	Spinach	0.256	11.0	0.035	1.245	0.029		0.007	0.239	0.004	0.009	0.021

[a] Ratio of TE contents in plant and soil

Boldface values exceed reference values

TE mobility. Even a simple correlation confirms the dependence of plant loads on pH
and the content of mobile species. The results were

Mn > Cd, Co, Zn and, to a lesser degree, Ni, Be, and Pb.

The influence of liming on plant loads (Figure 2.5) was most evident with Mn, Co,
Ni, Cd, Zn, Pb, and Be. Factor analysis of the entire standardized set of variables
(Table 2.9) shows that the first factor for Mn, Cd, Co, Zn, and Ni (and partly Be) is
again, in general, negatively pH dependent. The second factor is the mobilizable and
total pool of TEs, except for Mn (Be), where it is the third factor and the second is
clay. Communalities point to the lower role of humus. As far as the other elements are
concerned, the significant interrelations among TE mobilities and total content are
reflected, in general, to a much lesser degree, or there may even be no relation to plant
uptake.

On the basis of the pot experiments we have proposed prediction equations. They
are derived from TE contents, mobilities, and pH by means of multiple regression
analysis. In Table 2.10 we present examples of these regression equations for triticale.
The substitution of critical plant loads into the equation should result in critical soil
loads for the food chain or phytotoxicity. A more sophisticated statistical procedure
was also used, taking into account different levels of probability. However, the crucial problem remains the ecotoxicological relevance of critical plant loads, especially
from the viewpoint of zootoxicity and humanotoxicity. The present simplification of
this problem uses standards for fodder and food crops, which are based on statistical
data presented by Vollmer.[21]

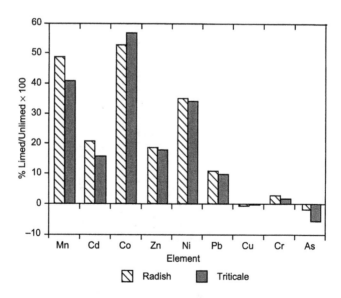

FIGURE 2.5   Influence of liming on trace element uptake by plants.

TABLE 2.9
Factor Analysis of the Standardized Set of Data Comprising TEs Contents, Mobilities in Soils, Transfers to Triticale (pot experiment), and Main Variables that Affect Them (n=97)

Loading of variables for the rotated factors: F1, F2, F3, F4

variables	Mn	Cd	Co	Zn	Ni	Be	Pb	Cu	As	Cr
pH	**-0.917**	**-0.832**	**-0.847**	**-0.851**	**-0.462** 0.781	**-0.892**	0.465 -0.696	**0.432** -0.705	0.690 0.526	**0.514** 0.642
triticale	**0.914**	**0.768**	**0.900**	**0.688** 0.478	0.712	**0.374** 0.543	0.839	0.396 0.531	**0.847**	0.958
MC	**0.852**	**0.900**	**0.885**	**0.878**	**0.838**	**0.715** 0.463	**0.624**	**0.868**	**0.707**	**0.727**
MN	**0.943**	**0.835**	**0.890**	**0.886**	**0.902**	**0.846**	0.754	**0.869**	**0.793**	**0.788**
ED	0.884	0.910	0.925	0.898	0.871	0.872	**0.849**	**0.956**	**0.863**	**0.849**
TO	0.834	0.911	0.920	0.839	0.776	0.587	**0.790**	**0.772**	**0.854**	**0.716**
<1_m	0.931	**-0.832**	**-0.505** 0.704	**-0.803**	0.868	-0.877	0.905	-0.915	0.912	0.936
<10_m	0.854	**-0.678**	0.782	**-0.662**	0.809	-0.828	0.892	-0.844	0.866	0.877
Cox	0.594	0.663	-0.742	0.669	0.439 -0.626	0.624	**0.784**	**0.610**	0.794	**0.435**
weight of factor loads										
F1	**39.2**	**44.0**	**39.1**	**43.7**	**28.4**	**29.1**	**27.4**	**41.8**	**37.8**	**31.8**
F2	22.1	27.0	22.5	27.2	26.8	21.8	22.6	27.0	25.6	25.4
F3	20.8		19.7		22.0	20.7	21.6		13.0	11.9

TABLE 2.10
**Prediction Equations of Trace Element Uptake
to Triticale (multiple regression analysis; sig T
of all coefficients <0.05, n = 150)**

Element	Equation
Mn	$\ln T = 0.342 \ln MN_{Mn} - 0.481 \ln TO_{Mn} + 1.400$
Cd	$\ln T = 0.297 \ln MN_{Cd} + 0.223 \ln TO_{Cd} + 0.966$
Co	$\ln T = 0.463 \ln MN_{Co} - 1.235 \ln TO_{Co} + 2.421$
Zn	$\ln T = 0.117 \ln MN_{Zn} + 0.259 \ln TO_{Zn} + 2.346$
Ni	$\ln T = 0.256 \ln MN_{Ni} + 0.155 \ln ED_{Ni} + 0.090$

MN Extractable TEs in 1-M NH_4NO_3

ED Extractable TEs in 0.025-M Na–EDTA

TO Total content ($HNO_3 + HF + HClO_4$)

The critical concentration of Cd, which is approximately 40 μg kg^{-1} of species soluble in 1-M NH_4NO_3, is specified in relation to total content and pH. Cadmium is the most harmful trace element because it has the highest plant uptake and mobility. Strict plant standards are in place because of zootoxicity and humanotoxicity. The other mobile elements are much less harmful. Some of them have high mobility (Zn, Mn, Co) but diminishing uptake rates (Zn > Mn > Co). Some have high values for critical plant loads (Zn, Mn, Cu), and some are more phytotoxic than zootoxic. For some TEs (e.g., Be, V), reference values for plant contaminations are scarce or missing. The other elements are characterized by low solubility, especially for the high geogenic contents of Cu, Pb, As, and Cr (Ni). The discrepancies mentioned are responsible for the behavior of less mobile elements. They are characterized by low solubilities, especially in the case of the high geogenic contents of Cu, Pb, As, and Cr (as well as Ni). Some have high critical plant loads (Pb, Cu).

For elements that are characterized by the lowest mobilities (Pb, Cu, As, and especially Cr), our conclusion is that only high contents (geogenic loads of Pb, As, and Cr > 400 and Cu > 150 mg kg^{-1}) or an extraordinary anthropogenic mobility may be critical. This statement is demonstrated in Table 2.11, which shows soils with different, mostly very high (even extreme) contents of trace elements. The TE contents presented rarely occur in agricultural soils. Very high mobility (ED/TO) can be found in severely polluted soils, especially in some Fluvisols. Critical plant loads of Pb, Cu, Cr, and As are surpassed only in soils characterized by a high potential solubility in anthropogenically (ED/TO × 100) polluted soils, in soils with extreme geogenic concentrations, and, in the case of Ni, in very acid soils.

CONCLUSION

The study of simulated pollution of soils with trace elements from soluble salts proved to be suitable for assessing soil vulnerability to pollution. The responses of

TABLE 2.11

Extreme Contents of Trace Elements and Their Potential Mobility in Relation to Uptake to Plants (elements with low effective mobilities) (mg kg^{-1} dry weight)

Elem.	Kinds of Extremes	TO	ED	MN	ED/TO ×100	MN/TO ×100	pH	Radish	Triticale	Reference Value — Value	Reference Value — Exceeding
Ni	Gb	2900	133	1.80	4.6	0.062	6.2	**12.3**	**6.9**		+
	Gb	920	31	0.16	3.4	0.017	6.8	4.3	1.2	5–10	–
	Ac	49	15	3.95	*31*	8.060	4.8	**8.82**	**12.1**		+
	F	72	31	0.44	43	0.610	6.6	1.15	1.0		–
	F	325	110	3.45	34	1.060	6.3	3.96	4.78		–
	Ac	87	28	0.60	32	0.690	3.9	1.50	1.05		–
	GAc	*118*	54	0.10	46	0.080	4.6	0.51	0.67		–
Pb	G	203	114	0.10	56	0.050	6.1	1.00	0.75		–
	An	*1226*	844	5.10	69	0.420	5.0	**39.0**	**12.0**	10	+
	F	*2340*	1909	1.85	82	0.080	5.5	**33.8**	27.7		+
	F	*175*	158	0.10	90	0.060	6.4	0.23	0.35		–
Cu	Gb	83	9	0.04	11	0.050	6.8	2.0	7.5		–
	Ga	108	20	0.29	18	0.270	4.5	6.9	4.9		–
	Ga	110	55	0.64	50	0.580	6.0	4.0	9.4	30	–
	An	115	86	0.14	75	0.120	5.5	**36.2**	28.1		+
	F	408	370	1.52	91	0.370	6.4	26.1	34.5		+
Cr	Gb	1846	0.87	0.018	0.047	0.0010	6.2	2.9	1.8		–
	Gb	670	0.40	0.010	0.059	0.0015	6.9	4.8	1.2	10	–
	GAc	207	0.31	0.012	0.149	0.0058	4.5	0.9	1.0		–
	F	539	6.50	0.03	1.200	0.0056	6.7	0.7	0.6		–
	F	372	18.20	0.01	4.890	0.0027	6.3	0.2	0.6		–
As	Ga	2789	13.10	1.02	0.470	0.0360	4.8	**13.0**	**8.0**		+
	Ga	1200	4.00	0.02	0.330	0.0017	4.8	1.7	0.3		–
	Ga	319	9.70	0.17	3.040	0.053	3.9	3.0	2.2	4–5	–
	An	68	41	0.12	*60*	0.176	5.0	3.8	5.3		+
	F	448	348	0.59	78	0.132	5.5	**5.8**	**4.8**		+

Ac Very acid soils   An Anthropogenic loads   F Fluvial anthropogenic loads
G   Geogenic loads   Gb Geogenic loads of mafic rocks   Ga Geogenic loads of acid rocks

Italic—Extreme values in soils

Boldface—Values in plants that exceed reference values

mobilities and transfers reflect soil- as well as element- specific immobilization capability. The results can be generalized for soil units and displayed in the form of vulnerability maps. Mobile species that are retained for a long time affect transfers into plants; this approach cannot be used for critical-load assessment with them.

Investigations of soils that were sampled in the field showed the possibility of distinguishing the kinds of loads by means of the ratio of potentially mobilizable content and total content. Factor analysis revealed the TEs that are characterized by negatively pH-dependent mobility. This group comprises Mn, Cd, Co, Zn, Ni, and Be (Pb). Especially for these TEs, we can derive prediction equations of mobility by means of multiple regression analysis.

The uptake of TEs by test plants showed that, for Mn, Cd, Co, Zn, and Ni (Be, Pb), it correlates with pH-dependent mobilities. The individual transfer factors, however, which are much lower than those found in trials with simulated pollution, follow the sequence

$$Cd, Zn > Mn, Cu > Co, Ni.$$

The derived prediction equations of transfer, after substitution for critical plant loads, reveal relevance for only the most mobile elements with high transfer factors, especially for Cd. For the immobile elements (As, Pb, Cu, and Cr), high total contents can occur with no harmful effects. For some of the more mobile elements, the transfers do not lead to harmful effects because of differences in uptake rates, critical plant loads, and minimal phytotoxicity.

REFERENCES

1. LABO, *Soil Background and Reference Values in Germany,* Bayerisches Staatsministerium für Landesentwicklung und Umweltfragen, Germany, 1995.
2. Suttner, Th., Aussendorf, M., and Martin, W., *Hintergrundwerte anorganischer Problemstoffe in Böden Bayerns,* GLA Fachberichte 16, Munich, Germany, 1998 (in German).
3. Danneberg, O.H, Hintergrundwerte von Spurenelementen in den landwirtschaftlich genutzten Böden Ostösterreichs, *Mitteil. d. Österr. Bodenkundl. Ges.,* 57, 7, 1999 (in German).
4. Utermann, J., Düwel, O., Fuchs, M., Gäbler, H.E., Gehrt, E., Hindel, R., and Schneider, J., *Methodische Anforderungen an die Flächenrepräsentanz von Hintergrundwerten in Oberböden,* BGR, Hannover, Germany, 1999 (in German).
5. Pejve, J.V., *Rukovodstvo po promeneniju mikroudobrenij,* Kolos, Moscow, 1963, 120 (in Russian).
6. Sillanpää, M., Micronutrients and the nutrient status of soils global study, *FAO Soils Bulletin,* 48, 1982.
7. Von Staiger, K., Machelett, B., and Podlesak, W., Vergleich von Aufschlu β- und Extraktionsverfahren zur Ermittlung des Gehaltes von Kupfer, Zink, Cadmium, Blei und Nickel in Böden unterschiedlicher anthropogener Belastung, *Bodenkultur,* 36/II, 99, 1985 (in German).
8. Adriano, D.C., *Trace Elements in the Terrestrial Environment,* Springer Verlag, New York, 1986, 533.

9. Kabata-Pendias, A. and Pendias, H., *Trace Elements in Soils and Plants,* CRC Press, Boca Raton, 1992, 365.

10. Thiele, S. and Brümmer, G.W., Bestimmung der mobilen Fraktionen ausgewählter Elemente (Cd, Ni, Co, Cr, As) in Oberböden durch $CaCl_2$ und NH_4NO_3 Extraktion, *Mitteil. d. Deutschen Bodenkundlichen Gesellschaft,* 72, 1313, 1993, (in German).

11. Hornburg, V., Welp, G., and Brümmer, G.W., Verhalten von Schwermetallen in Böden. 2. Extraktion mobiler Schwermetalle mittels $CaCl_2$ und NH_4NO_3, *Z. Pflanzenernähr. Bodenk.,* 158, 137, 1995, (in German).

12. Brümmer, G., Gerth, J., and Herms, V., Heavy metal species, mobility and availability in soils, *Z. Pflanzenernährung und Bodenkunde,* 149, 382, 1986.

13. Prüess, A., *Vorsorgewerte und Prüfwerte für Mobile und Mobilisierbare, Potentiell Ökotoxische Spurenelemente in Böden,* Verlag Ulrich E. Graner, Wendlingen, 1992, 145 (in German).

14. Zeilen, H. and Brümmer, G.W., Ermittlung der Mobilität und bindungsformen von Schwermetallen in Böden mittels sequentieller Extraktion, *Mitteil. d. Deutschen Bodenkundlichen Gesellschaft,* 66, 1991, 439 (in German).

15. Herms, U. and Brümmer, G.W., Einflussgrössen der Schwermetallöslichkeit und -bindung in Böden, *Zeitschrift für Pflanzenernährung und Bodenkunde,* 147, 1984, 400 (in German).

16. Hornburg, V. and Brümmer, G., Untersuchungen zur Mobilität und Verfügbarkeit von Schwermetallen in Böden, *Mitteil. d. Deutschen Bodenkundlichen Gesellschaft,* 59/II, 727, 1989 (in German).

17. Hornburg, V. and Brümmer, G.W., Einfluβ gröβen der Schwermetall-Mobilität und–Verfügbarkeit, in *Böden, in Arbeitstagung Univ. Jena, Leipzig,* B. 2, 415, 1990 (in German).

18. Hornburg, V. and Brümmer, G.W., Schwermetall-Verfügbarkeit in Böden und Gehalte in Weizenkorn und in anderen Pflanzen, in *Kongreβband,* Berlin, 1990. VDLUFA– Schriftenreihe 32, 1990, 821 (in German).

19. Hornburg, V. and Brümmer, G.W., Schwermetall-Verfügbarkeit und Transfer in Abhängigkeit von pH und Stoffbestand der Böden, *Mitteil. d. Deutschen Bodenkundlichen Gesellschaft,* 66/II, 1991, 661 (in German).

20. Magnicol, R.D. and Beckett, P.H.T., Critical tissue concentration of potentially toxic elements, *Plant and Soil,* 85, 107, 1985.

21. Vollmer, M.K., *Herleitung und Anwendung von Prüf- und Sanierungswerten für schwermetallbelastete Böden in der Schweiz,* Eidgenössische Forschungsanstalt für Agrikulturchemie und Umwelthygienen (FAC), Switzerland, 1995, 102 (in German).

22. Sauerbeck, D., Welche Schwermetallgehalte in Pflanzen dürfen nicht überschritten werden, um Wachstumsbeeinträchtigungen zu vermeiden, *Landwirtschaftliche Forsch.,* 39, 108, 1983 (in German).

23. Sauerbeck, D., Der Transfer von Schwermetallen in die Pflanze, in *Vorträge der DECHEMA-Arbeitsgruppe Bewertung von Gefähderungspotentialen in Bodenschutz,* 1989, Frankfurt M., 281, 1989 (in German).

24. Styperek, P. and Sauerbeck, D., Eignung von chemischen Extraktionsverfahren zur Abschätzung des pflanzenverfügbaren Cd und Zn in verschiedenen Böden und Substraten, *Landwirtschaftliche Forsch.,* 41, 471, 1985, (in German).

25. Podlesáková, E., Nemecek, J., and Vácha, R., Approaches to the assessment of soil vulnerability against contaminants and soil pollution, in *Proc. Int. Conf. Soil Conservation in Large-Scale Use,* 241, 1999.

3 Sequential Extraction of Metals from Artificially Contaminated Soils in the Presence of Various Composts

Luis Madrid, Encarnación Díaz-Barrientos, and Isabel Cardo

INTRODUCTION

Heavy metals in soils are in forms that are strongly determined by their origin and history. Although native metals are frequently in highly immobile forms,[1] anthropogenic forms are often more reactive and thus are more available to plants. In the latter case, however, a number of reactions with soil components[2] contribute to a progressive insolubilization of metals entering the soils. Tagami and Uchida[3] concluded that, although exchangeable and adsorbed fractions of several metals were observed after being freshly introduced in soil samples, the exchangeable fractions decreased to almost zero within a few days. Ma and Uren[4] also found that when water-soluble heavy metals were added to a soil they are rapidly retained by the soils, and the reactive forms then slowly transform into highly stable forms.

In the last decades, it has been well established that the presence of organic substances in soils causes various, and often contradictory, changes in the soil's capacity to retain metals.[5] Organic amendments are often added to soils in the form of waste materials that increase the soluble organic ligands in the soil. Li and Shuman[6] found that in a soil amended with soluble metal salts, EDTA treatment removed metals from the fractions bound to Mn and Fe oxides and redistributed them into the exchangeable fraction. In addition, the results of Nyamangara[7] showed that for soils amended with soluble metal salts and sewage sludge, a dramatic increase in Zn concentration was observed, and most of the Zn went into the more reactive exchangeable form.

The behavior observed for metals concerning their mobility as related with organic residues is far from uniform. Obrador and his colleagues[8] pointed out that when soils mixed with sewage sludge were aged, the bioavailable metal contents increased or decreased, depending on the cation. Hooda and Alloway[9] concluded that application of sewage sludge to soils caused increased Cd and Pb retention immediately after application, but a gradual decrease in sorption was observed for samples taken up to 450 days after application. Madrid and Díaz–Barrientos[10] found that application of olive mill wastewater to soils caused an increase in mobility of Cu and Zn, but not of Cd. Shuman[11] observed that certain organic amendments, such as spent mushroom compost, lower Cd and Pb availability by redistributing the exchangeable or organically-bound fractions to less available forms. Other amendments have little effect on metal distribution, and poultry litter causes an increase in the more available fractions.

In the work described below, the effect of the addition to a soil of several composted residues on the distribution of metals added to the soil in soluble form is studied for different times of contact. The possible effect of temperature is also tested. To distinguish between different metal forms in solid phase—what is called "speciation"—a number of sequential extraction procedures have been proposed. In this work, the method developed by the European Union's Community Bureau of Reference (BCR) was preferred, since it was proposed to harmonize analytical values between laboratories, and it is believed to give environmentally useful data.

MATERIALS AND METHODS

A sandy, surface sample of a Typic Endoquept soil from a strawberry-growing area in southwestern Spain was collected. At other sites on the same farm, this soil has been receiving amendments with three composts: composted urban solid residues (USR), composted wastes from paper industry (WPI), and composted residues from the olive oil industry (OI). Some characteristics of the soil and of the three composts are shown in Tables 3.1 and 3.2, respectively. The effects of the treatments with the composts on the total soil metal contents and on their sequential extraction were discussed in an earlier work.[12]

A sufficient amount of the unamended soil was thoroughly mixed with a solution containing several metals in NO_3 form in a proportion corresponding to the maximum permissible contents for soils receiving sewage sludge.[5] These amounts were 210, 450, 112, and 300 mg kg^{-1} for Cu, Zn, Ni, and Pb, respectively. The mixture was allowed to dry at room temperature (about two days).

Six 300-g triplicates of the artificially contaminated soil were then mixed in plastic containers with two proportions of each compost, corresponding to amendments of 65,000 and 200,000 kg ha^{-1} (called Levels 1 and 2 in the following text). Another compost-free (dose 0) triplicate was prepared, and the set of 21 containers was

TABLE 3.1
Soil Sample Properties

pH	Sand	CO$_3$ (%)	Organic Material	Cu	Zn	Ni	Pb (mg kg^{-1})	Mn	Fe
6.54	93	0	0.51	4.1	23.2	11.4	13.9	109	6090

TABLE 3.2
Compost Characteristics

Material	pH	% Organic Material	Cu	Zn	Ni (mg kg^{-1})	Pb	Mn	Fe
OI	9.4	31	70	67	107	19	240	10,000
USR	8.0	44	283	509	31	134	217	1,500
WPI	8.2	48	1280	2470	64	38	2750	474

subjected at 25°C to cycles in which they were alternately wetted at field capacity and allowed to dry, at the rate of about 2 cycles per week. A similar set of mixtures was handled in the same way in a ventilated oven at 37°C. Four-gram subsamples were collected from each container after 2, 8, 35, 52, and 78 weeks, and sequential extraction was carried out following the BCR procedure.[13] In this procedure, the metal forms with environmental significance can be grouped into three fractions, sequentially extracted by 0.11 M CH_3COOH (f_1), $NH_2OH \cdot HCl$ at pH 2 (f_2), and CH_3COONH_4 at pH 5 after digestion with H_2O_2 at 85 ± 5°C (f_3). More experimental details are described by Ure and his colleagues.[13] These fractions represent a progressively decreasing mobility in soils, although Ure and his colleagues were careful not to define any specific chemical nature for each fraction. However, the evaluation of the procedure by Coetzee et al.,[14] and the admitted specificity of the proposed reagents,[7,15] suggest that some wide assignments can be attributed to the three steps, as agreed by these authors, although some overlap is likely to exist. Thus, f_1 mainly includes adsorbed metals and some metals bound to $CaCO_3$, but some complexes with low complexation constants can be dissolved here; f_2 brings reducible species into solution, and thus includes metals bound to Fe and Mn oxides, but some oxides can be partially attacked in the first step; and, finally, f_3 includes oxidizable matter, particularly organic matter, but some metals forming weak complexes can already be released in the first step. Trivalent chromium in some barely soluble Cr^{3+} forms can be oxidized to chromate and dissolved in fraction f_3.

Concentrations of the metals added to the soil as well as Mn and Fe were determined in all the extracts by flame atomic absorption spectrophotometry (AAS). For each sample, the AAS result was obtained from the average of at least three readings. Sequential extraction data were also obtained for a BCR reference material;[16] the results are discussed elsewhere.[17]

RESULTS AND DISCUSSION

The amounts released in each fraction for each metal are summarized in Figures 3.1 through 3.6. Each point corresponds to the average of three replicates. Some erratic data suggest that the sequential extraction method is very sensitive to some uncontrolled errors, which is also suggested by the relatively high dispersion observed in the raw data for certification of the reference material given by Quevauviller and his colleagues.[16]

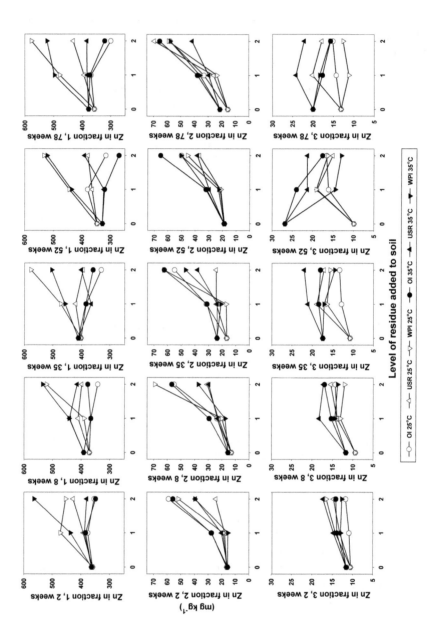

FIGURE 3.1 Results for sequential extraction of Zn (mg kg^{-1}) for each sampling.

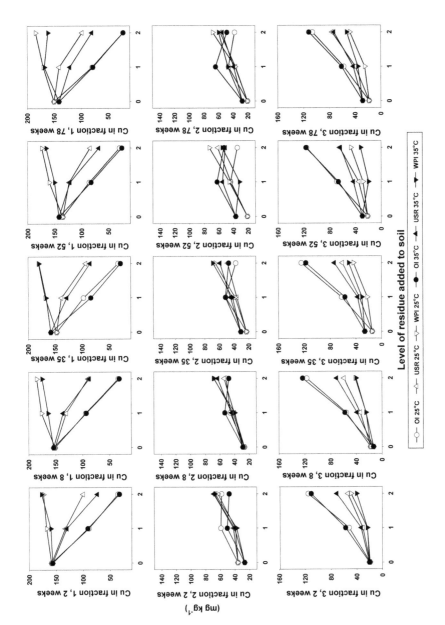

FIGURE 3.2 Results for sequential extraction of Cu (mg kg^{-1}) for each sampling.

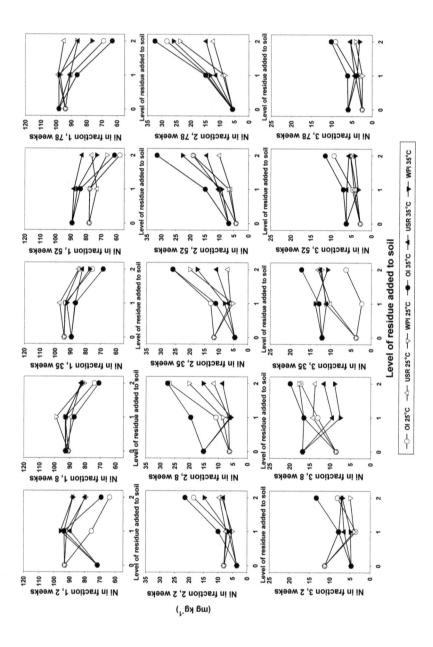

FIGURE 3.3 Results for sequential extraction of Ni (mg kg^{-1}) for each sampling.

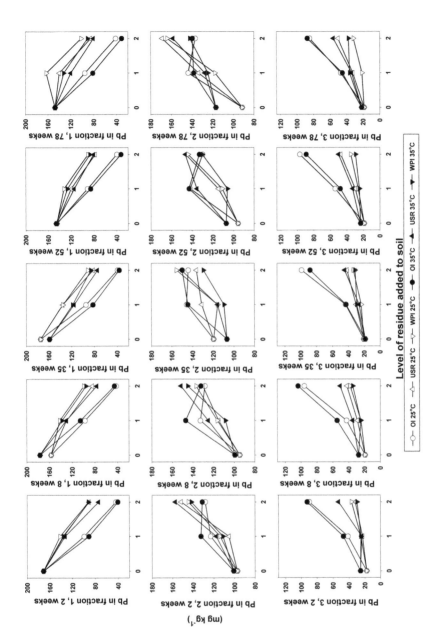

FIGURE 3.4 Results for sequential extraction of Pb (mg kg^{-1}) for each sampling.

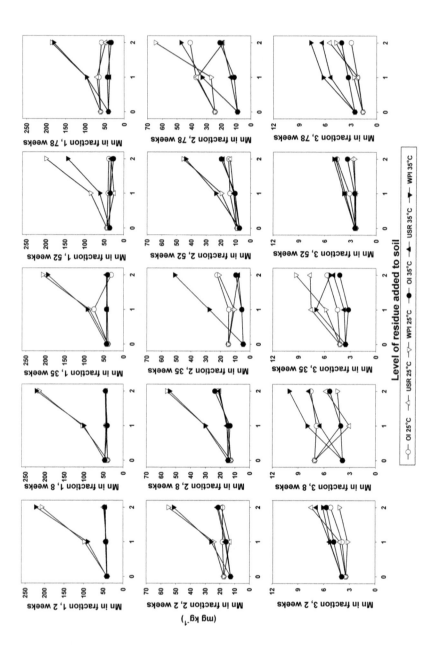

FIGURE 3.5 Results for sequential extraction of Mn (mg kg^{-1}) for each sampling.

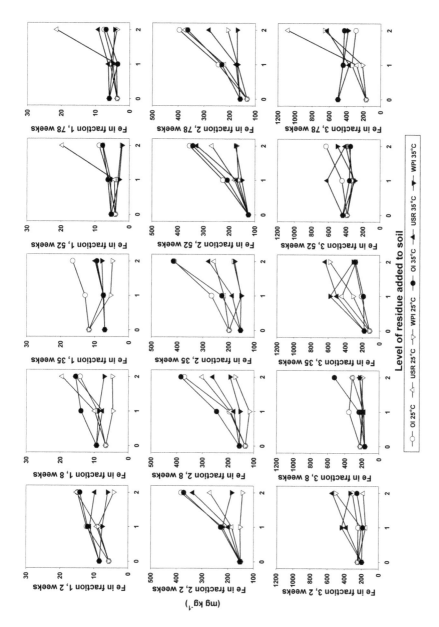

FIGURE 3.6 Results for sequential extraction of Fe (mg kg^{-1}) for each sampling.

Figure 3.1 shows that f_1 for Zn increases with the WPI doses decreases with the OI doses, and does not show any consistent relationship with time or temperature. The amounts in f_2 increase with the three composts and also show little or no dependence upon time or temperature. In contrast, f_3 shows a time-increasing difference between the two temperatures, even though the differences between doses of any compost are small.

The data for Cu are given in Figure 3.2. It is evident that the doses of OI and USR cause a strong decrease of f_1, while a slight increase is favored by WPI. Fraction f_2 shows a small but significant increase with the doses of USR and WPI, and f_3 clearly increases with all the composts, especially OI. No consistent effect of the temperature is observed in any case, and the systems clearly do not change over time.

Ni (Figure 3.3) shows erratic variations in some cases, but f_1 seems to be depressed by the presence of OI and, to a lesser degree, the other two composts. On the other hand, f_2 is significantly increased by OI and WPI, with a less evident effect of USR. A slight increase is observed in this fraction as the time of contact increases. The soil seems to lose part of fraction f_3 with time, but none of the composts has a clear effect upon this fraction.

The behavior of Pb (Figure 3.4) is very similar to that of Cu, except that f_1 decreases strongly with the doses of the three composts, while Cu showed this behavior only for USR and OI.

The behavior of these metals, which were added to the soil in soluble forms, can be compared with that of other metals that have been added only as components of the composts. This is the case with Mn and Fe (Figures 3.5 and 3.6). Addition of WPI causes significant increases of the more available fractions f_1 and f_2 of Mn, while f_3, which constitutes a minor part of the Mn sequentially extracted, shows very little variation with the doses of compost in quantitative terms. Although the low values of f_3 cause some of them to be erratic, some tendency to increase with the doses is suggested in some cases in Figure 3.5. On the other hand, most of the Fe that can be extracted from the soil is in the more strongly held fractions f_2 and f_3 (Figure 3.6), while f_1 accounts for a minor part, and OI and USR seem to favor f_2 and, to a much lesser degree, f_1.

To assess to what extent the treatments alter the relative distribution of each metal among the three fractions and the unextracted portion (what can be called the 'residual'), the sequential extraction data can be expressed in terms of percentage of the total contents, which should include the native content of the soil, the amount added in soluble form, and the metal contained in the compost mixed with the soil. Tables 3.3 through 3.8 compare these proportions in the last and (in parentheses) first samplings of each mixture for each treatment and temperature. The significance of the differences among treatments within each fraction and each sampling, estimated by the least significant differences (LSD), is expressed by the letters a, b, and c. Pairs of data within each group followed by the same letter are not significantly different at the P = 0.05 level.[18]

Table 3.3 shows that f_1 for Zn does not undergo any significant variation due to the presence of OI or USR when the data for the last sampling are considered. Some

TABLE 3.3
Relative distribution of fractions f_1 to f_3 and the sum Σ (per cent of total contents) of Zn for each treatment and temperature in the last (78 weeks) sampling, as compared with the first (2 weeks) sampling (in parentheses). For each sampling, pairs of data in the same line followed by the same letter are not significantly different at P = 0.05 level.

Residue	Temp.	Fraction	Doses 0		Doses 1		Doses 2	
OI	25	f_1	75 a	(90 b)	80 a	(83 ab)	67 a	(78 a)
		f_2	3 a	(2 a)	7 b	(5 b)	13 c	(9 c)
		f_3	3 a	(1 a)	3 a	(2 b)	3 a	(2 b)
		Σ	81 a	(93 a)	90 a	(90 a)	83 a	(89 a)
OI	37	f_1	80 a	(66 a)	80 a	(83 a)	72 a	(81 a)
		f_2	5 a	(2 a)	8 a	(6 b)	15 b	(11 c)
		f_3	4 b	(2 a)	4 a	(3 a)	4 a	(3 a)
		Σ	89 a	(70 a)	92 a	(92 b)	91 a	(95 b)
USR	25	f_1	75 a	(90 b)	83 a	(85 a)	91 a	(87 ab)
		f_2	3 a	(2 a)	5 a	(3 a)	15 b	(5 b)
		f_3	3 a	(1 a)	4 a	(2 ab)	4 a	(3 b)
		Σ	81 a	(93 a)	92 b	(90 a)	110 b	(95 a)
USR	37	f_1	80 a	(66 a)	81 a	(84 b)	81 a	(76 ab)
		f_2	5 a	(2 a)	6 b	(7 b)	9 c	(6 b)
		f_3	4 a	(2 a)	5 a	(3 ab)	5 ab	(3 b)
		Σ	89 a	(70 a)	92 a	(94 b)	95 a	(85 b)
WPI	25	f_1	75 a	(90 a)	93 b	(90 a)	97 b	(86 a)
		f_2	3 a	(2 a)	5 a	(3 a)	10 b	(5 b)
		f_3	3 a	(1 a)	2 a	(2 b)	2 a	(2 b)
		Σ	81 a	(93 a)	100 b	(95 a)	109 b	(93 a)
WPI	37	f_1	80 a	(66 a)	97 b	(95 b)	88 ab	(105 b)
		f_2	5 a	(2 a)	7 a	(4 ab)	10 b	(7 b)
		f_3	4 b	(2 a)	4 b	(2 a)	3 a	(2 a)
		Σ	89 a	(70 a)	108 b	(101 b)	101 b	(114 b)

OI—Olive oil industry; USR—Urban solid residues; WPI—Paper industry.

significant differences observed among the data for 2 weeks (in parentheses) tend to disappear after 78 weeks. On the contrary, the presence of WPI causes a significant increase in f_1, although the data for the two doses do not differ significantly. The presence of any compost also increases f_2, with somewhat higher values for the higher dose. The influence of the presence of any compost on f_3 is negligible in most cases. Fractions f_2 and f_3 are in most cases consistently higher for the 78-wk than for the 2-wk data. The relative pool of available Zn, estimated by the sum Σ of the three fractions, seems slightly increased in several cases. The Zn content of WPI causes a (calculated) increase in total Zn in the soil from about 470 to 600 mg kg^{-1}, which can account for the increase of Σ observed in the experiments with WPI as compared to the dose 0 experiment, but the metal of the other two composts is not high enough to account for the variations observed in some cases of the OI or USR experiments. The variations of f_2 with the doses are likely to be related to some mobilizing power of the composts, while the variation of f_2 and f_3 with time may be due to the evolution of the added Zn, e.g., bonding to oxides and/or organic matter. However, the low percentages of f_3 suggest that organic matter binding is not important for this metal. The behavior of Σ does not suggest any consistent variation with the doses of any compost.

The percentages of Cu extracted in the three fractions and their sum Σ are shown in Table 3.4. The most available fraction, f_1, undergoes a marked decrease with the doses of OI (more than 50% units) or USR (around 30% units), but it is not sensitive to WPI. On the contrary, f_2 and f_3 markedly increase with the doses of any compost. The differences between the 2-wk and 78-wk data are small and do not follow a consistent trend, suggesting that the time of contact is not important for the distribution of Cu. The variations observed in the WPI experiments could be attributed to the high Cu content of this compost, if the Cu is assumed to be incorporated mainly to the f_2 and f_3 fractions. However, the fact that the other two composts behave similarly (but with stronger variations), despite their much lower Cu contents, suggests that those two composts cause a transfer of Cu from the more soluble fraction f_1 to the less available f_2 and f_3, without significant variation of the sum of the three fractions. This latter fraction is predominant in the higher dose of OI and tends to be so for USR, suggesting that these two composts favor the formation of metal–organic bonds.

Table 3.5 shows the relative distribution of Ni among the fractions. The influence of the doses of any compost on f_1 seems to be small and not consistent. Only OI seems to cause some depressing effect on this fraction. On the contrary, f_2 shows a consistent increase with the doses of the three composts, and in most cases the values are greater for 78 wk than for 2 wk. Fraction f_3 accounts for a minor proportion of the total Ni content and is greater for the higher dose of OI. Thus, the organic fraction is favored only by OI, but a progressive transfer of Ni from the most soluble fraction f_1 to the reducible fraction f_2 occurs in the presence of the three composts. The presence of any compost does not seem to cause consistent variations in the sum Σ.

The relative contents in each fraction and Σ for Pb are shown in Table 3.6. The most soluble fraction f_1 is clearly depressed by the three composts, causing decreases of 13 to 37% from dose 0 to dose 3 (23 to 38% in the 2-wk samples). On the contrary, f_2 and f_3 are significantly increased, so that $f_2 + f_3$ increases in a similar proportion to the loss of f_1, and the sum Σ is not significantly varied. The behavior of this metal is thus similar to that of Cu, except that in the case of Pb, WPI also favors the formation of organic complexes, and fraction f_2 is predominant in all cases. Again, the time of contact has no influence on the process.

The behavior of a metal that has not been added in soluble form to the soil can be illustrated by the Mn data (Table 3.7). The proportion of the metal in the soil (native + that present in the composts) that can be mobilized in the three fractions (Σ) is initially clearly smaller than those observed for the other metals discussed

TABLE 3.4
Relative distribution of fractions f_1 to f_3 and the sum Σ (per cent of total contents) of Cu for each treatment and temperature in the last (78 weeks) sampling, as compared with the first (2 weeks) sampling (in parentheses). For each sampling, pairs of data in the same line followed by the same letter are not significantly different at P = 0.05 level.

Residue	Temp.	Fraction	Doses 0		Doses 1		Doses 2	
OI	25	f_1	70 c	(74 c)	38 b	(48 b)	11 a	(19 a)
		f_2	10 a	(9 a)	22 b	(21 b)	18 b	(21 b)
		f_3	8 a	(3 a)	27 b	(5 b)	52 c	(6 b)
		Σ	88 a	(86 b)	87 a	(72 b)	81 a	(46 a)
OI	37	f_1	66 c	(67 c)	37 b	(47 b)	11 a	(17 a)
		f_2	13 a	(10 a)	30 c	(25 b)	24 b	(22 b)
		f_3	13 a	(6 a)	29 b	(27 b)	55 c	(52 c)
		Σ	92 a	(93 a)	96 a	(99 a)	90 a	(91 a)
USR	25	f_1	70 c	(74 c)	65 b	(61 b)	45 a	(42 a)
		f_2	10 a	(9 a)	17 b	(17 b)	26 c	(25 c)
		f_3	8 a	(3 a)	18 b	(3 a)	35 c	(11 a)
		Σ	88 a	(86 a)	100 b	(81 a)	106 b	(78 a)
USR	37	f_1	66 b	(67 b)	56 b	(63 b)	37 a	(41 a)
		f_2	13 a	(10 a)	18 b	(19 b)	25 c	(24 c)
		f_3	13 a	(6 a)	20 b	(20 b)	34 c	(25 b)
		Σ	92 a	(93 a)	94 a	(102 a)	96 a	(90 a)
WPI	25	f_1	70 a	(74 b)	72 a	(73 b)	66 a	(61 a)
		f_2	10 a	(9 a)	16 b	(15 b)	24 c	(19 c)
		f_3	8 a	(3 a)	10 b	(9 b)	17 c	(12 c)
		Σ	88 a	(86 a)	98 a	(97 b)	107 a	(91 a)
WPI	37	f_1	66 a	(67 a)	71 a	(74 a)	58 a	(69 a)
		f_2	13 a	(10 a)	20 b	(17 b)	21 c	(23 c)
		f_3	13 a	(6 a)	15 a	(9 a)	19 b	(15 b)
		Σ	92 a	(93 a)	106 a	(100 a)	98 a	(107 a)

OI—Olive oil industry; USR—Urban solid residues; WPI—Paper industry.

above, except in the case of the higher dose of WPI. However, after 78 weeks, the value of Σ tends to increase in the OI and USR experiments as compared with the 2-wk data at 25°C, but not at the higher temperature. This increase in Σ with time seems to occur in fractions f_1 and f_2. Although these two composts do not have very high Mn contents, their organic matter can favor reducing conditions that can transfer Mn from initially insoluble to more available forms. At 37°C, the reducing conditions are probably not favored, and f_1 and f_2 are maintained close to the original situation. The Mn content of WPI is high, and the rise in f_1 with doses, only observed in the WPI experiments, is probably a consequence of it, suggesting that the Mn in

TABLE 3.5

Relative distribution of fractions f_1 to f_3 and the sum Σ (per cent of total contents) of Ni for each treatment and temperature in the last (78 weeks) sampling, as compared with the first (2 weeks) sampling (in parentheses). For each sampling, pairs of data in the same line followed by the same letter are not significantly different at P = 0.05 level.

Residue	Temp.	Fraction	Doses								
			0				**1**			**2**	
OI	25	f_1	75	b	(75	b)	73	b	(67 b)	56 a	(55 a)
		f_2	5	a	(6	a)	12	b	(11 b)	23 c	(16 c)
		f_3	2	a	(9	a)	3	a	(11 a)	7 b	(11 a)
		Σ	82	a	(90	a)	88	a	(89 a)	86 a	(82 a)
OI	37	f_1	79	b	(57	a)	70	b	(75 b)	51 a	(55 a)
		f_2	5	a	(3	a)	12	b	(8 b)	26 c	(17 c)
		f_3	5	a	(4	a)	5	a	(6 b)	8 b	(11 c)
		Σ	89	a	(64	a)	87	a	(89 b)	85 a	(83 b)
USR	25	f_1	75	a	(75	a)	80	a	(76 a)	80 a	(74 a)
		f_2	5	a	(6	a)	7	b	(6 a)	10 c	(8 a)
		f_3	2	a	(9	a)	3	a	(6 a)	5 b	(6 a)
		Σ	82	a	(90	a)	90	b	(88 a)	95 b	(88 a)
USR	37	f_1	79	a	(57	a)	74	a	(73 a)	72 a	(67 a)
		f_2	5	a	(3	a)	9	b	(5 b)	12 c	(7 c)
		f_3	5	a	(4	a)	4	a	(5 a)	5 a	(6 a)
		Σ	89	a	(64	a)	87	a	(83 b)	89 a	(80 b)
WPI	25	f_1	75	a	(75	b)	79	a	(75 b)	71 a	(66 a)
		f_2	5	a	(6	b)	7	a	(4 a)	20 b	(7 b)
		f_3	2	a	(9	b)	2	a	(3 a)	3 b	(4 a)
		Σ	82	a	(90	a)	88	a	(82 a)	94 a	(77 a)
WPI	37	f_1	79	b	(57	a)	80	b	(78 b)	63 a	(73 b)
		f_2	5	a	(3	a)	11	b	(5 b)	21 c	(12 c)
		f_3	5	b	(4	a)	3	a	(4 a)	3 a	(6 a)
		Σ	89	a	(64	a)	94	a	(87 b)	87 a	(91 b)

OI—Olive oil industry; USR—Urban solid residues; WPI—Paper industry.

this compost is in available form. Some increases with the doses of OI and USR are also observed for f_2 and, to a much lesser extent, f_3, suggesting that the Mn contained in these composts is mainly in reducible form and a small proportion is related with organic matter.

The proportion of Fe extracted, Σ (Table 3.8), is considerably lower than that of any of the other metals, and the distribution among the fractions is quite different: f_1 is negligible, and f_3 predominates with percentages somewhat greater than those found for Zn, Ni, or Mn in this fraction. Fractions f_2 and, sometimes, f_3 seem to be favored by increasing doses of the composts. However, it must be noted that these

TABLE 3.6
Relative distribution of fractions f_1 to f_3 and sum Σ (percent of total contents) of Pb for each treatment and temperature in last (78 wk) sampling, as compared with first (2 wk) sampling (in parentheses). For each sampling, pairs of data in same line followed by same letter are not significantly different at $P = 0.05$ level.

Residue	Temp. (°C)	Fraction	Doses 0		1		2	
OI	25	f_1	47 c	(53 c)	31 b	(33 b)	14 a	(15 a)
		f_2	29 a	(28 a)	47 b	(42 b)	46 b	(43 b)
		f_3	6 a	(5 a)	15 b	(12 b)	29 c	(20 c)
		Σ	82 a	(86 a)	93 a	(87 a)	89 a	(78 a)
	37	f_1	47 c	(53 c)	26 b	(33 b)	10 a	(16 a)
		f_2	37 a	(29 a)	45 a	(48 b)	47 a	(43 b)
		f_3	7 a	(6 a)	15 b	(14 b)	30 c	(29 c)
		Σ	91 a	(88 a)	86 a	(95 b)	87 a	(88 a)
USR	25	f_1	47 b	(53 c)	45 b	(43 b)	28 a	(27 a)
		f_2	29 a	(28 a)	43 b	(33 a)	56 c	(47 b)
		f_3	6 a	(5 a)	10 b	(7 a)	17 c	(13 b)
		Σ	82 a	(86 a)	98 b	(83 a)	101 b	(87 a)
	37	f_1	47 c	(53 b)	39 b	(43 b)	27 a	(27 a)
		f_2	37 a	(29 a)	41 a	(39 b)	52 b	(45 b)
		f_3	7 a	(6 a)	11 b	(11 b)	19 c	(14 c)
		Σ	91 a	(88 a)	91 a	(93 a)	89 a	(86 a)
WPI	25	f_1	47 a	(53 c)	53 a	(46 b)	34 a	(30 a)
		f_2	29 a	(28 a)	38 a	(36 b)	55 b	(42 b)
		f_3	6 a	(5 a)	7 a	(9 b)	11 b	(10 b)
		Σ	82 a	(86 a)	98 b	(91 a)	100 b	(82 a)
	37	f_1	47 c	(53 c)	43 b	(44 b)	30 a	(28 a)
		f_2	37 a	(29 a)	41 ab	(38 ab)	48 b	(45 b)
		f_3	7 a	(6 a)	12 b	(9 a)	13 b	(8 a)
		Σ	91 a	(88 b)	96 a	(91 b)	91 a	(81 a)

OI—Olive oil industry; USR—Urban solid residues; WPI—Paper industry.

increases are only of a few percentage units, which means that they are probably due to the more available status of the Fe contained in the composts, and it is not necessary to assume any special interaction of the soil Fe with other components of the composts.

The observations of Tagami and Uchida[3] and Ma and Uren,[4] concerning a fast decrease in the exchangeable or soluble forms of metals added to soils, are confirmed in the present case only for Pb, which shows that the most easily mobilizable fraction, f_1, is about 50% of the total contents in the absence of composts, although this proportion varies little (or not at all) up to 78 weeks. The other metals added in

TABLE 3.7
Relative distribution of fractions f_1 to f_3 and the sum Σ (per cent of total contents) of Mn for each treatment and temperature in the last (78 weeks) sampling, as compared with the first (2 weeks) sampling (in parentheses). For each sampling, pairs of data in the same line followed by the same letter are not significantly different at P = 0.05 level.

Residue	Temp.	Fraction	Dose 0			Dose 1			Dose 2		
OI	25	f_1	56	a	(38 a)	58	a	(38 a)	50	a	(38 a)
		f_2	22	a	(11 a)	33	b	(13 a)	35	c	(13 a)
		f_3	1	a	(5 a)	2	ab	(5 a)	3	b	(5 a)
		Σ	79	a	(54 a)	93	a	(56 a)	88	a	(56 a)
OI	37	f_1	37	ab	(30 a)	38	b	(43 b)	28	a	(39 b)
		f_2	8	a	(12 a)	10	a	(12 a)	18	b	(15 b)
		f_3	2	a	(5 a)	3	ab	(6 a)	3	b	(9 b)
		Σ	47	a	(47 a)	51	a	(61 b)	49	a	(63 b)
USR	25	f_1	56	b	(38 a)	63	c	(35 a)	41	a	(40 a)
		f_2	22	a	(11 a)	33	b	(12 a)	16	a	(15 a)
		f_3	1	a	(5 a)	2	a	(5 a)	5	b	(5 a)
		Σ	79	a	(54 a)	98	b	(52 a)	62	a	(60 a)
USR	37	f_1	37	a	(30 a)	34	a	(40 b)	34	a	(38 b)
		f_2	8	a	(12 a)	12	b	(15 a)	17	c	(15 a)
		f_3	2	a	(5 a)	5	b	(5 a)	5	b	(6 a)
		Σ	47	a	(47 a)	51	a	(60 b)	56	a	(59 b)
WPI	25	f_1	56	a	(38 a)	58	a	(59 b)	68	b	(72 c)
		f_2	22	a	(11 a)	16	a	(15 b)	24	b	(17 c)
		f_3	1	a	(5 a)	2	a	(4 a)	2	a	(2 a)
		Σ	79	a	(54 a)	76	a	(78 b)	94	b	(91 b)
WPI	37	f_1	37	a	(30 a)	58	b	(64 b)	66	c	(86 c)
		f_2	8	a	(12 a)	20	b	(16 b)	17	b	(19 b)
		f_3	2	a	(5 a)	4	b	(4 a)	3	b	(4 a)
		Σ	47	a	(47 a)	82	b	(84 b)	86	b	(109 c)

OI—Olive oil industry; USR—Urban solid residues; WPI—Paper industry.

soluble forms (Zn, Cu, Ni) showed proportions of f_1 for dose 0 close to 80% for Zn and Ni and around 70% for Cu. This discrepancy must be due to the different nature of the soils of those authors. Tagami and Uchida did not give data of clay or carbonate contents of their samples, so we cannot be sure of which parameters are responsible for the different behavior, but they described one of their samples as an Andosol, which probably contained variable-charge surfaces that can have a significant capacity to retain metals. Ma and Uren used a soil with 17% clay, which means that the available surface for metal retention or for catalyzing the formation of sparingly

TABLE 3.8

Relative distribution of fractions f_1 to f_3 and the sum Σ (per cent of total contents) of Fe for each treatment and temperature in the last (78 weeks) sampling, as compared with the first (2 weeks) sampling (in parentheses). For each sampling, pairs of data in the same line followed by the same letter are not significantly different at $P = 0.05$ level.

Residue	Temp.	Fraction	0		1		2	
OI	25	f_1	0 a	(0 a)	0 a	(0 a)	0 a	(1 b)
		f_2	2 a	(2 a)	4 b	(3 b)	6 c	(5 c)
		f_3	3 a	(3 a)	5 b	(4 a)	5 b	(3 a)
		Σ	5 a	(5 a)	9 b	(7 a)	11 b	(9 b)
OI	37	f_1	0 a	(0 a)	0 a	(0 a)	0 a	(0 a)
		f_2	3 a	(2 a)	4 b	(4 b)	6 c	(5 b)
		f_3	8 a	(2 a)	7 a	(3 b)	7 a	(5 c)
		Σ	11 a	(4 a)	11 a	(7 b)	13 a	(10 c)
USR	25	f_1	0 a	(0 a)	0 a	(0 a)	0 a	(0 a)
		f_2	2 a	(2 a)	4 b	(3 b)	6 c	(5 c)
		f_3	3 a	(3 a)	5 a	(2 a)	19 b	(4 a)
		Σ	5 a	(5 a)	9 b	(5 a)	25 c	(9 a)
USR	37	f_1	0 a	(0 a)	0 a	(0 a)	0 a	(0 a)
		f_2	3 a	(2 a)	3 a	(4 b)	5 b	(4 b)
		f_3	8 a	(2 a)	6 a	(5 b)	11 a	(5 b)
		Σ	11 a	(4 a)	9 a	(9 b)	16 a	(9 b)
WPI	25	f_1	0 a	(0 a)	0 a	(0 a)	0 a	(0 a)
		f_2	2 a	(2 a)	3 ab	(2 a)	4 b	(3 b)
		f_3	3 a	(3 a)	4 a	(5 c)	11 b	(4 b)
		Σ	5 a	(5 a)	7 a	(7 b)	15 b	(7 b)
WPI	37	f_1	0 a	(0 a)	0 a	(0 a)	0 a	(0 a)
		f_2	3 a	(2 a)	3 a	(2 a)	3 a	(3 b)
		f_3	8 a	(2 a)	7 a	(3 a)	7 a	(4 b)
		Σ	11 a	(4 a)	10 a	(5 a)	10 a	(7 b)

OI—Olive oil industry; USR—Urban solid residues; WPI—Paper industry.

soluble metal compounds is greater than in our case. The strong decrease in the most available fraction f_1 for Cu with the OI or USR doses and its transfer to fractions f_2 and f_3 agrees with the conclusions of Nyamangara,[7] who showed that the presence of sewage sludge favored that applied Cu were incorporated to less available forms. The same author found that most of the Zn applied to the soil was incorporated to the more bioavailable, exchangeable form, as found in the data presented here. In contrast, Obrador and his colleagues[8] found that Zn bioactivity diminished with the addition of sewage sludge, and that of Cu increased. However, Pb data from Obrador and his colleagues agreed with those found here, and they found that the bioavailability of this metal decreased with the doses of sludge. The results of Hooda and Alloway[9] and Shuman[11] also pointed to a decreased mobility of Pb by some organic amendments, although Shuman also pointed out that some other amendments may have a different behavior, e.g., poultry litter, which caused an increase in the more available forms. After two years of incubation of soils with sewage sludge, Hooda and Alloway[15] found that f_1 was predominant for Ni and Zn, while Cu and Pb were mainly in f_3. The results presented here point in the same direction.

When comparing the values of the first and last sampling for the dose 0 (compost-free) soil, some significant changes are observed in the proportions of Zn, Ni, and Mn released in fraction f_1. These changes can be related to changes in the nature of metal bindings during the wetting–drying cycles, especially in the case of those metals added in soluble form. Although some heterogeneity among subsamples taken from the containers cannot be excluded, in most cases the general trends of the absolute values with time and their dependence upon the doses (as observed in Figures 3.1 through 3.6) suggest that these changes are not important and are masked by the consistent effects of the compost additions, as described above.

In general, the differences between the two temperatures tested here are of little significance, except perhaps in the case of Mn. The rather large errors of the fractionation method probably mask any possible effect of temperature.

CONCLUDING REMARKS

It is evident from the data presented here that a common behavior of all heavy metals in soils cannot be expected as to the effect of organic amendments, nor can it be expected that all such amendments are uniform in their action on metal mobility. In some cases, the action does not seem to be due to metal–organic associations with the organic compounds present in the amendments, and the effect on the availability of a given metal may be caused mainly by relatively high contents of the metal in the amendment, as in the case of Zn. In some metals, immobilization is favored by the presence of some amendments by depressing the more available fraction and increasing the proportion of more strongly held metal forms, while other amendments have no significant effect. This is observed in the case of Cu or Ni, while Pb shows a similar behavior with the three composts tested here. In the case of native metals, although Mn availability seems favored by OI and USR, changes in Fe status can be attributed mainly to the metal present in the amendments themselves. Most of these effects are detectable within the first few weeks, with only minor changes afterwards, suggesting that any environmental consequence of the amendments can be of significance shortly after amending the soils.

Therefore, the use of organic amendments should be carried out carefully, as possible effects on metal availability may become noticeable if the metal contents prior to amending are relatively high. The use of composts with low contents of toxic metals is obviously preferable, but their contents in soluble organic components should also be taken into account, as it is likely that this parameter will favor any mobilizing effect on the soil metals. The greater effect of OI and, to a lesser extent, USR as compared with WPI, which is often observed, points to this direction.

ACKNOWLEDGMENTS

The authors are indebted to Dr. R. López for the analysis of the soil and compost samples, and to Miss P. Burgos for providing us with the compost samples. The financial help of the Comisión Interministerial de Ciencia y Tecnología (CICYT) of Spain under Project nos. AMB98-0888 and AMB97-0692 is also acknowledged.

REFERENCES

1. Ross, S.M., Sources and forms of potentially toxic metals in soil-plant systems, in *Toxic Metals in Soil-Plant Systems,* Ross, S.M., Ed., Wiley, Chichester, U.K., 1994, chap. 1.
2. McBride, M.B., Reactions controlling heavy metal solubility in soils, *Adv. Soil Sci.,* 10, 1, 1989.
3. Tagami, K. and Uchida, S., Aging effect on bioavailability of Mn, Co, Zn and Tc in Japanese agricultural soils under waterlogged conditions, *Geoderma,* 84, 3, 1998.
4. Ma, Y.B. and Uren, N.C., Transformations of heavy metals added to soil: application of a new sequential extraction procedure, *Geoderma,* 84, 157, 1998.
5. Madrid, L., Metal retention and mobility as influenced by some organic residues added to soils: a case study, in *Fate and Transport of Heavy Metals in the Vadose Zone,* Selim, H.M. and Iskandar, I.K., Eds., Lewis Publishers, Boca Raton, 1999, chap. 10.
6. Li, Z. and Shuman, L.M., Redistribution of forms of zinc, cadmium and nickel in soils treated with EDTA, *Science Total Environ.,* 191, 95, 1996.
7. Nyamangara, J., Use of sequential extraction to evaluate zinc and copper in a soil amended with sewage sludge and inorganic salts, *Agric. Ecosys. Environ.,* 69, 135, 1998.
8. Obrador, A., Rico, M.I., Álvarez, J.M., and Mingot, J., Mobility and extractability of heavy metals in contaminated sewage sludge-soil incubated mixtures, *Environ. Technol.,* 19, 307, 1998.
9. Hooda, P.S. and Alloway, B.J., Sorption of Cd and Pb by selected temperate and semi-arid soils: Effects of sludge application and ageing of sludged soils, *Water, Air Soil Poll.,* 74, 235, 1994.
10. Madrid, L. and Díaz-Barrientos, E., Retention of heavy metals by soils in the presence of a residue from the olive-oil industry, *Eur. J. Soil Sci.,* 45, 71, 1994.
11. Shuman, L.M., Effect of organic waste amendments on cadmium and lead in soil fractions of two soils, *Commun. Soil Sci. Plant Anal.,* 29, 2939, 1998.
12. Díaz-Barrientos, E., Madrid, L., and Cardo, I., Influence of the addition of organic wastes on the metal contents of a soil, *Fresenius J. Anal. Chem.,* 363, 558, 1999.
13. Ure, A., Quevauviller, Ph., Muntau, H., and Griepink, B., *Improvements in the Determination of Extractable Contents of Trace Metals in Soil and Sediment Prior to Certification,* BCR Report EUR 14763 EN, Commission of the European Communities, Brussels, Belgium, 1993.
14. Coetzee, P.P., Gouws, K., Plüddemann, S., Yacoby, M., Howell, S., and den Drijver, L., Evaluation of sequential extraction procedures for metal speciation in model sediments, *Water SA,* 21, 51, 1995.

15. Hooda, P.S. and Alloway, B.J., Changes in operational fractions of trace metals in two soils during two-years of reaction time following sewage sludge treatment, *Intern. J. Environ. Anal. Chem.,* 57, 289, 1994.

16. Quevauviller, Ph., Rauret, G., López-Sánchez, J.F., Rubio, R., Ure, A., and Muntau, H., *The Certification of the Extractable Contents (Mass Fractions) of Cd, Cr, Cu, Pb and Zn in Sediment Following a Three-Step Sequential Extraction Procedure,* BCR Report EUR 17554 EN, European Commission, Brussels, Belgium, 1997.

17. Díaz-Barrientos, E., Madrid, L., and Cardo, I., Effect of flood with mine wastes on metal extractability of some soils of the Guadiamar river basin (SW Spain), *Sci. Total Environ.,* 242, 149, 1999.

18. Snedecor, G.W. and Cochran, W.G., *Statistical Methods,* Iowa State University Press, Ames, IA, 1967.

4 Induced Hyperaccumulation: Metal Movement and Problems

Chris Anderson, Annabelle Deram, Daniel Petit, Robert Brooks, Robert (Bob) Stewart, and Robyn Simcock

ABSTRACT

Induced hyperaccumulation of lead was studied in field and pot trials in northern France (over Cu/Pb/Cd-rich soils polluted from a nearby smelter at Auby, France) and in New Zealand. The plants studied were Agrostis tenuis (1), Arrhenatherum elatius (2), Berkheya coddii (3), Brassica juncea (4), Cardaminopsis halleri (5), Silene humulis (6), and Thlaspi caerulescens (7).

INTRODUCTION

Phytoremediation technology has become firmly established in the literature. Until very recently, phytoremediation has relied on the use of plants known as hyperaccumulators, defined as plants that accumulate of the order of 100 times more metal than nonaccumulator plants growing in the same environment.[1,2,3] Hyperaccumulators are often species of low biomass and slow growth rate, leading to a slow time frame for metal uptake and soil decontamination. Hyperaccumulator plants are also limited in the range of metals that can be accumulated. There are many metals for which no hyperaccumulators have yet been recognized.

To combat these problems, a recent advance in phytoremediation technology has been induced hyperaccumulation. In this scenario, a chemical is applied to the soil that acts to solubilize a normally insoluble target metal. The mechanism for the subsequent uptake of the soluble metal complex is a matter of some debate. The entire metal–chelate complex may be taken up by the roots and translocated to shoots,[4] or the metal–chelate complex may dissociate at the root/soil interface, with the subsequent uptake of a free metal ion. Induced hyperaccumulation was first reported for uptake of the heavy metal Pb by corn (*Zea mays*) using a protonated form of the chelating agent ethylenediaminetetraacetic acid (EDTA).[5] The most recent addition to the list of metals for which induced hyperaccumulation has been reported is gold.[6]

Little or no attention, to our knowledge, has been paid to the potential importance of the specific metal mineral phase present in a heavy-metal-contaminated environment. The mineral phase of metals present at a polluted site is a function of the original source of metal contamination; different point sources for metal contamination generate different mineral phases of metal in soil.[7] This study outlines three environments that are examples of field settings where anthropogenic metal pollution is of concern. Each of these environments has been contaminated by metals present in different mineral phases. Hyperaccumulation was induced, with varying degrees of success, and is reported here for plants growing on soil from two of these three environments.

A model is also presented, based upon pot-trial results, where artificially contaminated soils were generated by adding several different Pb salts to commercial potting mix. The Pb salts used were chosen to represent the original mineral phase of metal that could contaminate different environments. This model explains relative differences in induced metal uptake as a function of the original polluting metal phase.

METAL UPTAKE—TRIALS ON POLLUTED SUBSTRATES

FIELD TRIALS IN NORTHERN FRANCE

During the Northern Hemisphere summer of 1998, a field trial run in collaboration with the University of Lille was initiated on an area of polluted land that has suffered considerable heavy-metal contamination due to industrial activity near the French town of Auby in the province of Nord Pas de Calais (Figure 4.1). The surface profile (0–10 cm) of the Auby soil is characterized as being extremely high in organic material (C% = 36, N% = 0.6). The high heavy metal concentration has inhibited the microbial breakdown of plant material.[8] Typical soil metal values for the surface profile are up to 4% Zn, 1% Pb, and 300 mg kg^{-1} Cd. The mean pH of these soils is 6.0. Chemical fractionation methods have been used to determine the distribution of the heavy metals Pb, Zn, Cd, and Cu in industrially polluted soils from northern France.[9]

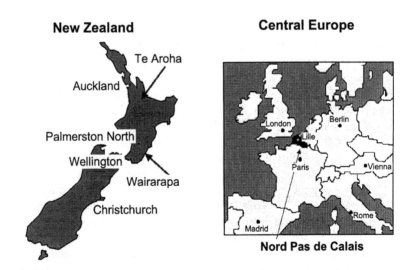

FIGURE 4.1   Map showing location of study sites in Europe and New Zealand.

The principal metal phases present in soils surrounding the Auby region have been shown to be heavy metal oxides and carbonates.

Two areas, each measuring 5 m^2, were selected at the Auby site and divided into nine 1-m^2 plots. The natural vegetation in each of these plots was a mixture of the Zn hyperaccumulator *Cardaminopsis halleri,* the metal-tolerant grasses *Arrhenatherum elatius* and *Agrostis tenius,* and the metal-tolerant herb *Silene humilis.* In each area, three replicate plots were treated with equal-volume solutions of EDTA (disodium salt) and citric acid at an application rate of 0.5 g per kg field weight soil, to a depth of 15 cm (75 g m^{-2} surface area). Vegetation was sampled each week for four weeks to determine the effect of the chemical amendments either on inducing Pb, Zn, and Cd uptake, or on increasing the already apparent level of uptake. In addition, soil cores to a depth of 0.6 m were taken each week during this period to monitor the possible movement of heavy metals down the soil profile. The plant metal concentrations at the end of the trial (day 21) are shown in Table 4.1.

The data show hyperaccumulation of Cd and Zn by *Cardaminopsis halleri,* but no significant differences between the treatments for each plant species. The addition of soil chemical amendments did not improve the uptake concentration of Pb, Zn, and Cd in either the hyperaccumulating or nonhyperaccumulating species studied. Similarly, no significant differences were observed between soil profiles that could be attributed to the addition of EDTA or citric acid (data not shown). Natural variation across each area was very large, showing the pollution at the site to be very heterogeneous. This natural variation may have masked any induced differences in metal distribution within the soil profile.

TRIALS ON MATERIAL FROM THE TUI BASE-METAL MINE

The Tui base-metal mine, located on the flanks of Mount Te Aroha in the Coromandel district of the North Island, New Zealand (Figure 4.1), has a long history of exploration and aborted attempts at active production. The most recent period of activity

---

TABLE 4.1
Metal concentrations (dry weight) in field experimental plants 21 days after EDTA treatment. Values are mean and standard ( ) deviation. Significance (ANOVA p <0.05) is for differences between the treatments for each metal

	Arrhenatherum elatius			Cardaminopsis halleri		
	Cd (mg kg^{-1})	Pb (mg kg^{-1})	Zn (mg kg^{-1})	Cd (mg kg^{-1})	Pb (mg kg^{-1})	Zn (%)
Control	7.05 (2.76)	17.10 (4.23)	609.5 (292.2)	167.6 (36.4)	50.0 (7.4)	2.47 (0.81)
Citric acid	7.03 (6.77)	16.68 (1.85)	733.0 (191.4)	123.4 (7.4)	49.7 (10.3)	1.64 (0.23)
EDTA	6.90 (1.30)	17.60 (1.30)	712.0 (222.3)	174.0 (24.3)	78.2 (29.7)	2.47 (0.51)
Significance	n.s.	n.s.	n.s.	n.s.	n.s.	n.s.

n.s.—Not significant

occurred between 1967 and 1974 with the extraction and processing of up to 100 tons of ore a day, yielding up to 10 tons of a Pb–Cu–Zn concentrate containing minor amounts of Cd–Ag–Au.[10] The cessation of mining activities in 1974 left a substantial tailings dam containing 100,000 m^2 of tailings with high levels of heavy metals.[11] The tailings continue to weather (oxidize), producing acid mine drainage. The principal metal-bearing minerals present in the Tui ore are galena (PbS), sphalerite (ZnS), and chalcopyrite ($CuFeS_2$).[12] The less abundant Cd minerals are the CdS minerals greenockite and hawleyite.[13] In the Tui mine tailings, Pb has been oxidized *in situ* to an anglesite ($PbSO_4$) mineral phase. The primary Zn phase in the tailings remains as sphalerite and, although not identified, it seems likely that CdS remains the primary mineral phase of Cd in the tailings due to the high solubility of the respective Zn- and Cd-oxidized sulphates.

Revegetation trials on this substrate have been ongoing at Massey University for a number of years.[12] More recently, hyperaccumulation, both natural and induced, has been tested as a potential answer to the problem of heavy-metal pollution at this particular location. The Tui mine tailings used in this study contain approximately 26 mg kg^{-1} Cd, 1.15% Pb, and 5400 mg kg^{-1} Zn. The pH of these tailings was 3.9.

Figure 4.2 outlines the results of an experiment in which the high-biomass nickel hyperaccumulator *Berkheya coddii* was grown in limed Tui tailings and subsequently treated with EDTA to induce uptake of heavy metals. Lime was applied at a rate of 2.5% w/w. This was necessary to raise the pH of the substrate to the point where plant growth could be sustained. EDTA induced a highly significant increase in the concentration of Pb in *Berkheya coddii,* although there was no difference between the three treatment levels. EDTA did not induce an increase in the Cd concentration of the plants relative to the control treatment.

MODEL FOR INDUCED METAL UPTAKE

The preceding sections have outlined trials conducted on two different substrates that are typical of two different polluted environments. The heavy-metal loading at each of these sites would preclude the use of phytoextraction for soil remediation because of the long time frame that would be associated with soil decontamination. However, these sites do represent ideal substrates to test the theories and mechanisms of natural and induced metal uptake.

Induced hyperaccumulation experiments on these substrates have produced surprising results. Enhanced accumulation of Pb was observed in *Berkheya coddii* once EDTA was applied to the Tui mine tailings, but no increase in Cd uptake was apparent. The concentration levels observed in both of these environments fall short of the Pb and Cd values reported in the literature for plants after treatment with EDTA. No increased uptake at all was observed in the plants growing at the northern French site after EDTA had been applied to the substrate.

Each of the sites can be characterized as being contaminated with heavy metals that exist in different mineral phases. Another site is introduced in Table 4.2 to illustrate a third possible environment where heavy metal pollution may be

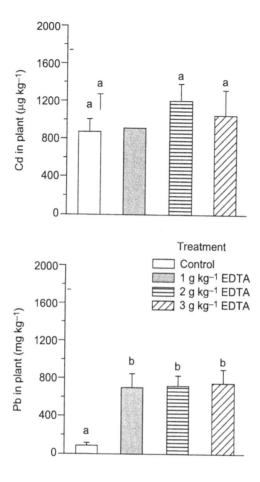

FIGURE 4.2 EDTA-induced uptake of Cd and Pb by *Berkheya coddii* growing on Tui mine tailings—mean + SE (n = 5). Means with same letter are not significantly different (ANOVA p >0.05). Data were normally distributed.

TABLE 4.2
The Dominant Mineral Phase of Metal Present at Three Contaminated Sites

Location	Dominant Mineral Phase of Contaminating Metal	Source of Contamination
Auby—northern France	Oxide, carbonate	Industrial fallout
Tui tailings—New Zealand	Sulphide, sulphate	Mine tailings
Wairarapa—New Zealand	Phosphate, carbonate	Superphosphate fertilizer

encountered. It is an area of pastoral land adjacent to a superphosphate storage shed, contaminated with Cd due to the high levels of this heavy metal found in some phosphatic fertilizers.

LEAD UPTAKE MODEL

A question raised by the summation presented in Table 4.2 is what effect do these different mineral phases have on the bioavailability of metals under conditions of natural uptake and under conditions of induced uptake? To answer and model this question, a greenhouse pot trial was carried out in conjunction with the 1998–1999 field trials. In this trial, plants were grown in potting mix spiked with Pb salts of different mineral phases.

Experimental Design

Six salts were chosen to represent Pb pollution that could occur in a wide range of environments: carbonate, nitrate, oxide, phosphate, sulphate, and sulphide. The soluble nitrate salt was used to model Pb, after dissolution, as part of the soil–organic phase. The justification for this statement is that soluble Pb would either leach out of the artificial soil or be adsorbed onto the organic phase. Each salt was added to commercial potting mix to give a dry-weight Pb concentration of 1% (w/w). A total of 280 pots (250 mL each) was planted with equal numbers with *Brassica juncea* and *Thlaspi caerulescens*. A control soil was used, where the two species were planted in pots containing potting mix with no added Pb. During the growing cycle, pot positions were randomly changed on a periodic basis to equalize light exposure. The ambient temperature of the greenhouse was set at 15–25 °C with no humidity control. Overhead watering was carried out each day with a hand-held hose.

After approximately 10 weeks' growth, 5 replicates of each plant species, for each mineral phase, were treated with one of 2 g kg^{-1}* EDTA, 2 g kg^{-1} citric acid, 2 g kg^{-1} acetic acid, and water was used as a control treatment. All treatments were applied as a solution (20 mL) and were randomly allocated to replicate pots. Two weeks after treatment, the above-ground portions were harvested, dried at 60°C until constant weight, and subsamples digested in concentrated nitric acid before analysis by flame atomic absorption spectroscopy (FAAS). As replicate specimens of *Brassica juncea* had reached different stages of maturity, and thus showed different weight ratios of stems, leaves, and flowers, only the leaves of this species were analyzed to minimize the variations in results that could be attributed to an uneven distribution of organs for an individual plant.

Potting-mix samples were cored from each of the control pots of each metal phase at the time of plant harvest, ground using a porcelain mortar and pestle, and subsamples digested in aqua regia to give the total metal concentration for each of the prepared soils. Ammonium acetate (1 M, pH 7) was used to estimate the concentration of plant-available Pb for each metal phase[14,15] by overnight shaking at a soil:liquid ratio of 1:10. Analysis of the filtrates was performed using FAAS. Measurement of soil pH was conducted in water using a 1:2.5 soil:liquid ratio.

---

* g kg^{-1} is grams of chemical applied per kilogram free weight of potting mix.

Results

Brassica juncea

For every mineral phase, EDTA caused a significant increase in the concentration of metal observed in the plant (Figure 4.3). There was no significant difference between the control and citric acid treatment for each phase. In the case of the carbonate and oxide salts, acetic acid caused a significant increase in the concentration of Pb observed in the plant. In these two cases, the increase in Pb uptake was greater than that for EDTA, although this result is not statistically significant.

Figure 4.4 summarizes the EDTA treatment response of Pb in *Brassica juncea* for each salt. The relative efficacy of EDTA in inducing Pb uptake was dependent upon the phase of the metal. The suitability of these phases to EDTA-induced hyperaccumulation can be ranked as follows:

phosphate > sulphide > sulphate ~ oxide = nitrate ~ carbonate > control.

EDTA appears to be the most effective chemical inducing agent for an environment polluted with Pb phosphate or sulphide. Reference to the individual species' phase combination plots (Figure 4.3) shows that for an environment polluted with Pb oxide or carbonate, acetic acid would be a better choice; its relative efficacy compared with EDTA in these cases was nominally greater. EDTA-induced hyperaccumulation from the nitrate phase was relatively low. The nitrate salt is completely soluble and, hence, after dissolution, would most likely be present in the potting mix as an organic phase. Soluble metal salts have often been used to model polluted soils in pot trials. This result suggests that results obtained from such experiments may be misleading and may actually underestimate the ability of EDTA to induce uptake for different metals.

Thlaspi caerulescens

An easily defined metal uptake pattern is less obvious for *Thlaspi caerulescens* (Figure 4.3). With respect to the Pb sulphate, sulphide, and phosphate phases, EDTA caused a significant increase in Pb uptake relative to the control treatment. In the case of the sulphate phase, acetic acid also caused a significant increase in Pb uptake. For each of these three phases, citric acid caused no significant increase in Pb uptake relative to the control treatment.

Acetic acid and EDTA did not cause a significant increase in Pb uptake relative to the control for Pb carbonate, but citric acid caused a significant decrease in plant Pb. The mean Pb concentration for control plants growing in the carbonate-contaminated soil was 1200 mg kg^{-1}, so natural hyperaccumulation was observed for *Thlaspi caerulescens* in this particular model environment.

Hyperaccumulation was similarly observed for Pb added as a nitrate salt (mean value of 1200 mg kg^{-1}). After treatment with EDTA, the increase in Pb uptake was significant and represented the greatest increase in Pb concentration observed in this experiment. The comparison of induced Pb uptake between the two plant species is an interesting one and may involve some physiological response of *Thlaspi caerulescens* to the Pb–organic phase and EDTA interaction. Pot trials conducted where Pb is added as a soluble metal salt may again be misleading and this

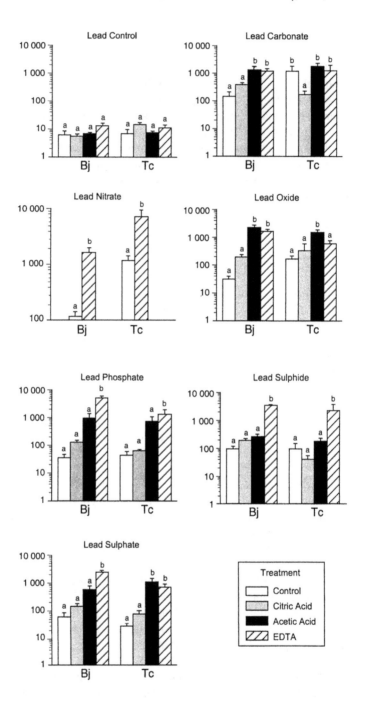

FIGURE 4.3 Natural uptake and induced uptake of Pb by citric acid, acetic acid, and EDTA in *Brassica juncea* (Bj) and *Thlaspi caerulescens* (Tc) growing on artificial 1% Pb soils of different mineral phases—mean + SE (n = 5). Means with same letter are not significantly different (ANOVA p >0.05). Data were normally distributed but plotted on log scale graph.

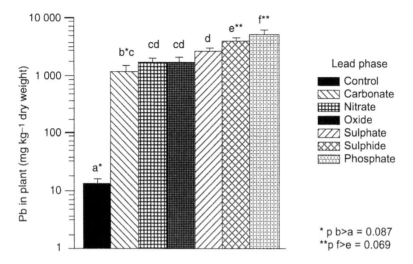

FIGURE 4.4 Summary of the relative efficacy of EDTA to induce Pb uptake in *Brassica juncea* growing on artificial 1% Pb soils of different mineral phases—mean + SE (n = 5). Means with same letter are not significantly different (ANOVA p > 0.05).

time overestimate the induced metal uptake potential of *Thlaspi caerulescens* from polluted sites. The only treatment to cause a significant increase in Pb uptake from the oxide phase was acetic acid. In this case, the increase in uptake due to EDTA was not significant.

Total soil lead

Digestion by aqua regia, and subsequent analysis of subsamples of the potting mix taken from the control pots, showed the final concentration of each mineral phase to agree with the target concentration of 1%.

Plant-available lead

Table 4.3 shows the concentration of plant-available Pb extracted by ammonium acetate (1 M, pH 7) from each of the mineral phases. The order of natural bioavailability for the seven soils was

oxide = nitrate = carbonate > phosphate ~ sulphate ~ sulphide > control.

This ordering is different from that observed for EDTA-induced uptake of Pb by *Brassica juncea*:

phosphate > sulphide > sulphate ~ oxide = nitrate ~ carbonate > control.

The difference can be attributed to the chelation effect of the ligand on Pb, i.e., the effect of EDTA to induce uptake was independent of the plant-available or ammonium acetate-soluble concentration of Pb in the soil.

TABLE 4.3
Total and plant-available Pb concentrations and pH for artificial Pb mineral-phase soils of this study. Mean Pb concentrations with same letter are not statistically different (ANOVA p > 0.05).

Pb phase	pH	Total metal (%)		Plant-available metal $(mg\ L^{-1})$	
Control	4.4	$(2.42 \pm 0.88) \times 10^{-4}$	a	0.28 ± 0.2	a
Sulphide	4.3	0.81 ± 0.28	b	33.1 ± 5.4	b
Oxide	4.8	0.92 ± 0.24	b	42.7 ± 4.1	bc
Sulphate	4.2	1.20 ± 0.22	c	55.2 ± 9.5	c
Carbonate	4.4	1.21 ± 0.10	c	92.6 ± 34.5	d
Phosphate	4.3	1.30 ± 0.20	c	106.6 ± 33.1	d
Nitrate	4.5	1.30 ± 0.20	c	115.5 ± 26.4	d

Ammonium acetate does not appear to model Pb uptake particularly well from each of the artificial Pb salts. The high plant-available metal concentrations of the contaminated soils did not translate into natural uptake. The anomalously high uptake of Pb by *Thlaspi caerulescens* from the nitrate and carbonate phase soils is not reflected in the data presented in Table 4.3. In general, both plant species appear to exclude Pb uptake.

SUMMARY—A MODEL FOR INDUCED METAL UPTAKE

The criteria that must be met for an induced hyperaccumulation operation to be successful are

- Use of a plant species with high biomass and rapid growth rate.
- Use of a chemical-inducing agent that is specific to the target metal.
- Minimization of secondary environmental problems such as metal leaching down the soil profile and high toxicity of the chemical agent.

The choice of a suitable plant is relatively easy. A species like *Brassica juncea* that has a high biomass and rapid growth rate is now regarded by many as an ideal choice for such an operation. It has been suggested that induced hyperaccumulation of a metal by a plant, which is not actively taking up the target metal, is relatively species independent.[16] This is a simplistic and broad statement and does not explain the large disparity in uptake results reported in this chapter for EDTA-induced uptake of Pb by *Thlaspi caerulescens,* relative to *Brassica juncea,* from Pb bound to an organic phase (Pb nitrate). However, the statement does appear substantiated by data for a number of plant species. In general, if the same concentration of Pb uptake could be induced into both *Brassica juncea* and *Thlaspi caerulescens,* then *Brassica juncea* would show greater phytoextraction potential due to the significantly larger biomass of this species (15 t ha^{-1} compared with 2 t ha^{-1}).

The choice of chemical-inducing agent is not so easily made. EDTA is known to complex well with Pb, but this chapter shows that more consideration needs to be given to the specific interaction between an individual metal phase and the inducing agent being used.

Trials conducted at the Auby site, where no further accumulation was induced in either a hyperaccumulator or nonhyperaccumulator species, may be explained by our data. Pb at this site is present in the carbonate and oxide phase, in an environment rich in organic material with a pH similar to that for the Pb-phase model. For both the carbonate and oxide phases, EDTA has been shown to be relatively ineffective at inducing uptake when compared with other sources of pollution. It appears that at this site acetic acid may be a better agent to use. Acetic acid is a naturally occurring organic acid; it seems likely that fewer environmental problems would be associated with the use of this chemical than have been encountered with the use of EDTA. In the future, the authors will test this theory and attempt to induce the hyperaccumulation of Pb in the field with acetic acid.

The Pb uptake results from the trial conducted on the Tui mine tailings do not fit so well with the Pb metal-phase uptake model presented in this chapter. EDTA-induced Pb uptake from a substrate polluted with metal present in a sulphide/sulphate phase could be expected to be very effective (Figure 4.2). However, the mean concentration of Pb in *Berkheya coddii* after treatment was less than 1000 mg kg^{-1} (Figure 4.3). Assuming species independence for Pb uptake, we could expect to see significantly higher concentrations in the plant after treatment.

An experiment similar to the one described in this chapter has been carried out for the metal Cd. This pot trial was run in tandem with a field test, examining the potential for the hyperaccumulation of Cd from a New Zealand pastoral environment contaminated with metal due to the loading of Cd to soil with superphosphate fertilizer.[16] The pot-trial model predicted that EDTA-induced uptake from a phosphate phase should be significant for *Brassica juncea,* but the field-trial data did not support the model; *Brassica juncea* was used in the field trial. Again, the pH at this site may be an important factor, as several years of lime storage near the field site had led to a pH increase across the trial area (pH 7.0). Mass balance calculations for the soil data at this site show a significant redistribution of Cd down the soil profile as a result of EDTA. The chemical thus solubilized Cd, but this soluble metal was not taken up by the plant species used.

The reason why EDTA-soluble Pb was not taken up from the Tui substrate, and why EDTA-soluble Cd was not taken up from the Wairarapa soil, remains unclear. In previous experiments, addition of lime to the Tui mine tailings has dramatically decreased the induced uptake of Pb by *Arrhenatherum elatius* above an addition rate of 0.125%.[17] The tailings upon which the data for Figure 4.2 are based were limed to 2.5%, so this could account for the discrepancy between the data set and the predicted model. Studies by some authors have shown that there is a strong pH dependency for EDTA-induced translocation of Pb from roots to shoots,[4] although the Pb–EDTA complex is stable at a wide range of pH values. We propose that there may be a pH constraint to the transfer of Pb–EDTA across the root cell wall membrane, a function of the geochemical conformation of the Pb–EDTA complex, that precludes Pb uptake into the roots at high pH.[16] This, of course, would preassume that Pb uptake occurred in the EDTA-complexed form. Further research needs to be conducted to examine this theory more fully.

CONCLUSION

Induced hyperaccumulation of heavy metals is strongly influenced by the chemical form of the metal present. Different sources of pollution generate contaminated environments where the pollution is present in different mineral phases. Three such environments have been illustrated in this chapter:

1. An area of mine tailings contaminated with metals present in a sulphide and sulphate phase.
2. An area of industrial pollution with metals present as oxide and carbonate phases.
3. An area of pastoral or agricultural pollution with metals present as a phosphate phase.

It appears that the choice of chemical-inducing agent to be used in such environments should be based upon a geochemical study of each specific area. EDTA may be well known as a complexing agent for metals such as Pb, but the efficacy of such a treatment will depend upon site-specific environmental geochemistry. Induced hyperaccumulation could be a powerful environmental tool, as it enables a dramatic increase in both the list of metals potentially viable for phytoremediation (and phytomining) and the list of plant species that could be used. If the site-specific limitations of induced phytoremediation technology can be realized, this technique could be used for the best potential gain.

An understanding of site-specific geochemistry is also relevant to fully appreciating natural hyperaccumulation. While *Thlaspi caerulescens* is not known as a hyperaccumulator of Pb, it appears that natural Pb uptake for this particular plant species is again dependent on the polluting metal phase. Natural hyperaccumulation of Pb was observed in two model environments: the carbonate phase and the organic (nitrate) phase. Specific hyperaccumulator plant species will probably be more effective than others in removing metals from different types of heavy-metal pollution. Again, if the site-specific species–phase interactions could be better understood, then hyperaccumulation technology could be used to the best possible advantage.

Future experiments will highlight these interactions for the metals Cd and Zn, using the hyperaccumulator species *Thlaspi caerulescens* and *Cardaminopsis halleri*, as well as the nonaccumulating species *Brassica juncea*.

Induced hyperaccumulation, where chemically induced metal movement from the soil to a plant system is the aim, appears to be a realistic objective. The inherent problems of this technology lie in the different geochemical interactions that ensue between target metals present in different mineral phases and the chemical inducing agent applied to the soil.

ACKNOWLEDGMENTS

The authors gratefully acknowledge financial support of the senior author by the Agricultural and Marketing Research and Development Trust of New Zealand (AGMARDT) through the award of a doctoral scholarship. We also thank the New Zealand Vice Chancellors' Committee for the award of a Claude McCarthy Fellowship to allow the senior author to travel to Vienna to present this paper at the 5th International Conference on the Biogeochemistry of Trace Elements.

REFERENCES

1. Chaney, R.L., Plant uptake of inorganic waste constituents, in *Land Treatment of Hazardous Wastes,* Parr, J.F., Marsh, P.B., and Kla, J.M., Eds., Noyes Data Corp, Park Ridge, NJ, 1983, 50.
2. Baker, A.J.M. and Brooks R.R., Terrestrial higher plants which hyperaccumulate metal elements—A review of their distribution, ecology and phytochemistry, *Biorecovery,* 1, 81, 1989.
3. Brooks, R.R., *Serpentine and Its Vegetation,* Dioscorides Press, Portland, 1987.
4. Blaylock, M.J., Salt, D.E., Dushenkov, S., Zakharova, O., Gussman, C., Kapulnik, Y., Ensley, B., and Raskin, I., Enhanced accumulation of lead in Indian mustard by soil-applied chelating agents, *Environ. Sci. Technol.,* 31, 860, 1997.
5. Huang, J.W. and Cunningham, S.D., Lead phytoextraction: species variation in lead uptake and translocation, *New Phytologist,* 134, 75, 1996.
6. Anderson, C.W.N., Brooks, R.R., Stewart, R.B., and Simcock, R., Harvesting a crop of gold in plants, *Nature,* 395, 553, 1998.
7. Tee Haar, G.L. and Bayard, M.A., Composition of airborne lead particles, *Nature,* 103, 135, 1971.
8. Labrune, L. and Douay, F., Métaux lourds (Pb, Zn et Cd) et microflore du sol, in *Programme de recherches concertées, étude d'un secteur pollué par les métaux—partie 1, Métaux polluants des sols (Cd, Pb et Zn) et organismes vivants,* unpublished report obtainable from the University of Lille, prepared for the Région Nord Pas de Calais, France, 1997 (in French).
9. Perdix, E., Gommy, C., Galloo, J.C. and Guillermo, R., Distribution du plomb, du zinc, du cadmium et du cuivre dans les sols par fractionnement chimique, in *Programme de recherches concertées, étude d'un secteur pollué par les métaux—partie 3, Distribution et spéciation des métaux,* unpublished report obtainable from the University of Lille, prepared for the Région Nord Pas de Calais, France, 1997 (in French).
10. Cochrane, R.H.A., Geology of Tui Mine, Mt. Te Aroha, M.Sc. Thesis, University of Auckland, New Zealand, 1969.
11. Morrell, W.J.M., Stewart, R.B., Gregg, P.E.H., Bolan, N.S., and Horne, D., An assessment of sulphide oxidation in abandoned base-metal tailings, Te Aroha, New Zealand, *Environ. Pollution,* 94, 217, 1996.
12. Morell, W.J.M., An assessment of the revegetation potential of base-metal tailings from the Tui Mine, Te Aroha, New Zealand, Ph.D. thesis, Massey University, Palmerston North, New Zealand, 1998.
13. Courtney, S.F., King P., and Rodgers K.A., A checklist of minerals from the Tui Mine, Te Aroha, New Zealand, *New Zealand Nat. Sci.,* 17, 95, 1990.

14. Ernst, W., Bioavailability of heavy metals and decontamination of soils by plants, *App. Geochem.,* 11, 163, 1997.
15. Robinson, B.H., The phytoextraction for heavy metals from metalliferous soils, Ph.D. thesis, Massey University, Palmerston North, New Zealand, 1997.
16. Anderson, C.W.N., Practical aspects of phytoextraction, Ph.D. thesis, Massey University, Palmerston North, New Zealand, 2000.
17. Deram, A., Petit, D., Robinson, B., Brooks, R., Gregg, P., and Van Halluwyn, C., Natural and induced heavy-metal hyperaccumulation *Arrhenatherum elatius:* Implications for phytoremediation, *Commun. Soil Sci. Plant Anal.,* 31, 413, 2000

5 Bioavailability of Cu, Zn, and Mn in Contaminated Soils and Speciation in Soil Solution

David L. Rimmer, Suzanne M. Reichman, and Neal W. Menzies

ABSTRACT

Greenhouse experiments were conducted in which five native Australian tree species were grown in soil spiked with copper, zinc, and manganese. The pots were equipped with samplers that allowed regular collection of the soil solution. The primary aim was to determine the critical concentration of each metal in soil solution that caused symptoms of toxicity in the trees. Preliminary experiments had shown the combination of added metal salt and lime that was required to achieve a given metal concentration in soil solution in the range 0.1–4 mM and a pH in the range 5–5.5. The Cu treatments at all concentrations caused the death of the plants. The critical concentration was therefore < 0.1 mM. Zinc and Mn were less toxic, and the treatments caused a reduction in growth with increasing solution concentration. The critical concentrations were at approximately 0.2 mM for Zn and 1 mM for Mn. Because solution concentrations were declining with time and were variable, there was considerable uncertainty in these values, and they should be used with caution. A secondary aim of the experiments was to assess the relative importance of inorganic and organically complexed forms of the metals by carrying out speciation of the collected solutions. In all cases it was found that the metals were only present in inorganic forms; possible explanations for this are given.

INTRODUCTION

The mining of metals can lead to sites that are contaminated through release of the metals from the tailings dams and rock waste dumps.[1] Many modern mines are carefully managed to minimize such contamination of the environment, but older mines, now often abandoned and derelict, are severely polluted. This is a worldwide problem, and examples can be found in many countries. In the UK there are metal-polluted mine sites in the Pennines in northern England[2] and in the southwest

1-56670-507-X/01/$0.00+$.50
© 2001 by CRC Press LLC

of the country.[3] In Australia sites such as Peelwood in New South Wales[4] and Rum Jungle in the Northern Territory[5] are well documented. There is a risk in any rehabilitation program that the metals present may cause toxicity in vegetation being established.

Assessment of the metal toxicity risk is usually based on the total metal content of the contaminated soil, because this has been found to correlate well with the soluble content, or soil solution concentration, that is available for plant uptake. A number of studies have quantified the link between the total metal content, associated with the solid fraction of the soil, and the concentration of soluble metal or that fraction of the metal in solution that is not complexed, i.e., the "free" metal ion.[6-8]

Sauvé et al.[7] found the following empirical relationships for copper (Cu) based on the analysis of a wide variety of urban, agricultural, and forest soils:

$$\text{soluble Cu} = 0.32 \text{ total Cu} + 13.2 \tag{5.1}$$

where soluble Cu is in $\mu g \ L^{-1}$ in the 0.01 M $CaCl_2$ extracts, and total Cu (HNO_3 digestion) is in mg kg^{-1} of dry soil, and

$$pCu^{2+} = 3.42 + 1.40 \ pH - 1.7 \ \log(\text{total Cu}) \tag{5.2}$$

where pCu^{2+} is the negative log of the free copper ion, measured by ion-selective electrode. McBride et al.[8] developed a similar generalized semiempirical equation for copper, zinc, cadmium, and lead based on metal complexation theory. The generalized equation, which also took into consideration the organic matter content (OM) of the contaminated soil, was

$$pM = a + b \ pH - c \ \log\left(\frac{\text{total M}}{\text{OM}}\right). \tag{5.3}$$

where M is the metal.

Our understanding of metal phytotoxicity risk has been based primarily on plant growth experiments carried out in solution culture in which it is possible to control the concentration of individual metals and obtain threshold concentrations for different species. Some workers, for example, Beckett and Davis,[9] have used the same approach to study the interaction between metals. In a study on barley growth, they found that Cu had a small antagonistic effect on the amount of Zn uptake, i.e., the presence of Cu in solution decreased Zn uptake.

When plants are grown in metal-contaminated soil, partitioning of the metals between the solid phases and the solution further complicates the situation. In a greenhouse experiment in which barley was grown on soil to which various combinations of Zn, Cu, Cd, and Pb salts had been added, Luo and Rimmer[10] found that the presence of added Cu increased the amount of Zn uptake (a synergistic effect) and led to a decrease in plant growth. This is the opposite of the effect reported by Beckett and Davis[9] and highlights the real differences that can occur when comparing soil and solution culture experiments aimed at assessing phytotoxicity risk.

Luo and Rimmer[10] hypothesized that the effect of the added Cu was to occupy the most energetically favorable adsorption sites, resulting in less Zn adsorption and greater plant availability of the added Zn. It was only possible to test this hypothesis indirectly,[11] because the soil solution was not sampled during the course of the greenhouse experiment. In the work reported here, the growth of a number of native Australian tree species in an artificially metal-contaminated soil was measured in a greenhouse experiment. At the same time, soil solutions were sampled and analyzed so that the effects of the metals on plant growth could be related to the concentration and form of the metals in solution.

In many cases, the rehabilitation of mine sites in Australia involves the reestablishment of native flora. There is very little reported work on metal toxicity in Australian native species in general and tree species in particular. Mitchell et al.[12] tested the tolerance of three native species (*Banksia ericifolia, Casuarina distyla,* and *Eucalyptus eximia*). Seeds were grown in the greenhouse in soil with added copper (as copper sulfate) at concentrations of 0–2000 mg Cu kg^{-1}. The EC50 was 205 mg kg^{-1} for the least tolerant *C. distyla* and approximately 600 mg kg^{-1} for the more tolerant *B. ericifolia* and *E. eximia*. The EC50 values for three crop plants (oat, cucumber, and soybean) were similar to the more tolerant native species. The only other study on Australian native species was the successful use of metal salts to reduce root growth around the edges of pots used to grow kangaroo paw (*Anigozanthus flavidus*).[13]

The primary aim of the work reported here was to establish the critical metal concentrations in soil solution that cause symptoms of toxicity or growth reduction in a number of native Australian tree species. This was achieved by growing seedlings in soil spiked with a range of Cu, Mn, and Zn concentrations in pots in a greenhouse and collecting and analyzing samples of soil solution from the pots on a regular basis. By carrying out metal speciation of the soil solution, it was hoped to achieve the secondary aim, which was to assess the relative toxicity of organically complexed metals compared with inorganic forms. The rationale for this was based on the extensive studies on aluminum toxicity in acid soils. Dissolved organic acids in soil solutions are important ligands for trace metals.[14] Complexation of Al with these ligands has been shown to reduce its toxicity.[15,16] It is assumed that the same will apply to the toxicity of other metals, but speciation studies have not been carried out to test this hypothesis.

MATERIALS AND METHODS

The soil used in the experiments was a Yellow Kurosol[17] (Albaquult[18]) with a pH (1:5 soil:water) of 4.8. It was collected from the University of Queensland's Mount Cotton Farm, at the sampling depth of 0–15 cm. The soil had a clay loam texture and an organic matter content of 22 mg kg^{-1}.

Copper, Mn, and Zn were added to the soil singly as sulfate salts in amounts ranging from 416 to 2257 mg kg^{-1} with the aim of producing soil solution concentrations in the range 0.1—4.0 mM. Sulfate salts were chosen to allow direct comparison with earlier work,[12] and because many contaminated mine sites are also sulfate-rich from the oxidation of iron pyrites. There were five target metal concentrations for each metal (0.1, 0.25, 0.5, 1.0, and 3.0 mM for Cu and Zn; 0.25, 0.5, 1.0, 2.0, and 4.0 mM for Mn), plus a control (no added metal), and each treatment was replicated four times. Lime was added to all of the soils and to the control samples to achieve a target pH in the range 5.0–5.5. The liming was necessary because adding soluble metal salts to soil causes acidification as a result of metal adsorption and the displacement of protons from soil particle surfaces.

The amounts of metal sulfate that had to be added to achieve the target solution concentrations, and the amounts of lime needed to reach the target pH, were determined by a preliminary incubation experiment. Good relationships (correlation coefficients of 75–85%) were obtained between added salt and solution metal concentration and added lime and final pH.

The soil was given a basal dressing of nitrogen, phosphorus, and sulfur fertilizer and placed in 20-cm-diameter plant pots (2.9 kg of soil in each). The pots were watered automatically at the base by means of an inverted long-neck bottle placed in the soil (adapted from Hunter[19]). This ensured that the soil at the base, from where the soil solution was sampled, was close to field capacity at all times.

After a period of one week for equilibration of the soil with the amendments, the pots were planted with seedlings of either *Acacia holosericea, Eucalyptus camaldulensis, Eucalyptus crebra, Casuarina cunninghamiana,* or *Melaleuca leucadendra.* Four plants were sown per pot; they were thinned to two plants after two weeks.

Each plant pot contained a small, hollow fiber soil solution sampler, which was ~1 mm in diameter and porous, with a 50,000 dalton cut-off.[20] A 15-cm length of fiber was placed toward the base of the pot and connected via Teflon® tubing to the surface. By attaching an evacuated glass sampling tube, a sample of soil solution (typically 5–10 mL) could be withdrawn from the soil. Analyses of the solutions included a pH determination and measurement of a range of element concentrations (Ca, Mg, Fe, S, P, Cu, Mn, Zn) by ICP–AES.

Speciation measurements on the soil solution allowed the partition of the soluble metals (Cu, Mn, or Zn) between inorganic and complexed forms. The methodology for metal speciation was based on that of Kerven et al.,[21] who used size exclusion chromatography to study the complexation of aluminum in acid soils. Solutions were passed through a Fractogel TSK HW-40(S) column, which has molecular size exclusion and only minimal ion exchange properties. The column was 30 cm in length, and the Fractogel was added as a slurry. The mobile phase was a degassed 0.02-M solution of $SrCl_2$ adjusted to pH 4.2 with HCl, and had a flow rate of 1 mL min^{-1}. Organic molecules passing through the column were separated according to molecular weight (MW), with the high-MW organics and any bound metals eluting first. Low-MW organics (<200 dalton) and inorganic anions and cations were retarded and eluted more slowly. The presence of organic molecules in the eluate was measured using a UV detector at 254 nm. The eluate was then collected in a fraction collector. The fractions were analyzed either by AAS for individual metals or by ICP–AES for all elements of interest, including carbon.[22]

For calibration, 1-mL samples of blue dextran, vitamin B12, metal sulfates, and copper citrate complexes were injected into the column. The calculated void volume was 8.75 mL, and the bed volume was 27 mL. Copper citrate complexes were eluted after ~10 min, whereas Cu and other metal sulfates were eluted after ~20 min. For analysis of soil solutions, 0.5-mL samples were injected into the column.

In the greenhouse experiment, the plants that survived were harvested after a period of 9 weeks. Dry weights of the shoots were obtained; the plant material was digested and the concentration of a range of elements (Ca, Mg, Fe, S, P, Cu, Mn, and Zn) was measured by ICP–AES.

RESULTS AND DISCUSSION

EFFECTS OF CU TREATMENTS

All of the added Cu treatments caused the plants to die within a week or two. Analyses of the soil solutions sampled after 3 and 5 weeks (Table 5.1) showed that the Cu concentrations in solution were initially above the target values for all but the largest addition. It was also found that the pH was well below the target of 5.0–5.5. This combination of low pH (~4) and high Cu concentration (>0.4 mM) was very toxic. At pH 4–4.5, Cu should be predominantly in inorganic forms, and the speciation measurements on the solution from these treatments showed this to be the case. It was also observed that many of the solutions were only slightly colored, suggesting that there was little dissolved organic carbon, and hence few ligands with which the Cu could be complexed.

TABLE 5.1
Soil solution copper concentrations and pH at 3 and 5 weeks after planting seedlings in Cu-treated soils

Target conc. (mM)	Metal added (mg kg^{-1})	20/4/98 (3 wk)		5/5/98 (5 wk)	
		Cu (mM)	pH	Cu (mM)	pH
Control	0	0.004	5.0	0.004	5.0
		(0.004)	(0.3)	(0.004)	(0.4)
0.1	1039	0.43	4.3	0.15	4.5
		(0.11)	(0.0)	(0.04)	(0.1)
0.25	1365	0.77	4.2	0.32	4.4
		(0.23)	(0.0)	(0.23)	(0.0)
0.5	1614	1.20	4.2	0.29	4.3
		(0.08)	(0.1)	(0.09)	(0.1)
1.0	1862	1.27	4.3	0.48	4.4
		(0.14)	(0.1)	(0.06)	(0.3)
3.0	2257	0.93	4.8	0.54	4.7
		(0.16)	(0.1)	(0.01)	(0.1)

Values are means of two subsamples. Standard deviations are given in parentheses.

Six weeks after planting, the Cu concentrations were at or below target values, and the pH of the solution was at 4.5 or above. At this point it was decided that the pots should be replanted with seedlings. For the second time, all plants, except the controls, died. It was confirmed that the toxicity was due to Cu and not Al arising from the low pH, as the concentration of Al in solution was similar for control treatments and Cu-treated pots and at nontoxic concentrations. Based on the observations in this experiment, the threshold solution concentration for Cu toxicity for these five species appears to be < 0.1 mM. The treatments with target concentrations of 0.1 mM were produced by an addition of 1039 mg Cu kg^{-1}; this can be compared with the EC50 values reported by Mitchell et al.[12] for three native Australian species, which ranged from 205 to 610 mg Cu kg^{-1}.

EFFECTS OF MN TREATMENTS

The manganese treatments produced Mn solution concentrations initially above target values, with pH values at or above target. With time, as slow equilibration took place, the solution concentrations decreased, and the pH increased slightly (Table 5.2).

The treatments produced toxicity symptoms and growth reduction of increasing severity with increasing amounts of added Mn (Table 5.3). In comparison with Cu and Zn, however, Mn was less toxic to the test species used in this study. The most sensitive species was A. *holosericea,* for which the critical addition was one designed

TABLE 5.2
Soil solution manganese concentrations and pH at 2, 4, and 9 weeks after planting seedlings in Mn-treated soils

Target conc. (mM)	Metal added (mg kg^{-1})	20/4/98 (2 wk)		5/5/98 (4 wk)		1/6/98 (9 wk)	
		Mn (mM)	pH	Mn (mM)	pH	Mn (mM)	pH
Control	0	0.03	4.8	0.01	4.8	0.004[a]	5.4[a]
		(0.01)	(0.2)	(0.00)	(0.2)	(0.003)	(0.3)
0.25	431	1.74	5.2	0.32	5.4	0.085[b]	5.5[b]
		(0.39)	(0.4)	(0.12)	(0.1)	(0.089)	(0.5)
0.5	672	2.37	5.2	1.11	5.1	0.149	5.7
		(0.85)	(0.0)	(0.46)	(0.1)	(0.122)	(0.4)
1.0	910	3.53	5.5	1.11	5.3	0.345	5.7
		(0.74)	(0.1)	(0.48)	(0.3)	(0.445)	(0.4)
2.0	1149	5.82	5.3	1.76	5.6	0.613[b]	5.9[b]
		(0.29)	(0.5)	(0.26)	(0.1)	(0.426)	(0.3)
4.0	1388	5.23	5.7	2.07	5.6	0.913	6.1
		(2.53)	(0.4)	(1.56)	(0.3)	(0.732)	(0.4)

Values are means of two subsamples for weeks 2 and 4 and pooled mean of 20 replicates for week 9, except where stated. Standard deviations are given in parentheses.

[a]One missing data point

[b]Two missing data points

to produce a soil solution concentration of 0.5 mM. This actually produced a solution concentration initially of 2.4 mM, which declined over the course of the experiment to ~0.15 mM. As a first approximation, a concentration of 2 mM can be considered to be the critical concentration for Mn for this species. Less sensitive were *E. crebra* and *M. leucandendra,* which declined from the target treatment of 2 mM (initial concentration of 5.8 mM). Least sensitive were *C. cunninghamiana* and *E. camaldulensis,* for which growth declined from a target treatment of 4.0 mM (initial soil solution concentration of 5.2 mM). There was considerable variation of the growth in the replicate treatments, as shown by the large standard deviations and the lack of many statistically significant differences.

Of the three metals studied, Mn is least likely to form complexes with dissolved organic ligands. Speciation measurements on the soil solution showed that in all cases the soluble manganese was in inorganic forms only.

EFFECTS OF ZN TREATMENTS

The added zinc behaved in a way similar to the manganese: initially high concentrations of Zn declined to below target values with time, while pH increased slightly (Table 5.4). Plant growth was more severely affected by the presence of elevated Zn concentrations in solution than it was for Mn (Table 5.5). All species, except

---

TABLE 5.3
Mean plant dry matter (grams per pot) for manganese-treated soils

Concentration (mM)		A. holosericea	C. cunninghamiana	E. camaldulensis	E. crebra	M. leucadendra
Target	Final					
Control	0.004[a]	1.77 a	0.46 ab	4.31 a	0.91 ab	2.59 a
	(0.003)	(0.41)	(0.10)	(2.34)	(0.43)	(0.34)
0.25	0.085[b]	1.18 ab	0.49 ab	4.88 a	1.34 a	1.42 ab
	(0.089)	(0.38)	(0.22)	(1.41)	(0.70)	(0.22)
0.50	0.149	1.26 ab	0.67 a	4.24 a	0.61 ab	0.71 ab
	(0.122)	(0.87)	(0.31)	(1.00)	(0.26)	(0.70)
1.0	0.345	0.68 b	0.48 ab	4.69 a	0.93 ab	1.35 ab
	(0.455)	(0.39)	(0.26)	(1.16)	(0.13)	(1.63)
2.0	0.613[b]	0.36 b	0.34 ab	4.97 a	0.35 b	0.18 b
	(0.426)	(0.47)	(0.17)	(1.39)	(0.15)	(0.09)
4.0	0.913	0.28 b	0.24 b	2.24 a	0.45 b	0.28 b
	(0.732)	(0.31)	(0.08)	(1.06)	(0.34)	(0.33)

Values of plant dry matter are means of four replicates, and final concentrations are pooled means of 20 replicates except where stated. Letters show dry-matter values that are significantly different at the 95% level. Standard deviations are given in parentheses.

[a] One missing data point

[b] Two missing data points

TABLE 5.4

Soil solution zinc concentrations and pH at 1, 3, and 9 weeks after planting seedlings in Zn-treated soils

Target conc. (mM)	Metal added (mg kg^{-1})	20/4/98 (1 wk)		5/5/98 (3 wk)		8/6/98 (9 wk)	
		Zn (mM)	pH	Zn (mM)	pH	Zn (mM)	pH
Control	0	0.005	5.3	0.005	4.7	0.0003	5.6
		(0.001)	(0.3)	(0.002)	(0.3)	(0.0001)	(0.2)
0.1	416	0.35	5.3	0.23	5.2	0.040	5.6
		(0.19)	(0.2)	(0.17)	(0.0)	(0.027)	(0.3)
0.25	726	0.22	5.7	0.29	5.0	0.060	5.8
		(0.08)	(0.3)	(0.07)	(0.1)	(0.033)	(0.3)
0.5	956	0.35	5.6	0.32	5.2	0.063	5.8
		(0.01)	(0.1)	(0.01)	(0.3)	(0.037)	(0.2)
1.0	1188	0.55	5.7	0.35	5.4	0.094	5.9
		(0.02)	(0.1)	(0.12)	(0.1)	(0.058)	(0.2)
3.0	1555	0.42	6.0	0.20	5.7	0.120	5.9
		(0.11)	(0.1)	(0.13)	(0.1)	(0.065)	(0.3)

Values are means of two subsamples for weeks 1 and 3 and pooled means of 20 replicates for week 9. Standard deviations are given in parentheses.

E. crebra, showed a significant growth reduction with all additions of Zn, and *Acacia* and *Casuarina* were the most sensitive. High concentrations of the metal (>0.2 mM) in the soil solutions sampled shortly after the start of the experiment (Table 5.4) were thought to be responsible for the growth reductions. Overall, the critical threshold soil solution concentration would appear to be approximately 0.2 mM. As with Mn, speciation measurements on the solutions showed that the zinc was only present in inorganic forms.

GENERAL DISCUSSION AND CONCLUSIONS

This novel approach to the phytotoxicity testing of metals in contaminated soils has clearly shown that the relative toxicity for the five native Australian tree species tested was Cu > Zn > Mn. There were experimental problems in maintaining a stable concentration of the metals in the soil solution and a stable pH. This highlights the need to allow much longer than one week for equilibrium to be established in the pots before planting the seedlings. The determination of exact threshold values for solution concentrations that cause toxicity symptoms has therefore proved difficult. Nevertheless, the approximate thresholds for the three metals are Cu < 0.1 mM, Zn 0.2 mM, and Mn > 1 mM. The Cu threshold agrees reasonably well with the only previously published metal toxicity threshold for native Australian species.[12] There was no clear pattern of species sensitivity, but for both Zn and Mn, *A. holosericea* was the most sensitive and the two *Eucalyptus* species least sensitive.

TABLE 5.5
Mean plant dry matter (grams per pot) for zinc-treated soils

Concentration (mM)		A. holosericea	C. cunninghamiana	E. camaldulensis	E. crebra	M. leucadendra
Target	Final					
Control	0.0003	1.13 a	0.53 a	5.98 a	1.14 a	1.43 a
	(0.0001)	(0.53)	(0.22)	(1.68)	(0.60)	(0.41)
0.1	0.040	0.09 b	0.07 b	2.66 b	1.01 a	0.64 b
	(0.027)	(0.05)	(0.04)	(1.37)	(0.93)	(0.30)
0.25	0.060	0.04 b	0.04 b	1.68 bc	0.75 a	0.27 b
	(0.033)	(0.01)	(0.01)	(0.48)	(0.70)	(0.19)
0.5	0.063	0.04 b	0.02 b	1.16 bc	0.18 a	0.35 b
	(0.037)	(0.03)	(0.01)	(1.16)	(0.26)	(0.54)
1.0	0.094	0.02 b	0.03 b	1.04 bc	0.15 a	0.14 b
	(0.058)	(0.02)	(0.01)	(0.36)	(0.13)	(0.15)
3.0	0.120	0.01 b	0.01 b	0.37 c	0.32 a	0.20 b
	(0.065)	(0.01)	(0.01)	(0.67)	(0.35)	(0.21)

Values of plant dry matter are means of four replicates, and final concentrations are pooled means of 20 replicates. Letters show dry-matter values that are significantly different at 95% level. Standard deviations are given in parentheses.

Speciation measurements on the soil solutions showed that in all cases the metals were only in inorganic forms. This is somewhat surprising, especially for the Zn treatments, where the pH was above 5, and because Zn is known to form complexes readily with organic ligands. The explanation could be that during separation on the gel column the metal complexes were unstable and the metals released from them were retarded as if they were inorganic ions. This seems unlikely, because the method has been used successfully to distinguish organic and inorganic forms of aluminum. An alternative explanation is that insufficient time had elapsed after application of the metal treatments to allow formation of the complexes. Again, this highlights the need for longer preequilibration. It also demonstrates a potential difficulty of using artificially contaminated soil compared with *in situ* and, hence, well-equilibrated, contaminated soil collected from the field.

ACKNOWLEDGMENTS

This research was carried out while David Rimmer was on study leave at the School of Land and Food at the University of Queensland. It was made possible by financial support from the University of Queensland, which provided a travel award, and the School of Land and Food, University of Queensland, which met the research costs. Graham Kerven, John Oweczkin, and David Appleton provided analytical support.

REFERENCES

1. Alloway, B.J., *Heavy Metals in Soils,* Blackie, London, 1995.
2. Colbourn, P. and Thornton, I., Lead pollution in agricultural soils, *J. Soil Sci.,* 29, 513, 1995.
3. Davies, B.E., Trace element content of soils affected by base metal mining in the west of England, *Oikos,* 22, 366, 1971.
4. Department of Mineral Resources NSW, Peelwood Rehabilitation Project, *MINFO New South Wales Mining and Exploration Quarterly,* 54, 69, 1997.
5. Menzies, N.W. and Mulligan, D.R., Vegetation dieback at the Rum Jungle uranium mine, in *Australian Society of Soil Science, National Conference: Environmental Benefits of Soil Management, Brisbane,* Mulvey, P., Ed., Australian Society of Soil Science, Sydney, 1998, 300.
6. Jopony, M. and Young, S. D., The solid-solution equilibria of lead and cadmium in polluted soils, *Eur. J. Soil Sci.,* 45, 59, 1994.
7. Sauvé, S., et al., Copper solubility and speciation of *in situ* contaminated soils: effects of copper level, pH and organic matter, *Water Air Soil Poll.,* 100, 133, 1997.
8. McBride, M., Sauvé, S., and Hendershot, W., Solubility control of Cu, Zn, Cd and Pb in contaminated soils, *Eur. J. Soil Sci.,* 48, 337, 1997.
9. Beckett, P.H.T. and Davis, R.D., The additivity of the toxic effects of Cu, Ni and Zn in young barley, *New Phytol.,* 81, 155, 1978.
10. Luo, Y. and Rimmer, D.L., Zinc–copper interaction affecting plant growth on a metal-contaminated soil, *Environ. Pollution,* 88, 79, 1995.
11. Rimmer, D.L. and Luo, Y., Zn–Cu interaction affecting Zn adsorption and plant availability in a metal contaminated soil, *Pedosphere,* 6, 335, 1996.
12. Mitchell, R.L., Burchett, M.D., Pulkownik, A., and McCluskey, L., Effects of environmentally hazardous chemicals on the emergence and early growth of selected Australian native plants, *Plant Soil,* 112, 195, 1988.
13. Baker, J.F., Burrows, N.L., Keohane, A.E., and de Filippis, L.F., Chemical root pruning of kangaroo paw (*Anigozanthos flavidus*) by selected heavy metal carbonates, *Sci. Hort.,* 62, 245, 1995.
14. Tam, S.C. and McColl, J.G., Aluminum and calcium binding affinities of some organic ligands in acidic conditions, *J. Environ. Qual.,* 19, 514, 1990.
15. Bartlett, R.J. and Riego, D.C., Effect of chelation on the toxicity of aluminum, *Plant Soil,* 37, 419, 1972.
16. Kerven, G.L., Asher, C.J., Edwards, D.G., and Ostatek-Boczynski, Z., Sterile solution culture techniques for aluminium toxicity studies involving organic acids, *J. Plant Nutrit.,* 14, 975, 1991.
17. Isbell, R.F., *Australian Soil Classification,* CSIRO Publishing, Melbourne, 1996.
18. Soil Survey Staff, *Keys to Soil Taxonomy.* U.S. Dept. of Agriculture, Natural Resources Conservation Service, Washington, D.C., 1992.
19. Hunter, M.N., Semi-automatic control of soil water in pot culture, *Plant Soil,* 62, 455, 1981.
20. Menzies, N.W. and Guppy, C.N., Soil solution extraction with polyacrylonitrile hollow-fibers, in *6th International Symposium on Soil and Plant Analysis, Brisbane,* Bruce, R.C., Ed., Australasian Soil and Plant Analysis Council, Kensington, 1999.
21. Kerven, G.L., Ostatek-Boczynski, Z., Edwards, D.G., Asher, C.J., and Oweczkin, J., Chromatographic techniques for the separation of Al and associated organic ligands

present in soil solution, in *Plant Soil Interactions at Low pH,* Date, R.A., Grundon, N.J., Rayment, G.E., and Probert, M.E., Eds., Kluwer Academic, Dortrecht, 1995, 47.
22. Oweczkin, I. J., Kerven, G.L., and Ostatek–Boczynski, Z., Determination of dissolved organic carbon by inductively coupled plasma atomic emission spectrometry, *Commun. Soil Sci. Plant Anal.,* 26, 2739, 1995.

Section II

Fluxes and Transfer Partitioning
of Trace Elements

6 Experimental and Theoretical Study on Equilibrium Partitioning of Heavy Metals

Willie J.G.M. Peijnenburg, Arthur C. de Groot, and Rens P.M. van Veen

ABSTRACT

Predicting the effects of metals on biotic species, communities, and ecosystems is at present seriously hampered by a lack of (quantitative) understanding of the factors that modulate metal bioavailability. Bioavailability needs to be dealt with as a dynamic process, comprising a physico–chemically-driven desorption process and a physiologically-driven uptake process. In this chapter, the physico–chemical aspects of bioavailability in a typical Dutch field situation are studied. Forty-nine Dutch soils, selected to cover a wide range of soil types occurring in the Netherlands, were sampled, and the partitioning of six metals (Cd, Cr, Cu, Ni, Pb, and Zn) and the metalloid As over the soil solid matrix and the pore water was studied. The main soil characteristics determining metal partitioning were quantified, and statistical models were derived to describe the partitioning process on the basis of a limited number of easily determinable soil properties. As there is evidence for predominant pore-water uptake of metals by organisms living in the soil, the models thus derived provide the first step in predicting the availability for uptake and, hence, prediction of the toxic effects of the metals studied.

INTRODUCTION

Environmental quality objectives for toxic substances are derived on the basis of risk considerations, where "risk" usually means the extent of an adverse effect. It is the purpose of ecotoxicological risk assessment to distinguish between soils or sediments that will or will not produce effects. In the case of metals, total concentrations in soils and sediments commonly span several orders of magnitude. Organisms, however, do not respond to total concentrations, so soil quality criteria that are based on total concentrations are unlikely to predict adverse biological effects. The total amount of a substance may not be toxicologically meaningful, as it may be partly nonavailable for uptake by organisms. This would not be important if availability were a constant factor. However, variations in some crucial soil properties result in substantially different availability for uptake of compounds by organisms in different soils. This variation should be taken into account to improve the accuracy in predicting (no) effects. For hydrophobic organic compounds this has, to a large extent, been achieved by developing and validating a procedure for normalization of the contaminant concentration of the amount of particulate organic carbon present in the system.[1,2]

In the case of soils there has been little consideration of the factors that modulate the bioavailability of metals. It is necessary to develop methods that contain qualitative and quantitative descriptions of differences in bioavailability

- Between soils typically used for laboratory testing and field soils
- Between contaminated and noncontaminated (natural background) soils
- Among contaminated field soils

It should be noted that (bio)availability must be dealt with as a dynamic process comprising at least two distinct phases: a physico–chemically-driven desorption process and a physiologically-driven uptake process requiring identification of specific biotic species as the endpoint.[3] Van Wensem et al.[4] and Van Straalen[5] have shown that it is eventually the body concentration that is critical in many organisms, as this is directly related to organ-effect levels. Soil organisms potentially have different uptake routes. It is thought that most organisms that live in the soil (including plants) are primarily exposed via pore water (e.g., Allen et al.[6] and references cited therein), but organisms that live *on* the soil are exposed indirectly via their food. There is evidence for predominant pore-water uptake of organic substances by soft-bodied animals, but due to their complex physico–chemical behavior such evidence is at present only circumstantial for metals.[7,8] Free metal ions in pore water are often considered to be the toxic species that can actually be taken up by organisms. Clearly, both abiotic (soil characteristics) and biotic (species-dependent) aspects determine "bioavailability."

In the Netherlands, maximum permissible and negligible concentrations (MPCs and NCs, respectively) for metals were first derived by Van de Meent and his coworkers[9] on the basis of the available ecotoxicological information, without taking into account that metals are naturally present in the environment. In most cases the methodology resulted in MPCs and/or NCs lower than what was considered to be the natural background situation. In those cases, the environmental quality objectives were set equal to the background concentration, but this was not considered an acceptable solution on the longer term. It was later realized that in the methodology proposed by Van de Meent and his colleagues,[9] some additional discrepancies were present because differences in bioavailability were insufficiently taken into account, and because no attention was paid to the fact that some metals are essential for the optimal functioning of living organisms.

As a first step toward solving the methodological problems identified, the so-called "added risk approach" was proposed by Struijs et al.[10] and implemented by

Crommentuijn and his colleagues[11] for the calculation of MPCs and NCs, taking existing background concentrations in the Netherlands into account. The starting point for this approach is the calculation of a maximum permissible addition (MPA) on the basis of data from laboratory toxicity tests. The MPA is considered to be the maximum concentration on top of the background concentration due to anthropogenic activities, taking the effects of the bioavailable fraction of the metals in the background into account. In the added-risk approach, fixed values are used to correct for differences in (bio)availability of the metal under consideration. It was, however, recognized that it is necessary to derive and validate methods for calculating and measuring potentially and actually bioavailable metal concentrations in soils to enable a more realistic estimate of the risks imposed by heavy metals in the near future. Among others, a definition study was carried out by De Rooij and his colleagues,[12] aimed at developing an improved methodology for the determination of heavy metal standards. In addition to a feasibility assessment for such a methodology, a research and development program was initiated.[13] This program encompasses the development of empirical models relating actually available metal concentrations to potentially available fractions, methods for measuring these concentrations, procedures for the extrapolation of data, and an uptake model that relates actually bioavailable metal concentrations to the concentrations in a number of test organisms. The program is anticipated to be based upon field soil samples selected to cover a wide range of soil types occurring in the Netherlands. The results of initial studies carried out by Van den Hoop et al.[14] and Janssen et al.[15] on metal partitioning in unpolluted (heavy metals at background levels) and moderately contaminated Dutch field soils were used in the design of the improved methodology for the determination of heavy metal standards.

EQUILIBRIUM PARTITIONING

A large fraction of soil organisms is directly or indirectly exposed via pore water. Metal concentrations and metal activities in the pore water are dependent upon the metal concentration in the solid phase as well as the composition of both the solid and the liquid phases. It is therefore of great practical interest and importance to have a quantitative understanding of the distribution of heavy metals over the solid phase and the pore water. A relatively simple approach for calculating the distribution of heavy metals in soils is the equilibrium partitioning (EP) concept.[16,17] The EP concept assumes that chemical concentrations among environmental compartments are at equilibrium, and that the partitioning of metals among environmental compartments can be predicted based on partition coefficients. The partition coefficient, K_p, used to calculate the distribution of heavy metals over solid phase and pore water is defined as

$$K_p = \frac{Metal_{solid\ phase}}{Metal_{pore\ water}}\ (L\ kg^{-1}) \tag{6.1}$$

K_p is not a constant and may vary by several orders of magnitude. It is affected by element properties and both solid-phase and pore-water characteristics. Knowledge of the relationship between soil characteristics and K_p values enables a calculation of the distribution of heavy metals over the solid phase and pore water for different soils. When coupled to an uptake model for metals by biota that are directly or indirectly exposed via the pore water, the relationships for predicting K_p values may be used to predict metal uptake for these organisms on the basis of the metal concentration in the solid phase, a property that is relatively easy to determine. It should be noted that, in the formula for calculating K_p, several expressions for metal concentrations in the pore water and the solid phase may be used. In this study, metal levels in the pore water are expressed in terms of total concentrations, so the calculated metal activities are used as the denominator in Eq. 6.1. Metal concentrations in the numerator are expressed in terms of total concentrations obtained after digestion of the soil matrix with either aqua regia or concentrated nitric acid. A 0.01-M $CaCl_2$-extraction was used as an expression of metal levels in the solid phase.

Finally, it should be noted that K_p is often referred to in the literature and text-books as the distribution coefficient (K_d). However, the difference between the two is often not clear, so in this chapter only the term K_p is used.

THE AIM OF THIS CHAPTER

The main aim of this chapter is to provide additional experimental data on *in situ* partitioning of six metals (Cd, Cr, Cu, Ni, Pb, and Zn) and As in Dutch field soils to be used in the validation and extrapolation stages of the research and development program proposed by De Rooij and his colleagues.[12] The data collected will in part supplement the database on metal partitioning in Dutch field soils that was established by Van den Hoop[14] and Janssen et al.,[15] and will provide the basis for deriving a bioavailability model for organisms predominantly exposed via pore water. In addition to data on metal partitioning, the study was aimed at determining a number of soil and pore water characteristics expected to influence metal partitioning. To that end, 46 sites were selected for sampling. To gain insight into possible changes of metal partitioning over time, the 46 sites selected encompass the 20 soils sampled by Janssen et al.,[15] as well as 3 of the 13 soils sampled by Van den Hoop.[14] OECD–artisoil[18] was included in the dataset for reference purposes, and duplicate sampling of one soil was carried out to gain insight into the reproducibility of the experimental procedures. OECD–artisoil is an artificial soil that is regularly used in toxicity studies. It was included in the dataset so metal partition in the artificial soil could be compared with partitioning in the field, with the aim of investigating possibilities to extrapolate results of toxicity and partitioning studies carried out in a typical lab setting to realistic field conditions. To gain insight into differences in metal partitioning of (aged) field soil samples and samples to which metals salts were added shortly before analysis, the final number of 49 soil samples was obtained by addition of an aqueous solution of metal salts to one of the field soil samples collected. It should be noted that although we often refer to the sampling of "Dutch" field soils, the dataset includes two samples taken from a highly polluted site in Belgium (Maatheide) and one sample taken in Germany. Because the German sample was expected to contain high natural lead levels, these three samples significantly broaden the range of metal levels included in this study.

A secondary aim of this study was to derive practical models for predicting metal partitioning in Dutch soils based upon easily determinable soil and pore water characteristics. The multivariate regression models are to be used to predict metal pore water levels or, preferably, metal activities in the pore water on the basis of easily determinable total metal concentrations in the solid matrix and a limited number of soil characteristics. The results obtained must be regarded as a first attempt toward deriving bioavailability models for organisms exposed via pore water, and will be further extended according to the research and development program mentioned above (De Rooij et al.[13]). It should be noted that, given the empirical and practical approach that was followed, the multivariate expressions require validation on the basis of the underlying mechanisms. Such a mechanistic approach was not the purpose of this contribution.

Within the concept of bioavailability, it is important to develop methods for calculating the nonavailable fraction of the metals present in the soil matrix. The difference in metal levels found in the digestion of the soil matrix with aqua regia or concentrated nitric acid and with the 0.01-M $CaCl_2$ extraction may give a first, practical indication of the immobile or nonavailable metal fraction. In this chapter we present models for calculating the nonavailable fraction thus defined.

EXPERIMENTAL METHODS AND MATERIALS

SOIL SAMPLING

Soil samples were collected at 46 different sites between September and December 1997. Of these, 43 were located in the Netherlands; 2 Belgian sites (both located at Maatheide, one of which was highly polluted) and one German site (Stolberg, a site expected to contain relatively high background levels of lead) were also sampled. The sampling procedure was then duplicated at the sites Veenoord and Winterswijk. To one of the latter samples (encoded as AR), an aqueous solution of a mixture of metal salts was added. Similarly, an aqueous solution of a mixture of metal salts was added to the OECD–artisoil (Sample X). Figure 6.1 shows the locations sampled, and the site codes are given in Table 6.1. The soils were classified according to the Dutch classification scheme[19] in which soils containing less than 8% clay are classified as sandy soils. Soils are classified as light clay if the clay content is between 8 and 35%, and as a heavy clay if the clay content exceeds 35%. A humus-poor soil contains less than 2.5% organic matter (OM), and a humic soil is between 2.5 and 15% OM.

The following criteria were used for the selection of sampling sites:

- Elevated metal concentrations, as well as metal levels in soils considered to contain metals at natural background concentrations, were to be included.
- The sites should have experienced little or no impact from agricultural practices.
- The sites should cover the Netherlands in a geographical sense.
- The soil characteristics had to vary among the sites.

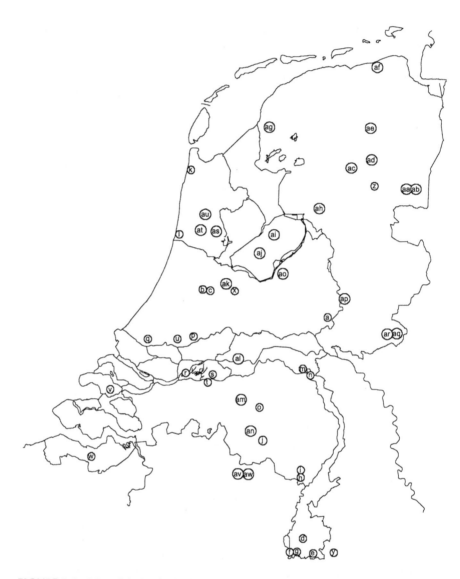

FIGURE 6.1 Map of the Netherlands showing locations of sampling sites. See Table 6.1 for site codes.

To gain insight into time-related changes of metal partitioning in undisturbed and slightly polluted soils, the 20 sites sampled by Janssen et al.[15] and 3 sites sampled by van den Hoop[14] [Lheebroekerzand (AC), Norgerholt (AE), and Eendenkooi (AL)] were resampled.

TABLE 6.1
Site Codes, Locations, and Main Sources of Metal Pollution (When Appropriate) of Sites Sampled

Site	Location	Main source of metals	Soil type
A	Rozendaal	Secondary lead smelter	Sandy soil, humus poor
B	Woerden	Motorway traffic	Heavy clay soil, humic
C	Woerden	Railway	Heavy clay soil, humic
D	Houthem, river bank (De Geul)	Lead/zinc mining	Light clay soil, humic
E	Epen, river bank (De Geul)	Lead/zinc mining	Light clay soil, humic
F	Eijsden	Zinc oxide factory	Light clay soil, humic
G	Eijsden, river bank (De Maas)	Sources upstream	Light clay soil, humic
H	Budel	Zinc factory	Sandy soil, humus poor
I	Budel	Zinc factory	Sandy soil, humic
J	Valkenswaard, river bank (De Dommel)	Sources upstream	Sandy soil, humic
K	Callantsoog	Former shooting range	Sandy soil, humus poor
L	Wijk aan Zee	Blast–furnace steelworks	Sandy soil, humic
M	Heumen	Power line pylon	Sandy soil, humus poor
N	Mook	Zinc plating factory	Sandy soil, humus poor
O	Boxtel, river bank (De Dommel)	Sources upstream	Sandy soil, humic
P	Bergambacht, river bank (De Lek)	Sources upstream	Sandy soil, humic
Q	Vlaardingen	Waste incinerator	Light clay soil, humic
R	Kop van het Land, river bank (Merwede)	Sources upstream	Light clay soil, humic
S	Hank, estuarine river bank (Biesbos)	Sources upstream	Light clay soil, humic
T	Drimmelen, river bank (Amer)	Sources upstream	Sandy soil, humic
U	Ouderkerk a/d ijssel		Light clay soil, humic
V	Nieuwerkerk		Light clay soil, humic
W	Sluiskil		Light clay soil, humic
X	OECD	Metals added in the laboratory	Light clay soil, humic
Y	Stolberg (D)	Natural lead background at elevated level	Light clay soil, humic
Z	Stuifzand		Sandy soil, humic
AA	Veenoord	Galvanization factory	Sandy soil, humic
AB	Veenoord	Galvanization factory, duplicate of AA	Sandy soil, humic
AC	Lheebroekerzand		Sandy soil, humic
AD	Westerbork		Sandy soil, humic
AE	Norgerholt		Sandy soil, humic
AF	Noord Polder		Sandy soil, humus poor
AG	Schraard		Heavy clay soil, humic
AH	Genemuiden		Light clay soil, humic
AI	Larserbos		Light clay soil,humic
AJ	Zenderpark		Sandy soil, humus poor
AK	Maarssen	Galvanization factory	Sandy soil, humic
AL	Eendenkooi	Natural zinc background	Heavy clay soil, humic
AM	Zandelei		Sandy soil humic
AN	Knegsel		Sandy soil, humic
AO	Ermelo		Sandy soil, humic
AP	Zutphen		Sandy soil, humic
AQ	Winterswijk geaddeerd	Metals added to sample AR	Sandy soil, humic
AR	Winterswijk		Sandy soil, humic
AS	Purmerend		Sandy soil, humic
AT	Krommenie		Sandy soil, humic
AU	De Rijp		Light clay soil, humic
AV	Maatheide (B)	Former zinc smelter	Sandy soil, humus poor
AW	Maatheide (B)	Former zinc smelter	Sandy soil, humic

PRETREATMENT OF SOILS

At each site, the upper litter or grass layer was removed and a total of 30 L of soil from the top layer (0–20 cm) was collected. The samples were transferred to the laboratory and stored in three 10-L polyethylene containers at 5°C for further handling and analysis.

All roots present in the samples were removed in the laboratory, and agglomerates were broken by hand or machine. Particles larger than 4 mm were removed by sieving, and the remaining soil was homogenized. One-third of the soil material collected was used for further pretreatment to enable chemical analyses; the remaining soil material was stored at 5°C.

The soil material to be used for chemical analyses was air-dried, particles larger than 2 mm were removed by sieving, and the remaining soil was stored at room temperature. The moisture content of air-dried soil was determined from the weight loss of approximately 10 g of soil heated at 105°C for about 12 hr.

COLLECTION OF PORE WATER

A somewhat modified procedure was applied to the collection of pore water, as compared with the procedure used by Janssen and his colleagues.[15] For practical reasons (collection of sufficient amounts of pore water, even for relatively dry soils), an electrolyte solution [2 mM $Ca(NO_3)_2$] was added to the soils prior to centrifugation: 2 kg of soil from each location was moistened with a 2-mM solution of $Ca(NO_3)_2$ to get a pF value of 2. Subsequently, the soils were stored for three weeks at 5°C.[20] After this equilibrium period, pore water was obtained by centrifuging the soils at 7500 rpm (6000 g of soil was used) at 5°C. Centrifugation was continued until about 150 mL of pore water was collected. If sufficient quantities of pore water could not be collected, another 2-kg portion of soil was centrifuged. After centrifugation, the pore water collected was filtered over a 0.45- μm pore size filter, and the pH was measured [pH(pw)]. The percentage of pore water collected, as related to the moisture content of the soil, ranged from 2 to 76% (w/w).

The pore water collected was divided into two polyethylene bottles: one bottle of 30 mL of pore water was acidified with concentrated nitric acid to set the pH to 2 and was used for metal analyses. The second bottle of 100 mL pore water was used to determine anions, pH, and dissolved organic carbon (DOC).

EXTRACTIONS AND DIGESTIONS

CaCl₂ Extraction

For each sample, approximately 10.0 g of air-dried soil was weighed into a 250-mL plastic bottle, and 100 mL of a 0.01-M $CaCl_2$ solution was added. The bottle was shaken for about 24 hr at 150 rpm. The supernatants were passed through a 0.45-μm filter. The pH in the extracts was measured [pH($CaCl_2$)], and the extracts were acidified with concentrated nitric acid to set the pH to 2 prior to metal analysis. Four blanks were prepared and treated in the same way as the soil samples.

Nitric Acid Digestion

Approximately 0.2 g of ground air-dried soil was weighed into a microwave digestion bomb, and 4 mL concentrated nitric acid was added. The soil samples were digested in a microwave oven (CEM Corporation MDS 2000) for 30 min at 180 psi. After the samples were cooled, the solution was quantitatively transferred into a volumetric flask, diluted to a final volume of 50 mL with Milli-Q water, and passed through a 0.45-μm filter. For reference purposes, seven blanks and seven standard soils were digested simultaneously.

Aqua Regia Digestion

About 1.0 g of ground air-dried soil was weighed into a microwave digestion bomb, and 4 mL concentrated nitric acid and 12 mL concentrated hydrochloric acid were added to each bomb. The soil samples were digested in a microwave oven (CEM Corporation MDS 2000) for 1 hr at 180 psi. After the samples had cooled, the solution was quantitatively transferred into a volumetric flask, diluted with Milli-Q water to a final volume of 100 mL, and passed through a 0.45-μm filter. For reference purposes, seven blanks and seven standard soils were digested simultaneously.

Ammonium Oxalate–Oxalic Acid Extraction

Approximately 1.0 g of air-dried soil was weighed into a 250-mL plastic bottle, and 60 mL of a 0.175-M ammonium oxalate–0.1-M oxalic acid solution was added to each bottle.[21] The bottles were shaken in the dark for 2 hr at room temperature. The supernatants were passed through a 0.45-μm filter. Three blanks were prepared and treated in the same way as the soil samples.

CHEMICAL ANALYSIS

Pore Water

Cations
The pore water-cation concentrations were obtained for several elements. Cd, Cr, Ni, and Pb were analyzed by graphite furnace atomic absorption spectroscopy (AAS) (Perkin Elmer 4100 and Zeeman background correction). Cu and Zn were analyzed by flame AAS (Perkin Elmer 2100 and deuterium background correction). Arsenic was analyzed by FI-AAS (Perkin Elmer 2100 with FIAS-200). Ca, Mg, Na, K, Zn, Fe, Mn, and Al were analyzed by ICP-AES (Spectro Analytical Instruments).

Anions
The pore water was analyzed for the following anions: Cl^-, NO_3^-, and SO_4^{3-} (all determined by ion chromatography), and PO_4^{3-} (continuous flow analysis).

Dissolved organic carbon
Dissolved organic carbon (DOC) was determined with a Dohrmann DC-190 TOC analyzer.

Extracts and digests

The $CaCl_2$ extracts (as a matter of course, with the exception of Ca), and the HNO_3 and aqua regia digests were analyzed for the same cations as the pore water. The ammonium oxalate–oxalic acid extracts were analyzed for Al and Fe by ICP-AES.

CHARACTERIZATION OF THE SOLID PHASE

The soils have been characterized in terms of pH(pw), pH(KCl), pH($CaCl_2$), loss-on-ignition (indicated as LOI, %), organic carbon content (OC, %), clay content (clay, %), granules between 2 and 38 μm (fraction, %), cation exchange capacity (CEC, cmol kg^{-1}), and amount of Al/Fe-oxyhydroxides (Al–ox and Fe–ox respectively, mmol kg^{-1}). Only for soils exceeding pH($CaCl_2$) = 5.5 was the carbonate content determined with an element analyzer (Model EA 1108, Fisons Instruments) after heating at 450°C for 3 hr.

Two expressions of LOI were obtained: LOI–1 and LOI–2, respectively. LOI–1 is considered representative of the organic matter content of the solid phase and was determined from the weight loss of approximately 5 g of dried soil (105°C) heated at 550°C for 3 hr. LOI–2 is considered representative of the inorganic matter content of the solid phase and was determined from the weight loss of the sample used to determine LOI–1 after additional heating at 900°C for 3 hr. The carbon and nitrogen content were determined with an element analyzer (Model EA 1108, Fisons Instruments). The organic matter content (OM, %) was calculated from the carbon content by multiplying with a value of 1.7. The pH(pw) was determined directly in the pore water; pH(KCl) was determined at a 2:5 soil:liquid ratio (w/v) with 1-M KCl, and pH($CaCl_2$) was determined in the 0.01-M $CaCl_2$ extract. The CEC of the soil was determined in an unbuffered $BaCl_2$ extract (based on NEN 5780). The amount of soil that was used for the determination of CEC was soil-type-dependent: 20 g of air-dried soil material for sandy soils and 7 g of air-dried material for the remaining soils was weighed into a centrifuge tube. Several portions of 100 mL of 0.5-M $BaCl_2$ were added, the suspension was shaken for 1 hr, and the cation exchange sites were loaded with Ba. Eventually, 100.0 mL 0.02-M $MgSO_4$ was added, upon which $BaSO_4$ precipitated and the cation exchange complex was fully occupied by Mg. The amount of Mg that remained in the supernatant was determined with capillary zone electrophoresis. The CEC was calculated from the difference between the added amount of Mg and the amount of Mg determined in the extract. The clay content was determined according to NEN 5753.[22]

CHEMICAL SPECIATION CALCULATIONS

The different forms (species) in which heavy metal ions (Me) and other ions may be present in the pore water (i.e., Me^{2+}, $MeCl^+$, $MeOH^+$, $Fe(OH)^{2+}$, $CaNO_3^+$, $MgHCO_3^+$, $Al(OH)^{2+}$, HCO_3^-, MeDOC, etc.) were calculated by chemical speciation calculations using the MINTEQ program.[23] Next to the measured pore water characteristics (DOC, cation and anion concentrations, and pH), the following provisions were made before the MINTEQ calculations were executed:

1. Chemical equilibrium constants between many of the possible inorganic species that may occur in water solutions were taken from the NIST database.[24]
2. According to program characteristics, MINTEQ was allowed to predict precipitation, which implies that finely dispersed particles may be formed under the prevailing conditions. In that case, pore water-metal concentrations are lower than under nonprecipitative conditions.
3. The average binding site concentrations of the DOC, relevant for the degree of metal complexation, were calculated by the RANDOM program[25,26] under the assumptions of Pretorius et al.[27] as to the composition and functional group content of the DOC:
   (a) All DOC is present as humic and fulvic type materials
   (b) Only R-COOH, R-OH, $\varnothing$-COOH, and $\varnothing$-OH are available to form binding sites
   (c) Average values for elemental composition and function contents are representative of the DOC in the samples.
   Ligands binding less than 1% metal at each pH value were disregarded. The eight remaining ligands were malic acid, acetylacetone, catechol, succinic acid, 2-hydroxy-2-methylpropionic acid, phthalic acid, and propionic acid.
4. The influence on metal speciation of the redox potential (pe), which was not measured in the pore water, was assumed to be negligible in view of simulation results obtained for each soil using the lower and upper limits for pe + pH.[28] The redox potential is important for Fe(II)/Fe(III) equilibria and for the formation and dissolution of ferric precipitates. The precipitation potential was found, especially for soils having a relatively high pH($CaCl_2$); among others, chloropyromorphite [$Pb_5(PO_4)_3Cl$], octavite ($CdCO_3$), and malachite [$Cu(OH)_2.CuCO_3$] were predicted to be formed. These precipitates could play a role in controlling the solubility of the cations involved.

Speciation calculations were carried out for all metals included in this study; insufficient data were available for As.

PRINCIPAL COMPONENT ANALYSIS (PCA) AND PARTIAL LEAST SQUARES (PLS)

Principal component analysis (PCA) was used to determine meaningful patterns among the soils and relations between the soil characteristics. Geladi and Kowalski[30] provide details of this method. Geometrically, the data points can be represented as points in a multidimensional space with the variables (in our case, the soil characteristics) as axes. Distances and clusterings of points can be interpreted as similarities and dissimilarities among the objects. PCA calculates vectors (principal components) that fit best through the multidimensional data points. The first principal component is the vector of best fit for the data points. Subsequently, principal components can be calculated orthogonal to each other, creating a plane or hyperplane with increasingly smaller R^2. To get an overview of the dataset, a few (two or three) principal components are often sufficient. Subsequently, identified principal components are characterized by a decreasing correlation coefficient, which usually becomes insignificant at the level of the third or fourth (or higher level) component. The number of significant components is determined via cross-validation criteria given within the program. A principal component consists of a score, which summarizes the X-variable (soil sample), and a loading, showing the influence of the variables (soil characteristics). In a score plot defined by two principal components, soil samples that have similar characteristics plot out near each other. This can give an indication of the similarity of soils. A loading plot defined by two principal components can give relationships among the soil characteristics. As in a score plot, soil characteristics that plot out near each other on the loading plot may be closely related.

Partition coefficients were related to soil characteristics using the partial least squares (PLS) projection to latent structures method. PLS is a multivariate projection method that finds relationships between predictor variables (here, soil characteristics) and a response variable (here, the partition coefficient) through regression modeling in latent variables in a way similar to PCA. As a measure of goodness-of-fit, we used the adjusted R^2. This is the variance of all the Ys explained by the principal components. By using the adjusted R^2 instead of a nonadjusted R^2, the R^2 values are corrected for the influence of the number of Xs entered in the model. R^2 has a maximum value of 1, and the higher this value, the better the model is considered to be. Adjusted R^2 values calculated with PLS are not sensitive for correlation between the descriptors when determining the relation between predictors and response. SIMCA-S 6.0 for Windows (Umetri AB[30]) was used for the PLS and PCA analyses.

Because the raw soil characteristics and K_p data showed a log-normal distribution, the datasets used in this study were log transformed (except pH) before PLS and PCA analyses were carried out to meet the assumption of homoscedasticity required for the regression models.[31] The final models were derived by means of stepwise multiple regression analysis.

EXPERIMENTAL FINDINGS

The results of the soil analyses are given here. Apart from the soil and pore water characteristics assumed to dominate metal partitioning, metal concentrations in the solid and liquid phase, calculated metal activities in the pore water and values of the partition coefficients of the metals included in this study are reported. Because three different expressions for the metal levels in the solid phase (aqua regia and concentrated HNO_3 digestion and $CaCl_2$ extraction) and two expressions for metal levels in the pore water (total concentrations and activities) were measured, six different partition coefficients could be calculated for each metal (three in the case of arsenic due to a lack of calculated arsenic activities in the pore water). In addition, the *differences* in metal levels, obtained by means of the two methods of digestion used in this study and by $CaCl_2$ extraction, were calculated. These differences might be indicative of the nonavailable metal fraction in the soil. In most cases, it turned out that, in an absolute sense, the values thus obtained did not deviate significantly from the total metal concentrations, so these differences are not reported. On the other hand, $CaCl_2$ extraction can be seen as an operationally defined expression of the available metal fraction present in the soil matrix. The operationally defined available metal fraction is expressed in this chapter as the percentage of the total metal content (aqua regia and HNO_3 digestion) that can be extracted by means of 0.01-M $CaCl_2$ extraction.

SOLID-PHASE CHARACTERISTICS

The main solid-phase characteristics assumed to influence metal partitioning for all soils studied are given in Table 6.2. In addition to the experimental findings for each soil, the minimum, maximum, and average values are also included. The large variation of soil properties reflects one of the main criteria used for selecting the sampling sites, i.e., that the soil samples should vary with respect to their physico–chemical composition.

For most of the parameters determined in these soils, the values found were well reproducible, and duplication of the whole procedure (soils AA and AB) showed that, in general, deviations of less than 2% were observable. However, in some cases (such as the OM content of the soils and the percentage of granules between 2 and 38 μm, for instance), large deviations of over 10% were detectable. This might be due to the fact that the data given represent point samples that do not necessarily reflect the pollution status of a larger area; even samples that are taken at the same site might deviate slightly.

METAL CONCENTRATIONS IN SOLID PHASE

The concentrations of some elements displaced by the aqua regia and HNO_3 digests, as well as by $CaCl_2$ extraction, are given in Tables 6.3 through 6.5. The data are expressed on a dry-weight basis, and the elements extracted by $CaCl_2$ have been corrected for the amount of each element that was supposed to be present in the pore water at the moment of sampling. $CaCl_2$ extraction releases only part of the metals from the solid phase; in particular, metals sorbed onto the oxide phases will not be released. As expected, much higher concentrations of elements were found in the digests than in the $CaCl_2$ extracts.

In addition to the total metal levels reported in Tables 6.3 and 6.4, a comparison is also made with the maximum permissible and negligible concentrations derived by Crommentuijn and his colleagues[11] using the added risk approach. This is done for metal concentrations determined by both aqua regia and HNO_3 digestion.

As can be deduced from Tables 6.3 and 6.4, in various samples the negligible and the maximum permissible metal concentrations are exceeded, and some soils contain more than one metal at levels exceeding the risk levels indicated. Again, it should be noted that the aim of this study was not to obtain soil samples representative for a

TABLE 6.2
Characterization of Solid Phase of Soils Sampled

Site	W_1 (%)	W_2 (%)	pH (CaCl$_2$)	pH (KCl)	Fe-ox (mM kg^{-1})	Al-ox (mM kg^{-1})	LOI$_1$ (%)	LOI$_2$ (%)	OM (%)	Clay (%)	Fraction (%)	CO$_3^{2-}$ (mM kg^{-1})
A	6.8	9.7	4.09	3.57	11.5	15.9	2.6	0.2	2.4	2.0	6.6	n.d.
B	27.7	56.5	5.60	4.71	136	70.1	15.1	1.1	15.0	39.1	19.5	n.d.
C	29.0	56.6	5.07	5.13	200	93.0	13.2	1.2	10.3	46.7	14.3	n.d.
D	17.2	29.2	7.26	6.93	41.8	18.3	4.5	2.9	4.7	11.0	35.2	0.7
E	26.6	33.6	6.65	6.30	100.1	36.6	6.0	0.9	5.2	8.9	25.2	0.2
F	18.2	36.0	7.38	6.92	43.1	28.1	6.2	3.2	8.7	13.3	40.7	1.2
G	15.1	41.0	7.24	7.20	184.8	37.9	4.8	5.9	6.8	10.0	30.3	1.5
H	2.9	8.2	3.97	3.98	3.9	11.3	2.0	0.1	2.2	0.5	1.7	n.d.
I	11.8	17.8	3.81	2.85	22.3	53.0	5.0	0.1	4.8	1.3	2.0	n.d.
J	2.8	18.7	4.55	3.98	8.2	20.7	2.1	0.2	2.6	1.3	1.2	n.d.
K	1.4	2.1	4.49	4.37	2.0	1.1	0.3	0.1	< 0.5	0.2	0.3	n.d.
L	6.3	6.8	7.12	7.80	10.0	4.2	1.6	1.9	2.7	0.8	0.6	0.6
M	6.7	7.7	3.99	3.42	27.0	22.1	1.9	0.2	1.9	3.0	2.1	n.d.
N	5.6	8.8	7.12	5.93	33.2	8.0	0.8	0.5	0.9	2.1	0.1	n.d.
O	18.9	25.1	6.09	n.d.	134	18.2	4.7	0.5	4.7	5.8	5.5	n.d.
P	22.9	30.9	7.22	7.34	146	26.1	3.5	5.6	6.0	4.4	7.9	1.4
Q	22.2	36.4	7.43	6.91	109	26.8	10.2	3.1	12.0	12.5	13.5	n.d.
R	12.8	19.2	7.36	7.16	46.7	15.1	4.1	2.6	4.7	8.2	10.6	0.5
S	33.8	65.7	7.08	6.52	214	66.1	15.0	1.9	16.8	24.6	35.9	0.7
T	21.3	29.0	7.36	7.00	48.7	10.5	3.2	0.7	3.8	6.7	5.5	0.2
U	48.8	68.1	4.88	4.36	234	248.0	35.2	1.9	32.3	27.3	22.9	n.d.
V	16.0	31.7	7.35	7.55	34.3	6.9	3.2	3.2	4.2	11.2	9.3	0.7
W	17.2	30.4	7.12	7.37	46.3	10.9	3.9	2.5	4.3	11.2	11.7	1.0
X	31.8	49.6	4.84	n.d.	1.7	7.0	11.1	0.5	6.7	11.8	8.1	n.d.
Y	22.9	36.4	4.19	n.d.	81.8	41.7	7.2	0.7	6.4	15.8	46.0	n.d.
Z	8.6	22.7	4.96	5.40	22.1	22.0	4.6	0.5	3.8	1.7	2.6	n.d.

Site	W_1 (%)	W_2 (%)	pH (CaCl$_2$)	pH (KCl)	Fe–ox (mM kg^{-1})	Al–ox (mM kg^{-1})	LOI$_1$ (%)	LOI$_2$ (%)	OM (%)	Clay (%)	Fraction (%)	CO$_3^{2-}$ (mM kg^{-1})
AA	11.0	21.6	5.50	6.00	16.6	40.5	4.3	0.2	5.2	1.7	3.0	0.2
AB	10.9	21.1	5.35	n.d.	16.0	40.4	4.4	0.2	3.8	1.9	2.3	n.d.
AC	37.3	60.4	3.59	3.06	24.1	80.0	10.5	0.2	9.3	1.4	1.3	n.d.
AD	11.8	30.5	3.65	3.22	14.8	28.6	4.9	0.1	5.3	1.1	1.9	n.d.
AE	23.5	40.8	3.09	2.36	19.0	11.0	6.8	0.2	5.8	2.3	3.7	n.d.
AF	11.3	29.6	7.30	7.63	18.7	4.2	1.1	3.1	1.9	5.4	5.7	0.6
AG	33.9	65.1	5.87	7.35	99.3	23.1	8.6	2.7	7.6	39.8	33.0	n.d.
AH	36.9	76.1	4.75	n.d.	389	51.6	26.7	1.6	23.4	29.3	30.8	n.d.
AI	22.4	50.3	7.23	n.d.	108	19.0	7.9	4.0	7.9	22.3	34.5	0.9
AJ	13.8	17.6	7.31	n.d.	27.6	5.8	3.0	0.6	2.1	4.7	4.3	n.d.
AK	17.8	25.3	6.60	5.64	64.8	36.0	5.8	0.5	5.8	3.3	4.8	n.d.
AL	30.3	68.4	5.59	n.d.	111	63.9	10.2	1.3	6.4	51.6	31.5	n.d.
AM	23.3	43.1	3.20	2.59	60.9	28.1	8.0	0.4	6.8	3.8	10.2	n.d.
AN	11.6	30.1	4.20	3.82	28.8	49.2	4.4	0.3	4.2	2.5	8.8	n.d.
AO	10.8	22.6	3.41	2.85	5.4	13.2	4.3	0.1	4.2	0.2	0.5	n.d.
AP	13.0	21.3	7.25	7.20	76.9	12.5	3.2	2.5	3.3	4.1	4.3	0.6
AQ	21.6	33.9	4.45	n.d.	34.7	41.1	6.7	0.3	6.4	2.2	3.1	n.d.
AR	11.9	29.6	4.36	3.92	40.7	47.5	6.4	0.3	6.1	2.3	3.4	n.d.
AS	20.7	25.5	7.21	7.67	57.2	16.2	4.5	1.9	5.0	6.2	3.6	0.5
AT	24.2	35.3	7.15	7.15	77.7	21.4	9.6	2.2	10.8	7.0	5.8	0.6
AU	37.5	73.8	7.16	7.45	132	23.4	14.8	2.1	16.2	15.0	9.0	0.4
AV	6.5	16.2	3.68	3.53	2.5	9.2	2.6	0.1	1.4	0.7	1.5	n.d.
AW	10.5	19.9	6.33	4.75	133.3	62.0	6.4	1.8	12.6	1.2	1.8	1.0
Min.	1.4	2.1	3.09	2.36	1.7	1.1	0.3	0.1	0.9	0.2	0.1	0.2
Max.	48.9	76.1	7.43	7.80	389.2	248.0	35.2	5.9	32.3	51.6	46.0	1.5
Avg.	18.5	33.4	5.68	5.47	71.5	33.6	6.9	1.4	7.0	10.0	11.6	0.7

W_1 = Moisture content of soil; W_2 = Moisture content of soil after centrifugation; Fe–ox/Al–ox = Amount of iron and aluminum extracted by ammonium oxalate/oxalic acid (assumed to be present as "active" or "amorphous" Fe and Al [oxyhydr]oxide[21]; OM = Organic matter; Clay = Soil particles < 2 mm; Fraction = Granules between 2 and 38 μm; CEC = Cation exchange capacity; n.d. = Not detectable; Min. = Minimum value; Max = Maximum value; Avg. = Average value.

TABLE 6.3

Total metal concentrations in solid phase (aqua regia–digestion). Boldface values represent samples containing metal levels between negligible metal concentration and maximum permissible metal concentration; italicized values represent samples exceeding maximum permissible metal concentration[11]

Site	Cd (μM kg⁻¹)	Cu	Cr	Ni	As	Pb	Zn	Al (mM kg⁻¹)	Fe	Mn	Mg	Ca	K	Na
A	0.60	0.07	0.67	**0.21**	0.07	0.17	0.26	159	89	1.85	18	9	13	1.73
B	2.98	0.43	1.51	0.72	0.16	0.18	1.52	1515	443	6.18	262	175	204	25.43
C	3.85	*1.65*	2.01	0.83	0.20	0.31	1.92	1868	560	8.28	289	128	247	25.25
D	**14.16**	0.20	1.12	**0.40**	0.09	**0.52**	7.07	464	257	6.75	92	600	71	4.89
E	*72.42*	0.41	0.76	*0.45*	0.23	4.28	*47.69*	628	318	10.70	118	120	107	35.58
F	23.98	0.48	1.26	**0.42**	0.29	0.47	*13.58*	685	330	10.75	108	503	110	12.50
G	*42.95*	**0.75**	1.28	*0.45*	0.18	n.d.	*9.14*	737	492	13.04	354	907	127	21.53
H	1.79	0.05	0.08	n.d.	0.02	0.09	0.20	37	17	0.45	3	2	11	3.79
I	3.13	0.05	0.47	0.16	0.05	0.15	0.23	88	37	0.27	5	2	7	0.84
J	**8.58**	0.07	0.19	0.03	0.07	0.11	0.84	159	50	1.86	14	14	32	6.34
K	n.d.	0.01	0.27	0.09	0.02	0.27	0.11	39	24	0.76	7	8	5	0.90
L	1.70	0.09	0.26	0.09	0.07	0.11	0.74	94	110	2.39	43	399	18	4.12
M	0.40	0.09	0.40	0.11	0.03	0.07	0.29	151	76	1.49	16	6	17	1.95
N	*2.72*	*1.61*	*1.37*	*0.45*	0.10	**0.66**	*3.05*	238	174	3.25	56	54	36	3.87
O	53.52	0.31	0.89	0.17	0.20	0.12	2.28	268	223	5.32	39	55	25	3.73
P	*37.61*	*0.84*	2.90	*0.42*	0.32	n.d.	8.92	548	394	11.29	264	1140	100	21.98
Q	5.57	**0.81**	1.17	*0.53*	0.11	*1.09*	2.29	613	370	10.46	129	594	109	63.78
R	**14.23**	0.54	*1.32*	*0.40*	0.26	0.44	5.06	424	232	7.78	169	542	69	14.84
S	*164.21*	*2.13*	3.06	0.98	**0.66**	*1.83*	*20.04*	1206	636	23.48	210	224	159	30.49
T	*17.83*	0.39	0.92	0.27	0.23	0.37	4.05	328	207	8.04	115	152	48	7.07
U	9.57	0.78	2.33	0.78	0.23	**0.90**	2.54	1259	394	3.69	168	193	104	14.28
V	1.92	0.10	0.70	0.17	0.18	0.06	0.69	492	234	n.d.	160	680	100	11.29
W	3.56	0.24	0.82	0.22	0.33	0.17	1.30	504	275	n.d.	176	437	105	42.95
X	59.45	0.61	0.65	**0.39**	0.17	**0.70**	7.97	980	21	0.23	16	52	56	4.21
Y	8.08	0.26	0.80	0.22	0.18	*1.27*	1.88	887	265	3.54	70	30	119	20.98

Site	Cd (μM kg⁻¹)	Cu	Cr	Ni	As	Pb	Zn	Al (mM kg⁻¹)	Fe	Mn	Mg	Ca	K	Na
Z	1.42	0.50	0.15	0.06	0.06	0.16	0.79	117	75	2.39	16	36	16	2.32
AA	2.31	0.28	0.23	0.19	0.02	0.18	6.09	184	69	1.60	22	40	27	6.23
AB	2.32	0.18	0.35	0.12	0.02	0.17	5.01	165	60	1.23	19	33	17	2.13
AC	2.15	0.03	0.12	0.04	0.02	0.11	0.16	104	30	0.58	7	8	10	2.05
AD	0.71	0.04	0.07	n.d.	0.02	0.04	0.12	77	29	0.47	5	6	6	1.43
AE	n.d.	0.02	0.15	0.02	0.02	0.07	0.10	91	44	0.74	9	11	16	3.01
AF	0.79	0.04	0.37	0.12	0.07	0.03	0.31	238	136	3.07	135	556	43	4.36
AG	3.21	0.22	2.65	1.10	0.24	0.14	1.24	1274	555	9.37	314	142	248	44.49
AH	6.47	0.43	1.27	0.55	0.25	0.25	1.47	911	506	12.28	180	143	121	23.60
AI	6.31	0.29	1.34	0.52	0.22	0.22	2.53	880	435	14.10	321	828	163	17.01
AJ	0.83	0.06	0.26	0.12	0.03	0.03	0.33	184	92	2.30	59	119	30	2.32
AK	10.22	0.54	0.25	0.17	0.07	0.62	19.14	224	157	3.31	46	135	27	5.71
AL	4.26	0.46	1.89	0.99	0.18	0.18	1.68	1768	595	9.48	356	145	220	51.26
AM	0.73	0.09	0.26	0.05	0.05	0.13	0.20	215	103	1.14	23	19	15	2.34
AN	4.67	0.17	0.23	0.04	0.05	0.15	0.56	146	52	3.00	12	17	11	1.62
AO	1.75	0.01	0.07	n.d.	0.01	0.04	0.09	32	12	0.20	2	3	4	1.16
AP	4.62	0.22	0.61	0.26	0.10	0.20	1.71	289	210	6.78	132	515	50	5.00
AQ*	27.74	0.47	0.56	0.24	0.14	0.58	3.92	227	118	6.67	26	50	23	3.87
AR	2.95	0.21	0.66	0.19	0.05	0.23	0.69	187	105	5.79	22	39	17	2.22
AS	0.84	0.26	0.31	0.16	0.10	1.79	0.77	267	135	3.38	83	381	42	6.11
AT	7.23	0.54	0.47	0.22	0.09	0.47	2.53	320	223	5.87	111	421	63	13.11
AU	4.28	0.37	0.83	0.27	0.20	0.54	2.24	584	241	5.30	112	291	107	25.11
AV	5.28	0.11	0.07	0.03	0.03	0.21	0.68	42	16	0.24	3	2	6	1.52
AW	188.83	5.13	0.39	0.48	0.67	7.11	115.67	142	201	3.82	18	28	14	17.91
Min.	0.40	0.01	0.07	0.02	0.01	0.03	0.09	32	12	0.20	2	2	4	0.84
Max.	188.83	5.13	3.06	1.10	0.67	7.11	115.67	1868	636	23.48	356	1140	248	63.78
Avg.	17.97	0.48	0.83	0.32	0.15	0.60	6.36	470	213	5.21	101	224	67	12.86

TABLE 6.4
Total metal concentrations in solid phase (HNO_3 digestion). Boldface values represent samples containing metal levels between negligible metal concentration and maximum permissible metal concentration; italicized values represent samples exceeding maximum permissible metal concentration[11]

Site	Cu (mM kg⁻¹)	Cr (mM kg⁻¹)	Ni (mM kg⁻¹)	As (mM kg⁻¹)	Cd (μM kg⁻¹)	Pb (mM kg⁻¹)	Zn (mM kg⁻¹)
A	0.04	0.66	**0.21**	0.07	n.d.	0.23	0.18
B	0.46	1.57	0.77	0.23	n.d.	0.15	1.73
C	*1.68*	1.87	0.75	0.30	0.69	0.34	2.69
D	0.19	1.24	0.32	0.13	**11.14**	0.48	7.95
E	0.39	0.58	*0.42*	0.23	*56.94*	*3.62*	*46.86*
F	0.50	1.34	*0.45*	0.39	**20.51**	0.44	*15.52*
G	**0.78**	1.39	*0.57*	0.20	**35.01**	n.d.	*10.90*
H	n.d.	0.24	n.d.	n.d.	1.41	0.09	0.27
I	n.d.	0.21	n.d.	0.06	2.48	0.13	0.40
J	n.d.	0.18	n.d.	0.06	6.23	0.08	0.73
K	n.d.	0.06	n.d.	0.04	n.d.	0.26	0.15
L	0.03	0.37	0.10	0.08	1.01	0.11	0.86
M	0.03	0.54	0.15	0.05	n.d.	0.06	0.15
N	*7.86*	*1.28*	*0.42*	0.28	2.56	*0.74*	*3.44*
O	0.30	0.63	0.13	0.25	*44.99*	0.17	2.31
P	*0.89*	*3.36*	*0.40*	0.37	*29.13*	n.d.	*9.01*
Q	0.70	0.91	**0.46**	0.15	3.36	**0.55**	2.63
R	0.51	*1.42*	*0.39*	0.31	**11.12**	0.39	*4.99*
S	*2.19*	*3.57*	*0.94*	*0.81*	*135*	*1.82*	*23.58*
T	0.35	0.90	**0.29**	0.25	**15.38**	0.33	*4.01*
U	0.84	*2.08*	0.77	0.36	7.12	**0.85**	2.77
V	0.06	0.49	0.16	0.21	n.d.	0.06	0.71
W	0.21	0.71	0.20	0.39	2.18	**0.15**	1.19
X*	0.48	0.46	0.32	0.22	*46.99*	**0.51**	*5.37*
Y	0.23	0.77	0.21	0.22	6.05	*1.09*	2.38
Z	0.07	0.17	n.d.	0.06	0.50	0.12	0.71
AA	0.10	0.15	n.d.	0.03	2.60	0.16	*5.50*
AB	0.11	0.14	n.d.	0.04	1.71	0.14	*5.37*
AC	n.d.	0.12	n.d.	0.05	1.70	0.10	0. 14
AD	n.d.	0.07	n.d.	0.02	n.d.	0.03	0.04
AE	n.d.	0.17	n.d.	0.02	n.d.	0.08	0.07
AF	n.d.	0.39	n.d.	0.08	n.d.	0.02	0.32
AG	0.16	*2.81*	*0.94*	0.31	n.d.	0.11	1.45
AH	0.54	0.98	0.49	0.37	3.48	0.23	2.35
AI	0.24	1.03	0.43	0.31	2.65	0.18	2.66
AJ	n.d.	0.32	0.13	0.07	n.d.	0.03	0.50
AK	0.50	0.48	0.17	0.11	**9.62**	**0.61**	*23.33*
AL	0.58	2.84	1.04	0.38	1.39	0.19	2.48
AM	0.04	0.33	n.d.	0.08	n.d.	0.12	0.20
AN	0.17	0.25	n.d.	0.07	3.76	0.13	0.65
AO	n.d.	0.09	n.d.	n.d.	1.61	0.03	0.12
AP	0.17	0.50	0.19	0.13	3.06	0.17	1.95
AQ*	0.27	0.38	0.15	0.11	*18.2*	0.37	*3.12*
AR	0.23	0.43	n.d.	0.09	3.10	0.27	1.11
AS	0.24	0.29	0.15	0.12	0.00	*1.77*	1.02
AT	0.54	0.51	0.19	0.14	5.41	0.44	*2.94*
AU	0.35	0.59	0.24	0.23	2.40	0.48	*3.27*
AV	0.06	0.10	0.01	n.d.	4.42	0.20	0.84
AW	*6.93*	0.57	*0.50*	*0.83*	*190*	*7.88*	*142*
Min.	0.01	0.06	0.01	0.01	0.00	0.02	0.04
Max.	7.86	3.57	1.04	0.83	189.85	7.88	142
Avg.	0.75	0.83	0.28	0.19	16.57	0.56	7.19

*Metals added in laboratory, so no comparison made to risk levels. n.d. = Not detectable;
Min. = Minimum value; Max. = Maximum value; Avg. = Average value.

TABLE 6.5
0.01-M CaCl$_2$-Extractable Metal Concentrations in Solid Phase

Site	Cu	Cr	Ni	As	Cd	Pb	Zn	Al	Fe	Mn	K	Mg	Na
				(μM kg⁻¹)								(mM kg⁻¹)	
A	0.99	0.20	3.67	0.06	0.19	5.33	17.0	1219	50.2	411	0.55	0.32	0.30
B	0.91	0.67	3.17	0.05	0.12	n.d.	6.58	41.1	14.0	141	0.61	8.25	3.54
C	3.74	0.18	8.59	0.03	0.23	0.06	28.3	87.2	14.8	167	14.5	12.81	1.87
D	0.49	n.d.	0.29	0.17	0.12	0.12	7.03	n.d.	n.d.	n.d.	2.48	2.10	0.44
E	0.86	n.d.	2.36	0.11	3.02	0.14	757	n.d.	2.00	11.4	0.70	3.44	0.44
F	0.87	n.d.	0.31	0.30	0.19	n.d.	11.9	n.d.	2.01	2.21	0.64	1.79	0.40
G	1.72	0.05	0.29	0.18	0.24	n.d.	5.73	8.03	5.02	1.49	0.49	1.97	0.26
H	0.29	0.09	2.42	0.08	0.95	3.41	68.0	1061	19.0	7.22	0.28	0.14	0.10
I	0.39	0.27	4.04	0.30	2.29	5.32	73.6	3219	108	15.9	0.43	0.28	0.12
J	0.19	n.d.	2.77	0.12	2.46	0.49	201	357	8.03	69.9	0.26	0.38	0.12
K	0.10	n.d.	0.42	0.09	0.02	63.23	3.38	146	35.2	4.77	0.13	0.16	0.17
L	0.29	n.d.	n.d.	0.18	0.02	n.d.	2.00	n.d.	2.99	1.40	0.33	1.02	0.34
M	2.30	0.17	3.84	0.08	0.19	1.29	26.5	1138	101	70.8	0.47	0.24	0.20
N	3.58	n.d.	0.27	0.02	0.07	0.06	2.71	n.d.	2.00	22.3	0.28	1.36	0.15
O	1.18	0.16	1.68	0.08	3.79	n.d.	82.5	9.03	5.02	136	3.37	5.97	1.63
P	1.95	0.07	0.64	0.44	0.26	n.d.	9.12	n.d.	n.d.	12.3	0.24	2.73	6.08
Q	1.20	n.d.	0.49	0.22	0.01	n.d.	2.10	n.d.	2.99	1.40	3.44	6.33	6.88
R	1.52	0.09	0.34	0.27	0.10	n.d.	2.50	n.d.	n.d.	5.61	3.51	3.03	2.59
S	1.22	0.11	0.95	0.21	0.97	n.d.	21.2	n.d.	n.d.	0.60	0.98	8.56	5.44
T	0.80	0.07	0.28	0.09	0.21	n.d.	4.22	n.d.	n.d.	4.42	0.62	3.33	4.09
U	2.05	0.55	4.68	0.12	0.55	1.39	49.7	500	61.3	744	1.51	9.88	2.27
V	0.27	n.d.	0.33	1.55	0.01	n.d.	n.d.	n.d.	2.01	n.d.	1.08	1.85	0.55
W	0.66	n.d.	0.27	2.71	0.02	n.d.	n.d.	n.d.	2.01	n.d.	1.57	1.57	0.74
X	4.56	8.40	131	14.0	27.6	3.79	3898	97.5	44.2	37.8	1.48	3.50	1.72
Y	1.35	0.15	3.82	0.01	4.06	22.6	231	605	20.2	209	1.45	2.73	0.27
Z	0.55	0.06	0.51	0.25	0.15	0.11	103	56.2	9.99	121	3.36	3.51	0.97

continued

TABLE 6.5 (continued)

Site	Cu	Cr	Ni	As	Cd	Pb	Zn	Al	Fe	Mn	K	Mg	Na
					(μM kg⁻¹)							(mM kg⁻¹)	
AA	0.29	n.d.	0.75	0.04	0.39	0.09	962	36.1	5.99	17.4	0.91	1.68	0.86
AB	0.48	0.06	0.91	0.04	0.44	0.11	989	26.0	9.01	21.3	1.07	1.68	0.96
AC	n.d.	0.08	1.90	0.06	1.31	4.41	40.5	2768	30.3	20.2	0.33	0.27	0.40
AD	0.18	0.11	1.19	0.05	0.37	0.70	33.8	1723	31.7	60.6	0.32	0.66	0.34
AE	0.17	0.12	1.18	0.13	0.10	3.51	14.5	761	42.0	52.7	0.63	0.73	0.22
AF	0.19	n.d.	n.d.	0.55	0.00	n.d.	n.d.	n.d.	1.85	0.60	1.27	1.13	0.43
AG	0.37	0.10	3.35	0.12	0.13	n.d.	2.71	10.0	8.04	200	5.56	21.89	4.37
AH	0.45	0.09	4.09	0.01	0.36	0.06	39.9	63.6	24.1	156	1.23	14.30	2.23
AI	0.48	n.d.	0.50	0.49	0.02	n.d.	1.81	n.d.	3.01	n.d.	6.85	6.52	2.06
AJ	0.10	n.d.	n.d.	0.09	0.00	n.d.	n.d.	n.d.	n.d.	n.d.	1.19	1.18	0.24
AK	0.70	n.d.	0.65	0.12	0.34	n.d.	435	13.0	3.98	2.01	0.88	2.56	1.34
AL	0.73	0.04	5.07	0.03	0.31	n.d.	3.82	14.8	2.85	372	0.55	13.16	1.92
AM	1.10	0.31	2.89	0.11	0.26	5.09	33.5	1901	98.5	75.7	0.79	0.95	0.23
AN	0.58	0.14	2.11	0.23	2.78	0.46	120	607	9.70	205	0.19	1.45	0.27
AO	0.08	n.d.	1.57	0.10	1.24	1.73	48.2	1110	5.99	6.92	0.25	0.52	0.21
AP	1.03	n.d.	0.20	0.12	0.02	n.d.	n.d.	n.d.	2.99	1.91	5.23	2.77	0.42
AQ	3.61	0.87	49.2	1.40	10.9	1.43	1851	445	29.7	791	1.32	1.66	0.67
AR	1.08	0.20	3.26	0.19	0.83	0.34	159	499	23.1	274	0.98	1.52	0.52
AS	0.51	0.08	n.d.	0.22	0.01	0.08	n.d.	n.d.	1.99	n.d.	0.80	1.67	0.81
AT	1.22	n.d.	0.59	0.51	0.02	0.09	3.72	n.d.	7.02	3.07	7.54	9.75	3.07
AU	0.59	n.d.	n.d.	0.13	0.01	n.d.	3.52	n.d.	2.01	8.09	0.60	8.19	8.09
AV	1.67	0.10	1.78	0.31	3.55	13.4	255	1436	25.2	0.15	0.33	0.14	0.15
AW	2.51	n.d.	8.77	0.06	28.9	5.51	6404	16.8	4.01	0.12	0.15	0.07	0.12
Min.	0.08	0.04	0.20	0.01	0.00	0.06	1.81	8.03	1.85	0.60	0.13	0.07	0.10
Max.	4.56	8.40	131	14.03	28.9	63.2	6404	3219	107.6	791	14.5	21.89	8.09
Avg.	1.09	0.47	6.07	0.55	2.05	4.98	396	666	20.6	107	1.71	3.71	1.46

n.d. = Not detectable; Min. = Minimum value; Max. = Maximum value; Avg. = Average value.

specific area. Instead, the data given represent point samples that do not necessarily reflect the pollution status of the whole area.

As can be seen from Table 6.3, 21 of the 46 soils sampled independently contain one or more metals at total concentrations (aqua regia digestion) exceeding the MPC; 25 of the 46 soils sampled independently contain one or more metals at total concentrations (aqua regia digestion) exceeding the negligible concentration (NC) (comparable numbers are derived for total metal concentrations obtained by means of HNO_3 destruction). Fourteen samples contain Zn levels above the maximum permissible concentration (MPC); in addition, one sample contains a Zn level in between the NC and the MPC. For Ni, these numbers are 10 and 5, respectively; for Cd, 9 and 4; for Pb, 6 and 5; for Cr, 6 and 0; and for Cu, 5 and 2. Only one sample contained As at levels exceeding the MPC (sample AW, a heavily polluted Belgian site), whereas one sample contained As at a total level in between the NC and the MPC. Thirteen samples contained more than one metal at levels exceeding the MPC, and, in addition, in four soils the NC for more than one metal was exceeded.

In Tables 6.6 and 6.7, the operationally defined available metal fraction is given, expressed as the percentage of the total metal concentration (aqua regia and HNO_3 digestion) that is extractable by means of 0.01-M $CaCl_2$ extraction. As can be deduced from these tables, the operationally defined available metal fraction expressed as the percentage of the total metal concentration that is extractable by means of 0.01-M $CaCl_2$ extraction varies greatly, both among metals and among soils. On average, this fraction is lowest for Cr (about 0.1%), followed by As and Cu (less than 1%), Pb (about 2%), Ni (about 3%), Zn (about 12%), and Cd (in between 18 and 27%). In addition, as illustrated in Figure 6.2 for Zn, there is no direct relationship between the operationally defined available metal fraction and the total metal concentration in the solid phase. This is despite the apparent trend of increasing extractable metal levels upon increasing total metal concentrations. As can be seen from Figure 6.2, deviations of extractable metal concentrations of over three orders of magnitude at similar total levels were measured. At first glance, a certain clustering might be deduced from the data presented in Figure 6.2. However, further analysis showed that this is not the case; the data are distributed randomly over the different soil types.

Because all soils sampled by Janssen and his colleagues[15] were resampled within the framework of this study, we compared the data on total metal concentrations reported in both studies. HNO_3 digestion was employed by Janssen et al.[15] as the sole method for liberating the metals from the solid phase. As a typical result of the comparison of both sets of data, in Figure 6.3 the log-transformed total Zn concentrations are plotted. As is obvious from this figure, and despite the fact that in some cases not exactly the same spot was resampled, in general the total metal concentrations that were reported correspond well.

PORE WATER CHARACTERISTICS

The composition of the pore water is given in Table 6.8 in terms of pH(pw), conductivity, DOC, anions, and major cations. MINTEQ-calculated (log-transformed) free

TABLE 6.6

Operationally defined available metal fraction, expressed as percentage of total metal concentration (aqua regia digestion) that can be extracted by 0.01-M CaCl$_2$ extraction

Site	Cu	Cr	Ni	As	Cd	Pb	Zn
A	1.42	0.03	1.78	0.09	31.87	3.12	6.41
B	0.21	0.04	0.44	0.03	4.16	n.d.	0.43
C	0.23	0.01	1.04	0.01	5.88	0.02	1.47
D	0.25	n.d.	0.07	0.19	0.87	0.02	0.10
E	0.21	n.d.	0.53	0.05	4.17	n.d.	1.59
F	0.18	n.d.	0.07	0.10	0.80	n.d.	0.09
G	0.23	0.00	0.06	0.10	0.55	n.d.	0.06
H	0.59	0.11	n.d.	0.49	53.28	3.64	33.38
I	0.77	0.06	2.58	0.62	73.35	3.66	32.24
J	0.27	n.d.	8.97	0.17	28.71	0.46	23.80
K	0.99	n.d.	0.45	0.50	n.d.	23.85	3.01
L	0.33	n.d.	n.d.	0.25	1.36	n.d.	0.27
M	2.58	0.04	3.49	0.24	47.40	1.94	9.25
N	0.22	n.d.	0.06	0.02	2.45	0.01	0.09
O	0.39	0.02	0.98	0.04	7.09	n.d.	3.61
P	0.23	0.00	0.15	0.14	0.69	n.d.	0.10
Q	0.15	n.d.	0.09	0.19	0.21	n.d.	0.09
R	0.28	0.01	0.09	0.10	0.70	n.d.	0.05
S	0.06	0.00	0.10	0.03	0.59	n.d.	0.11
T	0.20	0.01	0.10	0.04	1.17	n.d.	0.10
U	0.26	0.02	0.60	0.05	5.78	0.15	1.96
V	0.27	n.d.	0.19	0.88	0.32	n.d.	n.d.
W	0.27	n.d.	0.13	0.83	0.65	n.d.	n.d.
X	0.75	1.30	33.26	8.02	46.48	0.54	48.88
Y	0.52	0.02	1.76	0.00	50.21	1.78	12.29
Z	0.11	0.04	0.81	0.44	10.51	0.07	12.96
AA	0.10	n.d.	0.39	0.21	16.92	0.05	15.79
AB	0.27	0.02	0.76	0.21	18.85	0.06	19.74
AC	n.d.	0.07	5.41	0.29	60.91	3.89	25.40
AD	0.46	0.17	n.d.	0.31	52.07	1.64	28.33
AE	0.81	0.08	6.97	0.60	n.d.	4.89	14.83
AF	0.47	n.d.	n.d.	0.78	0.37	n.d.	n.d.
AG	0.17	0.00	0.31	0.05	3.91	n.d.	0.22
AH	0.11	0.01	0.74	0.00	5.55	0.02	2.71
AI	0.16	n.d.	0.10	0.22	0.27	n.d.	0.07
AJ	0.15	n.d.	n.d.	0.26	0.49	n.d.	n.d.
AK	0.13	n.d.	0.37	0.18	3.36	n.d.	2.27
AL	0.16	0.00	0.51	0.02	7.30	n.d.	0.23
AM	1.16	0.12	6.14	0.25	35.51	3.89	16.37
AN	0.33	0.06	5.46	0.42	59.66	0.31	21.70
AO	0.74	n.d.	n.d.	0.74	71.13	4.00	53.68
AP	0.46	n.d.	0.08	0.13	0.45	n.d.	n.d.
AQ	0.77	0.16	20.8	1.03	39.27	0.24	47.22
AR	0.51	0.03	1.74	0.37	28.18	0.15	23.08
AS	0.19	0.03	n.d.	0.23	0.62	0.00	n.d.
AT	0.22	n.d.	0.27	0.56	0.31	0.02	0.15
AU	0.16	n.d.	n.d.	0.06	0.30	n.d.	0.16
AV	1.52	0.14	6.86	1.23	67.25	6.38	37.65
AW	0.05	n.d.	1.81	0.01	15.32	0.08	5.54
Min.	0.05	0.00	0.06	0.00	0.21	0.00	0.05
Max.	2.58	1.30	33.26	8.02	73.35	23.85	53.68
Avg.	0.44	0.09	2.84	0.44	18.45	2.24	11.80

n.d. = Not detectable; Min. = Minimum value; Max. = Maximum value; Avg. = Average value.

TABLE 6.7
Operationally defined available metal fraction, expressed as percentage of total metal concentration (HNO_3 digestion) that can be extracted by 0.01-M $CaCl_2$ extraction

Site	Cu	Cr	Ni	As	Cd	Pb	Zn
A	2.25	0.03	1.79	0.09	n.d.	2.35	9.25
B	0.20	0.04	0.41	0.02	n.d.	n.d.	0.38
C	0.22	0.01	1.14	0.01	32.90	0.02	1.05
D	0.26	n.d.	0.09	0.14	1.11	0.03	0.09
E	0.22	n.d.	0.56	0.05	5.31	0.00	1.61
F	0.17	n.d.	0.07	0.08	0.94	n.d.	0.08
G	0.22	0.00	0.05	0.09	0.67	n.d.	0.05
H	n.d.	0.04	5.57	0.35	67.82	3.99	25.26
I	n.d.	0.13	8.62	0.50	92.55	4.06	18.63
J	3.54	n.d.	4.79	0.19	39.56	0.63	27.51
K	n.d.	n.d.	1.12	0.25	n.d.	24.44	2.23
L	0.93	n.d.	n.d.	0.22	2.28	n.d.	0.23
M	8.33	0.03	2.63	0.16	n.d.	2.25	17.33
N	0.05	n.d.	0.06	0.01	2.60	0.01	0.08
O	0.40	0.03	1.32	0.03	8.43	n.d.	3.58
P	0.22	0.00	0.16	0.12	0.89	n.d.	0.10
Q	0.17	n.d.	0.10	0.14	0.35	n.d.	0.08
R	0.30	0.01	0.09	0.09	0.89	n.d.	0.05
S	0.06	0.00	0.10	0.03	0.72	n.d.	0.09
T	0.23	0.01	0.10	0.04	1.36	n.d.	0.11
U	0.24	0.03	0.61	0.03	7.77	0.16	1.80
V	0.42	n.d.	0.21	0.75	1.64	n.d.	n.d.
W	0.32	n.d.	0.14	0.69	1.06	n.d.	n.d.
X	0.96	1.83	40.52	6.40	58.80	0.74	73.21
Y	0.59	0.02	1.86	0.00	67.10	2.08	9.73
Z	0.79	0.03	0.78	0.44	29.66	0.09	14.37
AA	0.29	n.d.	1.10	0.11	14.98	0.05	17.48
AB	0.43	0.04	1.60	0.10	25.60	0.08	18.41
AC	n.d.	0.07	8.28	0.12	77.16	4.58	29.10
AD	n.d.	0.16	9.13	0.21	87.93	2.05	83.49
AE	n.d.	0.07	2.97	0.52	n.d.	4.20	21.36
AF	n.d.	n.d.	n.d.	0.68	n.d.	n.d.	n.d.
AG	0.23	0.00	0.36	0.04	35.00	n.d.	0.19
AH	0.08	0.01	0.84	0.00	10.30	0.02	1.70
AI	0.20	n.d.	0.12	0.16	0.64	n.d.	0.07
AJ	n.d.	n.d.	n.d.	0.14	1.66	n.d.	n.d.
AK	0.14	n.d.	0.37	0.11	3.56	n.d.	1.87
AL	0.13	0.00	0.49	0.01	22.41	n.d.	0.15
AM	2.47	0.09	3.37	0.15	n.d.	4.26	17.06
AN	0.35	0.05	4.60	0.34	74.02	0.35	18.53
AO	n.d.	n.d.	5.81	0.89	77.43	5.15	41.88
AP	0.61	n.d.	0.11	0.09	0.67	n.d.	n.d.
AQ	1.34	0.23	32.95	1.31	59.85	0.38	59.31
AR	0.46	0.05	4.20	0.22	26.82	0.13	14.26
AS	0.21	0.03	n.d.	0.18	113.21	0.00	n.d.
AT	0.22	n.d.	0.30	0.38	0.42	0.02	0.13
AU	0.17	n.d.	n.d.	0.05	0.53	n.d.	0.11
AV	2.78	0.09	14.26	1.30	80.49	6.75	30.29
AW	0.04	n.d.	1.76	0.01	15.24	0.07	4.53
Min.	0.04	0.00	0.05	0.00	0.35	0.00	0.05
Max.	8.33	1.83	40.52	6.40	113.21	24.44	83.49
Avg.	0.78	0.11	3.76	0.37	27.44	2.38	13.18

n.d. = Not detectable; Min. = Minimum value; Max. = Maximum value; Avg. = Average value.

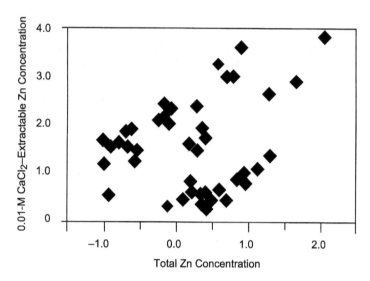

FIGURE 6.2   Plot of log-transformed 0.01-M CaCl$_2$-extractable Zn concentration (μmol kg^{-1}, aqua regia digestion) for each soil sample included in this study vs. log-transformed total metal concentration (mmol kg^{-1}).

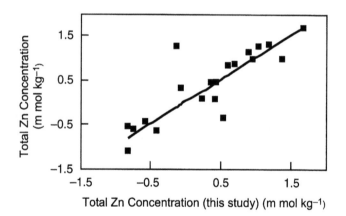

FIGURE 6.3   Plot of (log-transformed) total Zn levels (HNO$_3$-digestion) found in soil solid phase by Janssen et al.[15] vs. (log-transformed) total concentration of Zn in solid phase, as reported in Table 6.4 (soils A–T).

metal concentrations and metal activities for each of the soils included in this study are available on request.

As can be deduced from Table 6.8, the pore water composition varies strongly among soils, which is obvious given the variance among soil types and soil properties. Similar to total metal concentrations, the data reported in Table 6.8 were compared with the data reported by Janssen and his coworkers.[15] As a typical result of this

TABLE 6.8

Pore Water Composition of pH, EC (conductivity), DOC, Anions, and Major Cations

Site	pH (H₂O)	EC (µS cm⁻¹)	CO₃²⁻	DOC	NO₃⁻	SO₄²⁻	PO₄³⁻	Cl⁻ (mML⁻¹)	K	Mg	Mn	Na	Ca	Fe	Al	Cu	Cr	Ni (µML⁻¹)	As	Cd	Pb	Zn
A	4.23	431	0.02	8.38	2.67	0.27	1.3	0.43	0.40	0.19	89.4	0.63	0.46	15.5	65.6	1.67	0.11	0.83	0.02	0.03	0.11	5.9
B	6.10	1350	0.23	7.25	9.35	1.79	1.9	0.95	0.06	0.89	8.65	2.81	5.15	10.6	29.2	0.73	0.16	0.53	0.01	0.01	0.03	8.6
C	5.88	760	0.06	5.45	4.33	1.16	3.0	0.48	1.64	0.74	12.5	0.85	1.92	44.7	53.8	3.12	0.14	0.72	0.05	0.01	0.12	2
D	7.54	1750	1.45	4.86	15.48	0.48	0.7	0.36	0.95	0.54	0.07	0.73	7.83	1.21	2.3	0.79	0.03	0.35	0.02	0.01	0.01	1.4
E	6.96	1201	1.96	2.54	8.01	2.34	5.0	0.19	0.13	0.77	1.56	0.50	5.71	1.4	2.6	1.37	0.02	0.56	0.01	0.15	0.03	63
F	7.69	1460	3.78	3.52	11.08	0.26	1.2	0.11	0.05	0.39	0.32	0.33	7.53	0.6	3.6	1.02	0.02	0.33	0.04	0.01	0.01	1
G	7.33	909	2.74	2.52	3.90	0.81	4.5	0.10	0.06	0.39	2.98	0.25	4.95	1.1	2.5	1.64	0.03	0.39	0.03	0.01	n.d.	0.6
H	4.26	436	0.04	10.1	2.99	0.16	1.8	0.18	0.50	0.13	2.9	0.25	0.43	12.2	59	1.79	0.13	0.81	0.02	0.06	0.15	14.8
I	4.25	533	0.01	12.3	3.69	0.28	0.8	0.13	0.53	0.29	12.4	0.24	0.59	13.2	180.3	1.44	0.11	0.80	0.02	0.23	0.21	22.5
J	5.26	592	0.06	4.94	5.11	0.13	1.0	0.13	0.38	0.48	14.9	0.49	1.66	3.4	30.2	1.47	0.05	0.72	0.02	0.20	0.02	50.9
K	5.10	474	0.01	5.29	3.71	0.10	13.1	0.26	0.24	0.35	8.7	0.51	1.15	9.3	38.5	1.39	0.08	1.19	0.04	0.04	4.38	5.8
L	7.32	1440	1.57	3.23	11.74	0.39	6.2	0.78	0.23	0.70	0.98	1.15	6.25	4.2	5.3	1.54	0.03	0.40	0.05	0.01	0.05	1.2
M	4.63	538	0.01	10.2	3.78	0.20	11.2	0.20	0.45	0.23	27.0	0.43	0.76	21.85	78.75	1.97	0.11	0.81	0.02	0.04	0.06	12.85
N	7.16	1144	2.17	4.79	9.43	0.49	0.4	0.13	0.07	0.93	0.41	0.30	5.08	1.8	3.6	2.04	0.04	0.29	0.01	0.01	0.04	0.8
O	6.21	2710	0.14	4.74	15.97	7.43	1.6	1.65	2.18	3.77	2.4	3.33	10.39	1.3	3.4	2.03	0.08	0.77	0.01	0.27	0.01	16
P	7.12	2370	3.63	2.03	3.15	4.96	8.1	10.13	0.09	1.24	1.62	9.09	8.51	0.5	1.9	0.98	0.04	0.45	0.05	0.04	n.d.	5.4
Q	7.08	3700	0.91	4.23	10.70	14.83	7.9	7.71	1.17	2.89	0.59	8.21	18.43	3.5	4.8	1.22	0.02	0.32	0.02	0.02	0.01	1.5
R	7.16	3740	2.05	3.68	15.06	5.97	0.5	13.51	2.64	2.18	3.3	5.40	16.77	3.9	8.6	2.05	0.16	0.75	0.02	0.02	0.03	1.7
S	7.17	1095	2.71	2.23	4.53	1.36	0.6	2.27	0.12	0.75	0.15	2.79	3.84	1.3	2.2	0.97	0.07	0.28	0.03	0.02	0.01	1.1
T	7.09	2370	1.51	2.75	7.57	2.87	n.d.	9.99	0.26	1.39	1.17	7.11	8.27	4.2	4.3	0.84	0.07	0.47	0.02	0.02	0.02	1.3
U	5.60	2210	0.02	9.51	11.85	6.78	1.7	0.61	0.10	2.18	108	1.23	11.31	5.6	60.3	3.9	0.12	1.59	0.02	0.07	0.07	14.4
V	7.54	773	4.42	2.43	3.82	0.45	8.9	1.23	0.15	0.25	n.d.	0.62	3.70	0.4	1.4	2.02	0.04	0.72	0.38	0.01	0.01	1.4
W	7.50	663	3.17	3.32	1.37	0.50	13.0	31.71	0.21	0.16	n.d.	0.81	3.15	0.4	0.6	1.36	0.05	0.68	0.82	0.03	0.01	2.9
X	5.30	4000	0.12	44.6	7.08	0.53	1.4	0.52	0.84	2.07	12.5	1.43	17.83	4.5	17.2	8.18	5.15	44.1	28.8	9.62	1.00	1247
Y	5.08	369	0.02	4.80	1.87	0.41	2.5	2.52	0.27	0.35	41.8	0.41	1.03	136	176	1.84	0.58	0.48	0.18	0.18	1.59	18.1
Z	5.47	1390	0.04	6.91	9.48	0.75	166	0.51	2.84	1.24	1.99	2.05	3.42	10	27.6	3.96	0.07	0.62	0.10	0.02	0.07	9.5
AA	6.03	793	0.12	3.43	6.00	0.37	2.8	0.57	0.43	0.47	4.21	1.55	2.58	7.3	17.4	0.54	0.03	0.19	0.01	0.01	0.05	57.8
AB	6.04	860	0.45	3.13	6.71	0.40	3.1	0.57	0.50	0.52	2.04	1.63	2.75	7.1	17.6	0.46	0.05	0.18	0.01	0.01	0.05	67.7

continued

TABLE 6.8 (continued)

Site	pH (H₂O)	EC (µS cm⁻¹)	CO₃²⁻	DOC	NO₃⁻	SO₄²⁻	PO₄³⁻	Cl⁻ (mML⁻¹)	K	Mg	Mn	Na	Ca	Fe	Al	Cu	Cr	Ni (µM L⁻¹)	As	Cd	Pb	Zn
AC	4.05	165	0.05	4.52	0.27	0.20	5.5	0.39	0.14	0.04	1.86	0.29	0.16	131	238	1.31	0.30	0.34	0.14	0.03	0.44	2.7
AD	3.95	428	0.08	5.32	3.09	0.17	0.9	0.34	0.13	0.36	26.6	0.69	1.03	69.9	91.5	1.37	0.06	0.39	0.01	0.03	0.10	9.6
AE	3.49	482	0.03	10.0	1.89	0.18	136	0.35	0.41	0.21	11.0	0.34	0.38	41.4	79.7	2.49	0.09	0.41	0.10	0.04	0.32	5
AF	7.42	571	3.39	2.33	1.31	0.16	15.3	0.91	0.38	0.23	1.08	0.69	2.67	24	19.1	1.08	0.07	0.24	0.25	0.00	0.02	0.4
AG	6.28	2760	0.40	4.23	0.64	18.99	0.7	2.32	1.15	6.42	53.4	3.17	12.79	1.7	3.2	0.71	0.05	0.82	0.01	0.01	0.01	1.3
AH	5.35	321	0.27	8.45	0.07	0.87	4.6	0.62	0.08	0.47	18.7	0.75	1.23	147	114	1.79	0.30	0.95	0.07	0.02	0.13	2.3
AI	7.50	1181	5.12	3.99	1.34	0.72	23.1	2.41	1.64	0.88	0.24	1.77	4.99	1.2	2.4	0.83	0.03	0.28	0.15	0.00	0.01	0.2
AJ	7.34	819	2.03	2.29	4.25	0.27	4.6	0.43	0.52	0.29	0.06	0.41	3.81	1.3	3.1	0.22	0.01	0.16	0.03	0.00	0.01	1.8
AK	6.83	750	0.65	3.12	5.55	0.37	10.4	0.48	0.21	0.33	0.49	1.54	2.70	5.8	8.7	0.63	0.03	0.19	0.03	0.01	0.07	16.3
AL	6.44	364	0.25	4.85	0.24	0.78	1.2	0.72	0.04	0.38	23.9	0.73	1.31	23.9	47	1.35	0.11	0.64	0.02	0.01	0.01	0.6
AM	3.65	659	0.02	3.92	4.14	0.29	6.5	0.33	0.40	0.45	28.9	0.30	1.23	15.3	136	1.14	0.09	0.51	0.02	0.04	0.15	17.9
AN	5.58	126	0.03	3.84	0.03	0.26	16.7	0.23	0.05	0.12	39.0	0.33	0.34	52	113	1.69	0.27	0.30	0.27	0.02	0.23	4.4
AO	4.60	354	0.02	4.08	2.01	0.18	0.8	0.42	0.33	0.27	2.69	0.48	0.38	6.2	42.7	0.86	0.05	0.43	0.06	0.16	0.21	63.2
AP	7.07	1246	1.17	3.24	7.29	0.52	8.4	0.81	3.07	0.67	0.14	0.68	4.12	4	2.9	0.61	0.04	0.19	0.04	0.00	0.03	1.2
AQ	4.80	2310	0.11	22.6	0.32	0.53	15.4	19.67	1.08	1.17	605	1.22	5.95	505	80.6	1.34	0.46	9.07	0.30	1.59	0.36	355
AR	5.31	885	0.02	5.28	6.89	0.42	17.3	0.72	0.44	0.58	36.0	0.67	3.35	11.9	59	2.07	0.10	0.78	0.03	0.05	0.07	20.7
AS	7.30	760	2.18	2.87	1.36	1.28	3.0	0.91	0.23	0.26	0.19	0.94	3.77	3.1	5.4	0.28	0.04	0.11	0.07	0.00	0.13	0.4
AT	7.40	1790	2.42	6.74	6.63	3.34	10.9	3.54	2.65	1.80	0.57	2.99	5.98	3.5	4	0.66	0.04	0.26	0.10	0.00	0.03	0.9
AU	7.50	1460	4.12	4.76	0.07	4.88	0.6	3.12	0.12	1.13	0.34	5.42	5.66	0.8	1.3	0.7	0.01	0.29	0.04	0.00	0.01	0.2
AV	3.94	485	0.01	3.98	3.32	0.20	0.6	0.34	0.48	0.22	8.19	0.50	0.92	4.5	143	1.63	0.06	0.64	0.05	0.75	0.70	73.8
AW	6.16	395	0.12	1.32	2.41	0.54	0.4	0.22	0.04	0.18	0.42	0.29	0.93	2.1	44.6	0.65	0.01	1.06	0.01	1.34	0.25	837
Min.	3.49	126	0.01	1.32	0.03	0.10	0.4	0.10	0.04	0.04	0.06	0.24	0.16	0.40	0.6	0.22	0.01	0.11	0.01	0.00	0.01	0.20
Max.	7.69	4000	5.12	44.6	15.97	18.99	166	31.71	3.07	6.42	605	9.09	18.43	505	238	8.18	5.15	44.1	28.8	9.62	4.38	1247
Avg.	6.04	1191	1.14	6.02	5.17	1.87	11.5	2.60	0.63	0.88	26.0	1.62	4.59	28.2	43.6	1.55	0.20	1.60	0.67	0.31	0.24	62.3

n.d. = Not detectable; Min. = Minimum value; Max. = Maximum value; Avg. = Average value.

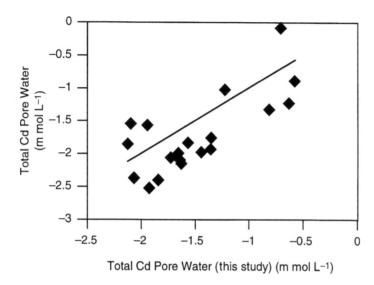

FIGURE 6.4 Plot of (log-transformed) total Cd levels found in pore water by Janssen et al.[15] vs. (log-transformed) total concentration of Cd, as reported in Table 6.8 (soils A–T).

comparison, in Figure 6.4 the (log-transformed) total Cd levels found in the pore water by Janssen et al.[15] are plotted as a function of the (log-transformed) total concentration of Cd in the pore water for soils A–T, as reported in Table 6.8. As can be deduced from Figure 6.4, Cd levels reported in this study systematically exceed the values reported by Janssen et al.[15] On the one hand, this finding is surprising in view of the fact that in this study the pore water was filtered over a 0.45-μm filter prior to analysis, whereas Janssen and his colleagues[15] used a 2.5-μm filter. On the other hand, however, this finding may well reflect the addition of a 2-mM aqueous solution of $Ca(NO_3)_2$ to the soils in order to generate sufficient quantities of pore water in all soils, as was done in this study.

PARTITION COEFFICIENTS

Calculated partition coefficients are available from the authors upon request. K_p values vary strongly, not only among metals, but also for a single metal among soils. In most cases, K_p values differ by several orders of magnitude. Consistently higher K_p values were found with the two digestion methods used in this study than with the 0.01-M $CaCl_2$ extraction. As the pore water concentrations in general are higher than the concentrations reported by Janssen et al.,[15] the K_p values reported here for soils A–T are systematically lower than the K_p values reported in the latter study.

From the values of the partition coefficients it may be deduced that addition of heavy metals leads, in general, to significantly lower K_p values. In other words, metals added to field soils show an increased bioavailability for organisms with predominant metal uptake via the pore water. When comparing the K_p values obtained for soils AR and AQ (soil AR after addition of a mixture of metal salts), it becomes clear that, with the exception of Cu (HNO_3 digestion) and As ($CaCl_2$ extraction) (although it should be noted that As was not present in the metal mixture that was added to the soil), the values obtained for soil AQ are significantly lower. In addition, from the data presented it can be deduced that K_p values for soil X (OECD artisoil, to which the same mixture of metal salts was added) for all metals studied are at the very low end of the range of partition coefficients found in this study, and in fact for Cr, Ni, and As, the values obtained for soil X are the minimum values found for all soils studied.

RELATIONSHIPS BETWEEN SOIL CHARACTERISTICS

Principal component analysis (PCA) was performed on all soil characteristics for which experimental data were available to study the correlation between the sampled soils. From the outcome of the score plot thus obtained, an indication of the similarity and diversity of the soils can be gained. Soils that plot out near each other on the score plot may be related. The first two principal components explained 73% of the variance. The third principal component explained only a small percentage of the remaining variance (8%) and did not add much more information. Figure 6.5 shows the scores of the soils for the first two principal components. As can be seen, soils AA and AB (which represent duplicate sampling) plot out near each other, clearly indicating the similarity of these soils. Figure 6.5 also shows that some clustering of the soil samples can be distinguished only vaguely. The soils are distributed over the score plot, which is a reflection of variation of the soil characteristics among sites

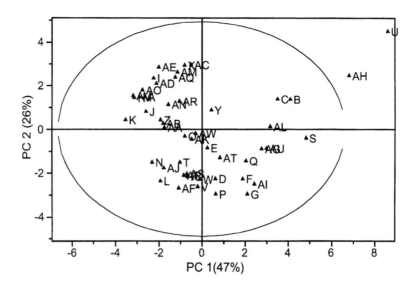

FIGURE 6.5 Principal component (PC) score plot of soils sampled. Ellipsoid indicates 95% confidence interval.

according to the criteria used for the selection of the sampling sites. Due to their high organic matter contents, soils U and AH do not fall within the 95% confidence interval. Nevertheless, these soils were included in all statistical analyses.

The relation between K_p values and soil characteristics depends strongly on the relationships between soil characteristics. The most important soil characteristics that need to be taken into account are:

- The adsorption phases: clay, organic matter (OM), and Fe– and Al– oxyhydroxides.
- pH, which was measured as pH(KCl), pH(pw), and pH(CaCl$_2$).
- Competitive sorbed ions, corresponding with the CEC, because CEC is determined as the amount of desorbed ions with a high concentration of cations. Ca is a particularly important cation at the exchange complex of soils in the Netherlands.

Among the pore water characteristics influencing metal adsorption, complex-forming anions are an important factor. In particular, dissolved organic carbon (DOC) is known to complex metals and keep them into solution. Therefore, DOC was taken into account as an important soil characteristic. Figure 6.6 shows the loadings of the soil characteristics for the first two principal components. The loading plot shows two clusters of soil characteristics. The first cluster contains most of the solid-phase characteristics: LOI–1, OM, N, Fe–ox, pF2, clay, and CEC. Within this cluster, a sub-cluster of Fe–ox, CEC, and clay seems to be present. To a lesser extent this also holds for Al hydroxide. CEC is correlated with Al–ox, OM, clay, and Fe–ox, because they provide the main adsorption phases in the soil.

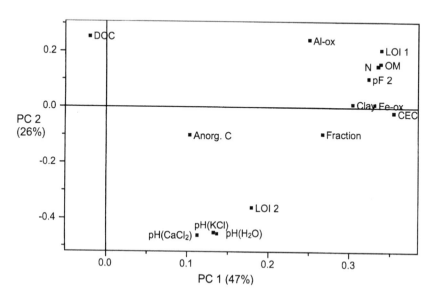

FIGURE 6.6 Principal component (PC) loading plot of soils sampled.

The second cluster shows that the pHs are closely related. This suggests that no further information is gained when the pHs are measured in several ways. To study the relationship between K_p values and soil characteristics, it is sufficient to use one type of pH measurement. For practical reasons the best choice seems $pH(CaCl_2)$ because the method to determine this pH is easy to carry out and is already commonly used. The correlations among the three are available from the authors upon request.

CORRELATIONS BETWEEN PARTITION COEFFICIENTS AND SOIL CHARACTERISTICS

As indicated above, one of the objectives of this study was to derive models for predicting metal partitioning in Dutch field soils based upon a limited number of easily determinable soil and pore water characteristics. As soils X (OECD artisoil) and AQ (a soil sample to which metal salts were added in the laboratory) do not represent field soils, they were left out of the dataset used for the multivariate data analyses. Soil AB is a duplication of sample AA. To guarantee an unbiased dataset, soil AB also was not included in the data analysis, which left a database of 46 field soils.

As indicated above, six different sets of K_p values were obtained by combining three expressions for the metal concentrations in the solid phase and either total metal concentrations in the pore water or pore water activities. In addition, the nonavailable metal fraction in the solid matrix (in this study defined as the difference in metal levels found between either aqua regia or concentrated nitric acid destruction and 0.01-M $CaCl_2$ extraction) was correlated to soil characteristics. Finally, models are derived for calculating these nonavailable metal fractions.

MULTIVARIATE REGRESSION MODELS FOR 46 SOILS

The results of the multivariate regression analyses for each of the metals included in this study are available from the authors upon request. Apart from the actual models, the relevant statistical information is also provided. In all cases, the relevant soil and pore water characteristics are given in decreasing order of importance.

Several conclusions may be drawn from the regression formulae for predicting K_p. In general, formulae are obtained that are a combination of one or more of the sorption phases present in the soil (metal sinks) and the soil-related factors that modulate metal sorption (pH). Generally speaking, soil pH is the dominant factor regulating metal partitioning in soil, and pH explained a high percentage of the variation in K_p values for nearly all metals. Other researchers reported similar results.[32,33] This can be explained by the fact that H^+ ions compete for binding sites. Further, the soil pH affects the surface charge and is important in regulating metal speciation in the pore water.

With regard to the elements included in this study, especially for the partition coefficients based upon the two digestion techniques employed, generally good correlations are obtained for the heavy metals studied. The standard errors of prediction are typically in the range of 0.2–0.4 log units. This corresponds to a standard error of prediction of K_p values of a factor of 2–3 in an absolute sense. As this error is in line with experimental uncertainties in determining K_p, this is quite acceptable. The models for predicting partition coefficients that are based upon $CaCl_2$- extraction clearly perform less well. The poor R^2_{adj} values observed for the $CaCl_2$ extraction suggest that K_p values based on this extraction method cannot explain the variation in K_p values by taking into account the determined soil characteristics. One reason for this may be that $CaCl_2$ is not successful in desorbing the metals from the adsorption phases. This is in line with the earlier observations by Janssen et al.[15] for Ni and Zn, that the amount of metal extracted by 0.01-M $CaCl_2$ corresponds to metal pore water concentrations. The latter observation would also suggest that K_p values for Ni and Zn can also be calculated as the ratio of the amount of metal digested by HNO_3 to the amount of metal extracted by 0.01-M $CaCl_2$.

Arsenic deviates from these general observations, and no satisfactory models could be derived for this element. Because arsenic is the only element studied that will be predominantly present in the field soils as a negatively charged species (arsenate), this finding most likely is a reflection of this deviating behavior. Clearly, more research is needed to better understand the partitioning of As in Dutch field soils. This research is now in progress.[13]

In general, the best models are obtained for partition coefficients based upon metal activities in the pore water. All models based on metal activities in the pore water have $pH(CaCl_2)$ as the dominant property determining K_p values. With respect to this finding, it should be noted that calculated metal activities were used to derive the underlying K_p values. It will be clear that the pH of the pore water may have strongly affected the values of the activities thus obtained, which implies that the present covariation may result from autocorrelation. At present, methods for measuring the activity of free metal ions in pore water are grossly lacking. To confirm the correlations found, it is essential that methods for actually measuring metal activities are developed. Such research is currently in progress.[13]

Finally, it may be deduced that the nonavailable metal fraction is associated with a limited number of sorption phases in the soil. Especially Fe–ox, Al–ox, and LOI–1 (which is a reflection of the organic matter content of the soil) are important factors in this respect. Again, pH appears to have an impact on the nonavailable metal fraction. In general, the best models are derived for data based upon aqua regia digestion, as compared with data based upon HNO_3 digestion.

MULTIVARIATE REGRESSION MODELS FOR NON-CARBONATE-CONTAINING SOILS

The dominant property modulating metal partitioning in soils is pH. To illustrate the impact of pH on metal partitioning, log-transformed K_p values for Zn are plotted in Figure 6.7 as a function of $pH(CaCl_2)$.

From a close inspection of the data shown in Figure 6.7, it became clear that metal partitioning in soils containing carbonate as an additional sorption phase deviates from partitioning in soils that do not contain carbonate, in the sense that

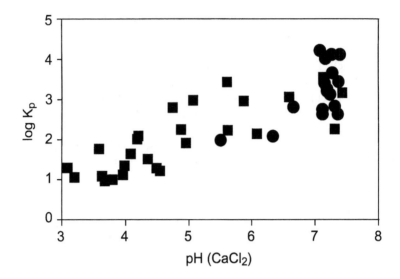

FIGURE 6.7   Plot of log-transformed K_p-values of Zn in 46 Dutch field soils vs. pH(CaCl$_2$). K_p values (L kg^{-1}) were derived on basis of aqua regia digestion and total zinc concentrations in pore water. Triangles represent K_p values measured in carbonate-containing soils.

K_p values obtained for these soils appeared not to be related to pH(CaCl$_2$). This may be due both to metal sorption to the carbonate phase and to precipitation. We therefore decided to split the database of 46 soils into a set of 18 soils that contain carbonate, and a set of 28 soils that do not. Subsequently, multivariate regression analysis was carried out on both datasets.

For the carbonate-containing soils, no models could be derived for predicting metal partitioning. This not only implies that additional research on this type of soil is required, but these findings may also be a consequence of the precipitation processes taking place in these soils. Precipitation may give rise to relatively high metal levels in the solid phase that are not related to any of the soil constituents. In addition, it should be noted that a majority of the sediments underlying Dutch surface water are also known to contain considerable amounts of carbonate, which implies that additional research should also take metal partitioning in these sediments into account.

The results of the multivariate regression, as carried out for the limited dataset of all soils that do not contain detectable carbonate levels in the solid phase, are available from the authors upon request. Again, in all cases the relevant soil and pore water characteristics are given in decreasing order of importance.

Basically, the same conclusions may be drawn from the models for the soils that do not contain detectable carbonate levels in the solid phase as for the data set containing all soils. Despite the obvious reduction in the number of data points, in most cases higher values for the statistical parameters are found, indicating an improvement of the significance of the correlations obtained.

CONCLUSIONS AND RECOMMENDATIONS

The aim of this study (i.e., providing a first set of data and models to be used in the validation and extrapolation stages of the research and development program proposed by De Rooij et al.[12]) was met. *In situ* partitioning of six metals (Cd, Cr, Cu, Ni, Pb, and Zn) and As in Dutch field soils was studied at 46 sites. For reference purposes, OECD–artisoil was included in the dataset, and duplicate sampling was carried out for one soil to gain insight into the reproducibility of the experimental procedures. In addition, the impact of freshly-added metal salts on the soil matrix was studied and, finally, models were derived that enable the prediction of metal partitioning in Dutch soils on the basis of a limited number of easily obtainable soil characteristics. The models may be used to predict pore water concentrations in field soils on the basis of measured metal concentrations in the solid phase.

Based upon these findings, the following conclusions may be drawn:

- Partition coefficients of heavy metals and As vary strongly among soils and among metals.
- Addition of metal salts to one of the soils leads to a decrease in the magnitude of the partition coefficients, which implies that the bioavailability of the metal salts added is increased for soil organisms whose dominant uptake route is via pore water. This finding could be related to nonlinearity of the sorption isotherm in these soil systems.
- Metal partitioning in soils containing carbonate differs from partitioning in which no detectable amounts of carbonate are present as an additional sorption phase.
- With the exception of arsenic, metal partitioning can be quantified by models that combine one or more of the metal-binding soil phases, such as amorphous iron- and aluminum-oxyhydroxide, organic matter and clay, and one of the soil characteristics that modulate metal partitioning. Soil pH is an especially dominant factor in this respect, as it was found that pH explained a high percentage of the variation in the values of the partition coefficients (K_p) for all metals.
- The "best" models in terms of statistical significance are obtained for K_p values that are based upon total metal concentrations in both the soil solid phase and the pore water; in general, between 70 and 90% of the data variance is explained in this case. The standard errors of prediction of K_p are typically in the range of 0.2–0.4 log units, or a factor of 2–3 in an absolute sense. This is in line with experimental uncertainties in determining K_p.
- Taking the activity of the free metal ion into account significantly increases the explained variance.
- Exclusion of the carbonate-containing soils from the data analysis further increases the predictive capability of the models thus derived.

As indicated above, the results given in this contribution will provide the basis for additional research that is aimed at developing empirical models that relate actually available metal concentrations to potentially available fractions.[13] In light of these future activities, the following recommendations can be made on the basis of the results presented in this report:

1. Further investigate the factors that modulate metal partitioning in carbonate-containing soils.
2. Further investigate the partitioning of negatively-charged species (such as arsenate) in soil–pore water systems.
3. Investigate the possibilities of using mild extraction methods to replace methods of pore water collection, which, as shown above [2 mM Ca(NO$_3$)$_2$], do not always yield comparable results.

REFERENCES

1. Belfroid, A.C., Sijm, D.T.H.M., and van Gestel, C.A.M., Bioavailability and toxicokinetics of hydrophobic aromatic compounds in benthic and terrestrial invertebrates, *Environ. Rev.,* 4, 276, 1996.
2. Di Toro, D.M., Zarba, C.S., Hansen, D.J., Berry, W.J., Swartz, W.J., Cowan, C.E., Pavlou, S.P., Allen, H.E., Thomas, N.A., and Paquin, P.R., Technical basis for the equilibrium partitioning method for establishing sediment quality criteria, *Environ. Toxicol. Chem.,* 11, 1541, 1991.
3. Peijnenburg, W.J.G.M., Posthuma, L., Eijsackers, H.J.P., and Allen, H.E., A conceptual framework for implementation of bioavailability of metals for environmental management purposes, *Ecotox. Environ. Safety,* 37, 163, 1997.
4. Van Wensem, J., Vegter, J.J., and Van Straalen, N.M., Soil quality criteria derived from critical body concentrations of metals in soil invertebrates, *Appl. Soil Ecol.,* 1, 185, 1994.
5. Van Straalen, N.M., Critical body concentrations: their use in bioindication, in *Bioindicator Systems for Soil Pollution,* Van Straalen, N.M. and Krivolutsky, D.A., Eds., Kluwer Academic Publishers, Dordrecht, The Netherlands, 1996, 5.
6. Allen, H.E., Fu, G., and Deng, B., Analysis of acid-volatile sulfide (AVS) and simultaneously extracted metals (SEM) for the estimation of potential toxicity in aquatic sediments, *Environ. Toxicol. Chem.,* 12, 1441, 1993.
7. Spurgeon, D.J. and Hopkin, S.P., Effects of variation of the organic matter content and pH of soils on the availability and toxicity of zinc for the earthworm *Eisenia Fetida, Pedobiologia,* 40, 80, 1996.
8. Belfroid, A.C. and Van Gestel, C.A.M., Blootstellingsroutes van toxische stoffen voor terrestrische invertebraten, Report no. E-99-06, Free University, Amsterdam, The Netherlands, 1999 (in Dutch).
9. Van de Meent, D., Aldenberg, T., Canton, J.H., Van Gestel, C.A.M., and Slooff, W., Desire for levels. Background study for the policy document, "Setting environmental quality standards for water and soil," RIVM Report no. 670101002, Bilthoven, The Netherlands, 1990.
10. Struijs, J., Van de Meent, D., Peijnenburg, W.J.G.M., Van den Hoop, M.A.G.T., and Crommentuijn, T., Added risk approach to derive maximum permissible concentrations for heavy metals: how to take natural background concentrations into account, *Ecotoxicology and Environ. Safety,* 37, 112, 1997.

11. Crommentuijn, T., Polder, M.D., and Van de Plassche, E.J., Maximum permissible concentrations and negligible concentrations for metals, taking background concentrations into account, RIVM Report no. 601501001, Bilthoven, The Netherlands, 1997.
12. De Rooij, N.M., Smits, J.G.C., Bril, J., Plette, A.C.C., and Van Riemsdijk, W.H., Methodology for determination of heavy metal standards for soil, WL/Delft Hydraulics, Report no. T2004, 1997.
13. De Rooij, N.M., Smits, J.G.C., Bril, J., Plette, A.C.C., and Van Riemsdijk, W.H., Methodology for determination of heavy metal standards for soil, phase 2a: development of models and measuring techniques, WL/Delft Hydraulics, Report no. T2117, 1998.
14. Van den Hoop, M.A.G.T., Metal speciation in Dutch soils: field-based partition coefficients for heavy metals at background levels, RIVM Report no. 719101013, Bilthoven, The Netherlands, 1995.
15. Janssen, R.P.T., Pretorius, P.J., Peijnenburg, W.J.G.M., and Van den Hoop, M.A.G.T., Determination of field-based partition coefficients for heavy metals in Dutch soils and the relationships of these coefficients with soil characteristics, RIVM Report no. 719101023, Bilthoven, The Netherlands, 1996.
16. Shea, D., Deriving sediment quality criteria, *Environ. Sci. Technol.*, 22, 1256, 1988.
17. Van der Kooij, L.A., Van de Meent, D., Van Leeuwen, C.J., and Bruggeman, W.A., Deriving quality criteria for water and sediment from the results of aquatic toxicity tests and product standards: application of the equilibrium partitioning method, *Wat. Res.*, 25, 697, 1991.
18. OECD, Guideline for testing chemicals no. 207, Earthworm acute toxicity tests, Organization for Economic Cooperation and Development, Adopted April 4, 1984, 1984.
19. Kuipers, S.F., *Bodemkunde.* Educaboek, Culemborg, The Netherlands, 1984 (in Dutch).
20. Houba, V.J.G. and Novozamsky, I., Influence of storage time and temperature of air-dried soils on pH and extractable nutrients using 0.01 mol/L $CaCl_2$, *Fresinius J. Anal. Chem.*, 34, 511, 1997.
21. Sparks, D.L., Page, A.L., Helmke, P.A., Loeppert, R.H., Soltanpour, P.N., Tabatabai, M.A., Johnston, C.T., and Sumner, M.E., *Methods of soil analysis,* Part 3, Chemical Methods, Soil Science Society of America, Inc., Madison, WI, 1996.
22. Dutch Normalization Institute, *Determination of clay content and size distribution by means of sieving and pipeting,* NEN 5753, Delft, The Netherlands, 1991.
23. Allison, J.D., Brown, D. S., and NovoGradac, K.J., MINTEQA2/PRODEFA2, a geochemical assessment model for environmental systems, U.S. Environmental Protection Agency, Athens, GA, 1991.
24. NIST—National Institute of Standards and Technology, Critical stability constants of metal complexes database, Version 1.0, U.S. Department of Commerce, Gaithersburg, MD, 1993.
25. Murray, K. and Linder P.W., Fulvic acids: Structure and metal binding. I. A random molecular model, *J. Soil Sci.,* 34, 511, 1983.
26. Woollard, C.D., A computer simulation of the trace metal speciation in seawater, Ph.D. thesis, University of Cape Town, South Africa, 1995.
27. Pretorius, P.T., Janssen, R.P.T., Peijnenburg, W.J.G.M., and Van den Hoop, M.A.G.T., Chemical equilibrium modelling of metal partitioning in soils, RIVM Report no. 719101024, National Institute for Public Health and the Environment, Bilthoven, The Netherlands, 1996.
28. Lindsay, W.L., *Chemical Equilibria in Soils,* J. Wiley and Sons, New York, 1979.
29. Geladi, P. and Kowalski, B.R., Partial least-squares regression: a tutorial, *Anal. Chim. Acta,* 185, 1, 1986.

30. Umetri AB, Simca-S for Windows, Version 5.1. Umeå, Sweden, 1994.
31. Draper, N.R. and Smith, H., *Applied Regression Analysis,* 2nd Edn., John Wiley and Sons, New York, 1981.
32. Anderson, P.R. and Christensen, T.H., Distribution coefficients of Cd, Co, Ni, and Zn in soils, *J. Soil Sci.,* 39, 15, 1988.
33. Buchter, B., Davidoff, B., Amacher, M.C., Hinz, C., Iskandar, I.K., and Selim, H.M., Correlation of Freundlich Kd and n retention parameters with soils and elements, *Soil Science,* 148, 370, 1989.

7 Isotopic Exchange Kinetics Method for Assessing Cadmium Availability in Soils

Emilie Gérard, Guillaume Echevarria, Christian Morel, Thibault Sterckeman, and Jean Louis Morel

ABSTRACT

Measurement of the availability of metals is essential for risk assessment of contaminated soils. The isotopic exchange kinetics method was adapted and evaluated to characterize the availability of cadmium in soils. Three soils were sampled in a Zn-, Cd-, and Pb-contaminated area, and a noncontaminated soil was collected in an agricultural field. The isotopic exchange kinetics method was derived from that developed for major elements and nickel. Various conditions were tested to adapt the method to cadmium: soil-to-solution ratio, equilibration time, filtration (pore size, sorption, nature of the filter), sampling pattern of the soil solution, and sorption on the surface materials. The speciation of Cd in solution was also determined. Results showed that optimal conditions to assess the kinetics of exchange for cadmium are a high soil-to-solution ratio and an equilibration time of 18 h. Results were independent of the sampling time (from 1 to 100 min). To limit the sorption of the radioisotope on the filter and to remove colloidal forms from the solution, cellulose nitrate filters of a 0.025- μm porosity were used. Ninety percent of the metal in soil extracts was Cd^{2+}. To account for the sorption of the radionuclide on the surface materials and the filtration apparatus, a reference control isotopic exchange kinetic was run. This procedure allowed comparison of the isotopic composition of Cd in the soil solution and that in plants grown on the same soil in a previous experiment, showing that the isotopically exchangeable pool was highly related to that of the metal taken up by plants.

INTRODUCTION

Toxic metals are present in various forms in soils, and risk assessment of food-chain contamination requires characterization of the available fractions. The total concentration of soil metals is not sufficient to assess the risk of soil–plant transfer of the element. The size of the available pool of Cd is closely dependent on the soil properties, because the metal is mostly associated with the solid phase and is present in a wide variety of chemical forms.[1] In addition, Cd is rather labile in soils[2], and the form of Cd taken up by plants is the free ion, Cd^{2+}, even though data suggest that chelated forms such as CdCl could be taken up by the plant.[3,4]

Many methods have been developed to evaluate available soil metals. Tests using plants predict quantities of available metal, but they are time-consuming and only give quantities effectively absorbed by a given plant under given conditions. Microbiological tests have been developed[5,6] that use enzyme activities inhibition by metals, and chemical extraction is widely used to evaluate the available quantities of trace elements. It allows metal fractions present in the solid phase to be solubilized, simulating the solubilization of the element during the growth period of any plants. Various extractants such as DTPA (diethylenetriaminepentaacetic acid), $CaCl_2$, and $NaNO_3$ give correlations between extracted Cd and Cd absorbed by plants, but for a specific plant type and soil type. For instance, available Cd for lettuce grown on a calcareous soils was well correlated with DTPA-extracted Cd,[7] but Brown et al.[8] did not obtain any correlation between extracted Cd and rye grass uptake for different pH-soils, nor did Hooda and Alloway[9] for rye grass on acidic soils. The DTPA extraction[10] should only be used on basic or neutral soils.[11] All methods only give indicative predictions on the pollution status of the soils and do not allow the measurement of the phytoavailable pool.

Isotopic methods, such as isotopic dilution and the isotopic exchange kinetics method (IEK), have been shown to measure truly phytoavailable elements in soils. Isotopic dilution consists of spiking the metal in soils and, after the growth of the plant tested, measuring the specific activity in the plant and the soil solution.[12,13] When the specific activity of the element in the plant equals the specific activity of the element in the soil solution, it means that the plant takes up the element from the isotopically exchangeable pool. Consequently, determining the specific activity of the soil solution allows a good prediction of the amount of phytoavailable element. The determination of the soil solution-specific activity is not easy, however, and this method does not give any dynamic indication about the element availability. IEK, a more rapid technique, is based on the same principle; it is used in the case of macronutrients (e.g., P)[14] and for trace elements such as Ni.[15] In IEK, the specific activity (or isotopic composition, IC) of the soil solution is calculated on the basis of a short experiment of isotopic exchange in a reconstituted batch system. This method indicates the dynamics of the element in soils and allows time-related parameters to be calculated that define their availability.

This work was undertaken to determine whether IEK methods could be used to determine the dynamic parameters and the phytoavailable amounts of Cd in soil. A series of trials were conducted to determine the optimal conditions for a standardized experimental procedure adapted to the specificity of the element. Then, the calculated isotopic compositions of Cd in four soil solutions (IC_S) at 90 d were compared with the isotopic compositions of Cd (IC_P) in three plants collected from a previous experiment.[16]

MATERIALS AND METHODS

SOILS

The A_p horizons of four soils were sampled from a series of silt loam brown soils (Hapludalf) located in northern France (Région Nord-Pas de Calais) and showing a gradient of distance from a lead and zinc smelter. Soils from different cultivated and limed plots (S0, S1, S2, and S3) were used for the experiments. Samples from the different plots had similar chemical properties, but the soils displayed a gradient of contamination by atmospheric industrial particles bearing Pb, Zn, and Cd, exhibiting total Cd content of 0.6, 8.9, 15.1, and 25.4 mg kg^{-1} and total Zn contents of 76, 515, 1010, and 1520 mg kg^{-1} (Table 7.1). Their slightly alkaline pH was due to liming. Soil S0 was sampled from a cultivated plot, far away from any source of severe metallic pollution.

IEK PROCEDURE

The method described in this study was first developed for potassium[12] and nickel[15] and is here adapted for cadmium.

TABLE 7.1
Characterization of Soil Samples*

Characteristic	Soil 0	Soil 1	Soil 2	Soil 3
Clay,[a] g kg^{-1}	149	178	209	154
Organic matter, g kg^{-1}	15.5	16.8	16.1	16.1
C/N[b]	9.4	11.1	11.7	11.4
pH$_{H2O}$[c]	8.2	7.9	8.1	7.8
Total CaCO$_3$,[d] g kg^{-1}	32	7	13	15
P. Olsen,[e] g kg^{-1}	0.17	0.09	0.07	0.11
CEC,[f] cmol$^+$ kg^{-1}	9.2	11.9	16.9	11.5
Total Cd,[g] mg kg^{-1}	0.6	8.9	15.1	25.4
Total Zn,[g] mg kg^{-1}	76	515	1010	1520
Total Pb,[g] mg kg^{-1}	65	440	766	1196

*Results Are Mean Values of Three Replicates

[a]Granulometric analysis (Standard NF X 31–107[17]).

[b]C/N–Organic carbon determined by sulfochromic oxidation (Standard NF X 31–109[18]), N by dry combustion (Standard NF ISO 13878[19]).

[c]pH–Standard NF ISO 10390.[20]

[d]CaCO$_3$–Standard NF ISO 10693.[21]

[e]P. Olsen–Standard NF X 31–161.[18]

[f]CEC–Cationic exchange capacity (Standard NF X 31–130[18]).

[g]Cd, Zn, Pb–See paragraph "Chemical Analysis and Statistics."

Fifty grams of soil in 99 mL of water (5:10 soil-to-solution ratio) were shaken for 18 h on an end-over-end shaker to reach a constant Cd concentration in the solution. One mL of a ^{109}CdCl$_2$ solution was injected into the suspension under constant stirring. Three aliquots of the suspension were sampled with a syringe after exchange times of 1, 4, and 10 min, and immediately filtered on cellulose nitrate (porosity 0.025 μm; Schleicher & Schuell GmbH, Germany) allowing more than 1 mL of soil solution to be collected. The radioactivity in the solution was measured by gamma spectrometry. The concentration of stable Cd in the filtered extracts (C_{cd}, mg L^{-1}) was also measured. Total radioactivity introduced at time = 0 was corrected by taking into account the amount of ^{109}Cd sorbed onto the surface of vials and filtration apparatus. Therefore, a blank solution was obtained from an air-dried soil suspended in deionized water (5:10 ratio) in a 250-mL polyethylene vial (Nalgene), shaken for 18 h, centrifuged at 10,000 g for 20 min, and filtered on cellulose nitrate (0.025-μm-porosity membrane) to obtain 99 mL of solution. One mL of the ^{109}Cd solution was injected in this solution, and three replicates of 1 mL each were sampled and filtered after 1, 4, and 10 min of exchange.

Isotopic Composition of Cd in Soil Solution

Isotopic composition of Cd in the soil solutions (IC_S) extrapolated to 90 d was calculated using a mechanistic model that already successfully describes the isotopic exchange of mineral ions in soils.[15,23–25] IC_S is expressed as follows:

$$IC_s = \frac{\frac{r_t}{R_s}}{Cd_s} \tag{7.1}$$

where r_t is the radioactivity in the soil–solution system, R_S is the total corrected radioactivity introduced into the system at time t = 0, and Cd_S (in μg g^{-1}) is the quantity of free Cd^2 in the soil solution per 1 g of soil when suspended in 10 mL of water, that is, for the 5:10 soil-to-solution ratio, $Cd_S = 10/5\ C_{Cd}$. C_{cd} is the concentration of stable Cd in the filtered extracts (mg L^{-1}). Parameters that describe the isotopic exchange, i.e., radioactivity remaining in the soil solution after 1 min (r_1/R_S) and n are determined by a nonlinear regression and are used in the theoretical equation[23] to calculate r_t/R_S over time:

$$\frac{r_t}{R_S} = \frac{r_1}{R_S}\left[t + \left(\frac{r_1}{R_S}\right)^{\frac{1}{n}}\right]^{-n} + \frac{Cd_S}{Cd_T} \tag{7.2}$$

where Cd_T is the total concentration of Cd in the soil (mg/kg^{-1}).

This equation is derived from a mathematical model[26] that describes the decrease of radioactivity in the soil solution with a sum of an infinite number of exponential terms:

$$r_t = A_0 + A_1 e^{-k1t} + \ldots + A_i e^{-kit} \tag{7.3}$$

where $A_0, A_1, \ldots A_i$; and $k_1 \ldots, k_i$ are constants.

It was mathematically demonstrated that this sum of an infinite number of exponential terms equals the exponential function[22] that is given in Eq. 7.2.

The optimal choice of the IEK parameters such as the soil-to-solution ratio, the period of time needed for a constant C_{Cd}, the minimal number of samplings, and the filtration, was made after a series of preliminary tests that were conducted primarily with the most polluted soil, S3.

The effect of the soil-to-solution ratio was investigated for the four soils, using soil-to-solution ratios of 1:10, 3:10, 5:10, and 7:10, equilibrated as described above (18 h) and 0.025-μm filtration porosity (cellulose nitrate). For each ratio, the IEK procedure was run to calculate the IC_S and parameters r_1/R_S and n. Conversely, equilibration times of 7 h, 18 h, and 48 to 72 h were tested on suspensions of soil 3 with a soil-to-solution ratio of 1:10, and at each equilibration time the soil suspensions were filtered on cellulose nitrate (0.025 μm). Cd concentration was determined in each case.

The effect of the sampling pattern on the IEK parameters was investigated. The IEK procedure was run on a 1:10 soil solution (soil 3), and the soil solution was sampled at 1, 4, and 10 min; 1, 4, 10, and 40 min; or 1, 4, 10, 40, and 100 min. The parameters r_1/R_S and n were then calculated.

The effect of filter pore size on solution Cd was tested using three different pore sizes: 0.45, 0.2, and 0.025 μm (cellulose nitrate). Suspensions (1:10, soil 3) were shaken for 18 h and centrifuged at 10,000 g for 20 min. The Cd was spiked with ^{109}Cd, and after 2 h of equilibration the suspension was filtered and analyzed for Cd and ^{109}Cd. The sorption of Cd on the filters was also assessed. A suspension (ratio 1:10, soil 3) was shaken for 18 h, centrifuged, and Cd was spiked with ^{109}Cd. This suspension was filtered at 0.45, 0.2, or 0.025 μm and then filtered a second time at the same porosity. The filtrates were analyzed for ^{109}Cd. The effect of the filter material was also tested using Teflon®,* cellulose nitrate, and cellulose acetate (0.2 μm). Suspensions (1:10, soil 3) were shaken for 18 h, centrifuged, and Cd was spiked with carrier-free ^{109}Cd. Then the suspensions were filtered and analyzed for ^{109}Cd. In addition to this, the capacity for sorption of Cd on the vials was tested. Ninety-nine milliliters of a soil solution (1:10, soil 3, centrifuged) was introduced in a Nalgene vial (250 mL), and the Cd was spiked with 1 mL ^{109}Cd. After an equilibration time of 2 h, the solution was analyzed for ^{109}Cd. For comparison, two aqueous solutions (deionized water and tap water) were introduced in the same manner in Nalgene vials.

To calculate the free Cd^{2+} concentration (C_{Cd}) in the filtrates obtained while running IEK on the four studied soils, the speciation software program GEOCHEM was used.[27] Soil suspensions were prepared in the same manner as for IEK (1:10, 18 h equilibration time) and filtered at 0.2 or 0.025 μm. A total of eight Cd species was considered to exist in the soil solution. The GEOCHEM program contains a database with complexation constants calculated for an ionic strength of zero preexists, and the activity coefficients were calculated using Davies equation. Major elements (Ca, Mg, K, Na, Si, Al, Fe, Mn), trace elements (Cd, Pb, Zn, Cu), and P, NO_3, NH_4, and organic matter were measured in the soil solutions. The partial pressure of CO_2 was assumed

---

* Registered trademark of E.I. du Pont de Nemours and Company, Inc., Wilmington, DE.

to be $10^{-3.5}$ atm; the ionic strength was computed as 0.012 mol L^{-1}. The soluble organic matter was assimilated to two categories of fulvic acids for which complexation constants have been established at pH = 5.

Chemical Analyses and Statistics

Total Cd in the soil was determined after dissolution by HF and $HClO_4$, using induced-coupled plasma atomic emission spectrometry (ICP–AES) or electrothermal atomic absorption spectrometry (ETAAS). Radioactive Cd was measured by gamma spectroscopy (Packard Cobra Auto-Gamma Counting Systems). The concentration of Cd in the filtered extracts was also determined by ETAAS. Aliquots were acidified with HNO_3 to prevent Cd precipitation before measurement. Analysis of variance was performed with the STATITCF software, Version 5.[28]

RESULTS

SOIL SOLUTION EQUILIBRATION CONDITIONS

Tables 7.2 and 7.3 show the effect of the soil-to-solution ratio on C_{Cd}, n, r_I/R_S, and IC_S of Cd in the four soils for an exchange period extrapolated at 90 d. The parameters were dependent on the soil-to-solution ratio. C_{Cd} decreased with the dilution of the soil in the suspension between ratios 7:10 and 1:10 with reduced or no differences between ratio 5:10 and 7:10 on soils 0 and 1. A decrease in the soil-to-solution ratio led to an increase in n values with no significant differences between ratios 5:10 and 7:10. The r_I/R_S values increased from ratio 7:10 up to ratio 1:10. IC_S decreased from ratio 5:10 to 1:10 on the contaminated soils. For these three contaminated soils, the IC_S for ratio 7:10 showed large standard deviations and was not statistically different from ratio 5:10. For soil 0, ratio 7:10 showed greater IC_S values than ratio 5:10. Figure 7.1 shows the effect of the equilibration time on C_{Cd}. There were no significant differences in C_{Cd} between 7 and 72 h.

SOIL SOLUTION SAMPLING

IEK parameters n and r_I/R_S were independent on the number of samplings (Table 7.4). After 10 min on soil 3, r_I/R_S reached 7×10^{-3} and n was 0.34.

FILTRATION

The effect of the filter pore size on the concentration of ^{109}Cd and Cd in solution from a centrifuged soil suspension (soil 3) is shown in Figures 7.2 and 7.3. The filters retained significant amounts of Cd (stable and radioactive), and stable Cd was retained more than ^{109}Cd. Up to 37% of ^{109}Cd in the centrifuged supernatant was retained at porosity 0.025 μm and up to 49% of stable Cd. The concentration of Cd in the 0.2-μm filtrates was lower than that obtained for 0.45 μm, and equivalent to that obtained at 0.025 μm. The phenomenon also occurred for ^{109}Cd, but differences

TABLE 7.2
Comparison of Four Soil–Solution Ratios in Measured C_{Cd} ($\mu g\ L^{-1}$), in Calculated n and r_1/R_S

Studied Value	Soils	Soil–Solution Ratio			
		1:10	3:10	5:10	7:10
C_{Cd}	S0	0.064 (0.02)	0.07 (0.03)	0.11 (0.01)	0.1 (0.03)
	S1	0.84 (0.08)	2.1 (0.6)	2.5 (0.3)	2.9 (0.5)
	S2	1.2 (0.13)	3.1 (0.1)	3.7 (0.2)	4.5 (0.1)
	S3	1.9 (0.09)	3.1 (0.1)	3.8 (0.4)	4.5 (0.1)
n	S0	0.34 (0.04)	0.32 (0.02)	0.31 (0.04)	0.21 (0.04)
	S1	0.35 (0.01)	0.36 (0.04)	0.27 (0.04)	0.24 (0.11)
	S2	0.35 (0.02)	0.26 (0.05)	0.19 (0.03)	0.22 (0.03)
	S3	0.29 (0.05)	0.22 (0.029)	0.19 (0.01)	0.20 (0.078)
$r_1/R_S \times 10^{23}$	S0	8.9 (0.7)	6.8 (0.2)	4.7 (0.30)	3.3 (0.2)
	S1	9.8 (0.2)	5.7 (0.6)	4.4 (0.10)	3.6 (0.2)
	S2	8.0 (0.4)	5 (0.5)	3.0 (0.10)	3.1 (0.3)
	S3	12 (1.4)	8.5 (0.1)	5.3 (0.02)	4.2 (0.1)

( ) Standard deviation associated with mean value of three replicates.

TABLE 7.3
Comparison between Four Soil-to-Solution Ratios in Calculated IC_S (90-d exchange period)

Soils	Soil-to-Solution Ratio			
	1:10	3:10	5:10	7:10
Soil 0	2.06 (0.27)	2.16 (0.23)	2.33 (0.27)	4.28 (1.57)
Soil 1	0.133 (0.05)	0.122 (0.002)	0.157 (0.03)	0.187 (0.068)
Soil 2	0.080 (0.02)	0.081 (0.005)	0.122 (0.02)	0.108 (0.01)
Soil 3	0.067 (0.015)	0.079 (0.013)	0.086 (0.008)	0.093 (0.051)

(IC unit in mg21 Cd kg soil)
( ) Standard deviation associated with mean value of three replicates.

were not significant. Table 7.5 shows the effect of a second filtration on Cd and ^{109}Cd measured in a centrifuged and prefiltrated soil suspension. A reduction of the measured Cd and ^{109}Cd was observed for porosity 0.45 μm (10% of Cd retained, 15% of ^{109}Cd) and for porosity 0.025 μm (31%), with, for both treatments, no significant difference between stable and radioactive Cd. Negligible quantities of Cd were retained at porosity 0.2 μm.

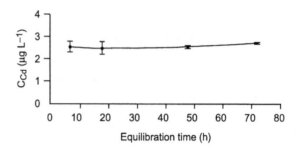

FIGURE 7.1   Effect of equilibration time on Cd concentration in solution (C_{Cd}).

TABLE 7.4
Effect of Sampling Times on IEK Parameters r_1/R_S and n, on Soil 3, Soil-to-Solution Ratio 1:10

Sampling Number	n	r_1/R_s ($\times 10^{-3}$)
1, 4, 10	0.338 (0.044)	6.98 (0.5)
1, 4, 10, 40	0.348 (0.047)	7.01 (0.5)
1, 4, 10, 40, 100	0.341 (0.055)	6.99 (0.5)

( ) Standard deviation associated with mean value of three replicates.

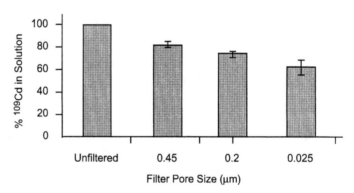

FIGURE 7.2   Effect of filter pore size on [109]Cd in soil solution.

Table 7. 6 shows the effect of the filter material on measured [109]Cd in a centrifuged soil suspension. Filters retained 26% to 35.5% of [109]Cd, and the retained [109]Cd was dependent on the filter material. Teflon filters retained up to 35.5% of [109]Cd, and cellulose nitrate and acetate filters retained 26% of [109]Cd. There was no significant difference between cellulose nitrate and cellulose acetate. The results of

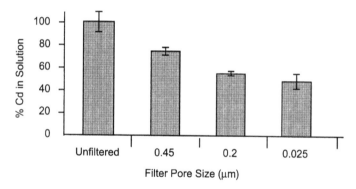

FIGURE 7.3   Effect of filter pore size on Cd concentration in solution (C_{Cd}).

TABLE 7.5
Effect of Filtration on Cd and ^{109}Cd Concentration in Liquid Phase of Soil Suspension, Soil 3, Soil-to-Solution Ratio 1:10

| | Cd in Solution (%)[a] | | ^{109}Cd in Solution (%)[a] | |
Porosity	Single Filtration	Double Filtration	Single Filtration	Double Filtration
0.45 μm	100 (4)	89.5 (5)	100 (2.5)	85 (4.5)
0.2 μm	100 (4)	104 (0.3)	100 (3.7)	97.2 (4.2)
0.025 μm	100 (14)	69.4 (1.5)	100 (9.9)	69.5 (8.5)

[a] Results are expressed in % of Cd or ^{109}Cd in solution after single filtration at a given membrane porosity.
( ) Standard deviation associated with mean value of three replicates.

TABLE 7.6
Effect of Filter Material on ^{109}Cd in Solution

Filtration Type	^{109}Cd in Solution[a]	
None	100	(2.5)
Teflon 0.2 μm	64	(5)
Cellulose nitrate 0.2 μm	74	(0.8)
Cellulose acetate 0.2 μm	73	(5.3)

[a] Results are expressed in % of ^{109}Cd in initial solution.
( ) Standard deviation associated with mean value of three replicates.

the tests on [109]Cd adsorption on Nalgene vials are shown in Table 7.7. For all solutions tested, [109]Cd was retained on the vials. [109]Cd was more retained on Nalgene vials when present in aqueous solutions (38% for deionized H_2O and 11% for tap H_2O). In the case of the supernatant of a centrifuged soil suspension (soil 3), 6% of [109]Cd was retained.

Cd SPECIATION IN THE SOIL SOLUTION

The GEOCHEM speciation software results showed that Cd was mostly present under Cd^{2+} forms in the soil solutions centrifuged and filtered at 0.2 or 0.025 μm (Table 7.8). From 88 to 91% of the Cd was under the free species in the 0.2-μm filtered solution, and from 82 to 91% for the 0.025-μm filtered solution, showing a decrease of the values in the latter case for soils 0 and 2.

ISOTOPIC COMPOSITIONS OF Cd IN SOIL SOLUTIONS AT 90 d
AND IN PLANTS

In general, the IC_S values for the four soils were in the same range as the isotopic compositions (IC_P) of Cd measured at 90 d on three plants varying in Cd accumulation pattern (data given by a previous experiment[16]). They ranged from 0.85 to 3.97 mg^{-1}Cd kg soil in soil 0, and from 0.09 to 0.26 mg^{-1} Cd kg soil in the

TABLE 7.7
Tests of [109]Cd Adsorption on Nalgene Vials

	[109]Cd Counts min^{-1}mL^{-1}		
Solution Type	Theoretically Introduced	Remaining in Solution	% Sorbed [109]Cd
Deionized H_2O	1481	912 (16)	38%
Tap H_2O	1481	1322 (48)	11%
Soil 3	12352	11868 (191)	6%

() Standard deviation associated with mean value of three replicates.

TABLE 7.8
Free Cd^{2+} in Solution Estimated with GEOCHEM Software

	% of Free Cd^{2+} in Solution			
Porosity of Filtration	Soil 0	Soil 1	Soil 2	Soil 3
0.2 μm	88	90	91	90
0.025 μm	82	91	85	89

contaminated soils. For a given soil, IC_S values were in general similar to IC_P, but significant differences were also observed, ranging from 0.99 to 1.64 mg^{-1} Cd kg soil in soil 0, and from 0.03 to 0.1 mg^{-1} Cd kg soil in the contaminated soils. In the contaminated soils, IC_S was smaller than IC_P of rye grass, and higher in the noncontaminated soil. After 90 d, the IC_S were identical to the IC_P in lettuce and *T. caerulescens* in soil 0, soil 1, and soil 3.

TIME-RELATED KINETICS PARAMETERS

Three factors describing the dynamics of phytoavailable Cd in soils (labile Cd) were characterized, as has already been done for P[23,29] and Ni[15]: quantity, E(t); intensity, C_{Cd}; and capacity $1/(r_f/R_S)^{16}$ (Table 7.10).

TABLE 7.9
Comparison between IC_S of Cd in Soil Solution (90 d) and IC_P of Cd in Rye Grass (*Lolium perenne* L.), lettuce (*Lactuca sativa* L.), and *Thlaspi caerulescens*[16]

Soil Type	Rye grass IC_P[a]	Lettuce IC_P[a]	Thlaspi caerulescens IC_P[a]	Soil solution IC_S[a] (5:10)
Soil 0	0.85 c	3.32 ab	3.97 a	2.33 b
Soil 0	0.26 a	0.18 b	0.18 b	0.16 b
Soil 2	0.20 a	0.10 c	0.15 b	0.12 c
Soil 3	0.16 a	0.09 b	0.11 b	0.09 b

[a] In mg^{-1} Cd kg soil. Values in same line followed by same letter are not significantly different (P = 0.05, four replicates).

TABLE 7.10
Cd Isotopic Exchange Kinetics Parameters for Four Soils (Ratio 5:10)[16]

Studied Value	Control	Soil 1	Soil 2	Soil 3
C_{Cd}[a]	0.1 (0.01)	2.5 (0.3)	3.7 (0.2)	3.8 (0.4)
r_f/R_S	0.0047 (0.0003)	0.0044 (0.0001)	0.003 (0.0001)	0.0053 (2×10^{-5})
E (0–1 min)[b]	0.04 (0)	0.93 (0.13)	1.7 (11)	2.01 (0.01)
E (1 min–30 d)[b]	0.34 (0.05)	4.99 (0.95)	5.81 (1.23)	8.29 (0.56)
E (30 d–60 d)[b]	0.03 (0)	0.35 (0.04)	0.5 (0.09)	0.83 (0.05)
E (60 d–90 d)[b]	0.02 (0.01)	0.2 (0.03)	0.29 (0.05)	0.5 (0.47)
E (>90 d)[b]	0.17 (0.05)	2.43 (0.98)	6.8 (1.28)	13.8 (0.63)

[a] C_{Cd} is Cd concentration in filtered extracts (μg L^{-1}).
[b] E(t) unit in mg Cd kg^{-1} soil.
( ) Standard deviation associated with mean value of three replicates.

The E(t) values represent the quantity of Cd isotopically exchanged during time interval t (t is related to the plant growth):

$$E(t) = 1/IC_S. \tag{7.4}$$

The intensity factor is the instantaneously available soil Cd and corresponds to C_{Cd}. The capacity factor represents the ability of the soil to maintain the intensity factor constant:

$$\frac{1}{\frac{r_\perp}{R_S}} = \frac{E(0 - 1 \text{ min})}{C_{Cd}}. \tag{7.5}$$

Most of the soil Cd was isotopically exchangeable within 30 d (from 40% on soil 3 to 66% on soil 1). Within 90 d, between 0.43 and 11.6 mg Cd kg^{-1} soil (soil 0 and soil 3) were exchangeable. During this period, therefore, the pool of nonisotopically exchangeable Cd ranged from 28% of total Cd in soil 0 to 54% in the most contaminated soil. The more the soil was contaminated, the less its Cd was exchangeable within 90 d. The Cd concentration of the filtered extracts (C_{Cd}) ranged from 0.1 to 3.8 µg L^{-1} for ratio 5:10. The $r_\perp/R_S$ values for the four soils ranged from 0.003 (soil 2) to 0.0053 (soil 3), and the n parameter exhibited values between 0.19 (soils 2 and 3) and 0.31 (soil 0).

DISCUSSION

The series of trials conducted on the soil solution equilibration procedure—sampling, filtration, and Cd speciation—allowed us to determine the optimal experimental procedure for running the Cd IEK experiment on the four calcareous soils. The range of soil-to-solution ratios was chosen on the basis that 7:10 was the highest ratio that was experimentally practicable on these soils, while 1:10 was the ratio used earlier for studying P^{22} and Ni.[15,30] Results showed that the kinetic parameters n and $r_\perp/R_S$ and then C_{Cd} and IC_S, were dependent on soil-to-solution ratios up to 5:10. However, differences between ratio 5:10 and 7:10 were mostly unsignificant, especially for n values and IC_S values in the contaminated soils. This dependence on the soil-to-solution ratio would be due to either an effect of dilution on ion exchange or a higher soil dispersion when using decreasing amounts of soil in the suspensions. In this case, the exchange surfaces between the solution and the solid phase might be increased. Hence, the soil capacity to maintain C_{Cd} a constant would be overrun when the soil solution ratio decreased from 5:10 to 1:10, i.e., the quantity of instantaneously exchangeable Cd, indicated by E(0–1 min), would not be sufficient to replenish the soil solution Cd when decreasing the soil-to-solution ratio. However, the strong buffering capacity of soils, indicated by $1/(r_\perp/R_S)$, allows a reduced and nonlinear decrease of C_{Cd} and IC_S values with the soil-to-solution ratio. As the concentration of Cd fixed in the solid phase is divided by 7 between ratios 7:10 and 1:10, Cd concentration in solution is divided by 2.4 for soil 3. The distribution coefficient of Cd between soil solution and solid phase increases from 0.25×10^{-3} for ratio 7:10 up to 0.74×10^{-3} for ratio 1:10. A stable C_{Cd} value appears to have been attained at ratio 5:10 for soils 0 and 1, but not for soils 2 and 3. For soils 2 and 3, a higher soil-to-solution ratio should be used. On the uncontaminated soil, IC_S was significantly higher for ratio 7:10 compared with ratio 5:10. In this case, concentrations of Cd in the soil solutions ranged from 0.06 to 0.1 $\mu g \, L^{-1}$, suggesting that these results could be due to analytical difficulties when measuring C_{Cd}. This implies that, on these noncontaminated high-pH soils, IEK methods could be limited when analytical techniques are not suitable to measure such concentrations.

Because differences in the IC_S values between ratios 5:10 and 7:10 were not significant on the contaminated soils and, moreover, because filtration of soil suspension with a ratio of 7:10 was not easy to perform and led to great standard deviations, the ratio 5:10 was used. The optimal ratio to run IEK on a given soil type should be chosen to give maximum C_{Cd} and stable n and r_t/R_S, taking into account the feasibility of the filtration. In addition, as C_{Cd} showed stable values after 7 h of equilibration, a standard equilibration time of 18 h (one night) was adopted for the study, which is consistent with the range of times generally used in other experiments on Cd.[31,32] As the experimental values of r_t/R_S and n did not show significant differences when sampling up to 10 or 100 min, the chosen sampling pattern was 1, 4, and 10 min, as for Ni.[33] Significant amounts of Cd were retained by the filters (0.45 to 0.025 μm), showing a possible effect of either Cd sorption or the presence of colloidal forms of Cd in the supernatant of this slightly alkaline soil. The differences in [109]Cd and Cd retention showed that [109]Cd was mostly present in noncolloidal forms in the soil solution, but stable Cd was present in both ionic and colloidal forms. This was not surprising because [109]Cd was introduced in the soil solution as $CdCl_2$. Hence, sorption is not the only phenomenon to explain the retention by filters; colloidal forms were present in the filtered extracts. The concentration of Cd was lower when using 0.2-μm and 0.025-μm filters than when using 0.45-μm filters. This shows that colloidal forms of Cd with diameter <0.45 μm exist in the soil solution and could occur even after a 0.2- μm filtration. This has also been demonstrated for Pb.[34] The use of 0.45-μm filters to obtain a colloidal-free soil solution must hence be avoided.

To prevent measuring Cd ions adsorbed on colloidal material in soil solutions, the porosity 0.025 μm was adopted for the experiment. The speciation GEOCHEM software allowed evaluation of the proportion of free Cd in soil solutions filtered at 0.2 or 0.025 μm. Results showed that even after 0.2- or 0.025-μm filtration, 10% of Cd was not in the Cd^{2+} form. The free Cd found in solution was, however, greater than the amount found in a 0.45-μm filtered extract from sludge-treated or $Cd(NO_3)$-amended silty loam soil solution (51%).[35] It was also greater than that of soil leachates (40 to 60%)[36] and extracts of forest soils to which Cd was added as $Cd(NO_3)_2$ (35%).[37] This difference might be due to the presence of large amounts of $CaCO_3$ in the four studied soils, and consequently to the competition between Ca^{2+} and Cd^{2+} at the exchange surface sites of the solid phase. Cd speciation in the soil–solution filtrates was taken into account when calculating IC_S values by subtracting 10% of the total Cd measured in solution. When studying possible sorption of Cd on filters, filter material tests led to the choice of cellulose nitrate. Results from a double filtration experiment showed that most sorption occurred on 0.45-μm and 0.025-μm porosity filters (10% and 30%). This can be due to sorption, but it may also be due to eventual further removal of colloidal material. As there seemed to be no sorption at 0.2 μm, we suggest that the phenomenon of sorption was minimal and that the differences were due to colloids in solution. The study of the sorption of [109]Cd on vials showed that, in the case of a centrifuged soil suspension, the retained quantities were small but statistically significant (6%). Sorption was greater when using H_2O (tap and deionized water) because there were small (or none, in deionized water) quantities of competing ions to exchange at the surface sites of sorption. The sorption of Cd and [109]Cd on vials, and eventually on the filtration apparatus, was taken into account while running the IEK method. A reference control IEK procedure was run on a prefiltered soil solution to evaluate the differences between the theoretically introduced [109]Cd in the soil solution and what remained in the solution after a period in the Nalgene vial and a filtration.

Results from this methodological study showed that IEK methods on Cd should be run with soil suspensions shaken for 18 h, using a soil-to-solution ratio as high as possible, and, to remove colloidal material from the solution, the smallest practicable porosity filtration. A reference control IEK procedure must also be run to take into account sorption of Cd and [109]Cd on the apparatus.

When comparing the IC_S values (ratio 5:10) for the four studied soils after 90 d with the IC_P values of three plants grown in another experiment on the same soils,[16] IC_S values were shown to be similar to IC_P for a given soil, especially in the contaminated soils. This suggests that the isotopically exchangeable pool assessed by short-term IEK experiments without plants was similar to the phytoavailable pool. Significant differences were, however, observed between the IC_S and IC_P of rye grass on the three contaminated soils. This could be explained by an early accumulation of Cd in the roots of rye grass during the first 30 d of growth and transfer of the metal, which had a greater IC, to the shoots during the following period. IC_P values were also in some cases slightly higher than IC_S, which is theoretically impossible but could be explained if the IEK experiments underestimated IC_S.

The IEK experiments also allowed measurement of the amount of the isotopically exchangeable pools (E(t)), thus permitting the phytoavailable pools of soil Cd to be estimated as a function of time.[16] Despite high pH values, Cd was easily exchangeable in the four soils. Most Cd (from 40.5 to 66.5%) was exchangeable within 30 d, and more than 50% was exchangeable within 90 d for all soils.

If the IEK method can be used on other types of soils, it will be a very useful tool for studying Cd phytoavailability. Whether plants access the same pools of soil Cd can be determined and, if not, their influence on the mobility of soil Cd can be evaluated. The method also describes the exchangeable pools of Cd as a function of time, allowing the risk of transfer of soil Cd to terrestrial plants to be predicted. This can be very useful when studying the quality of sewage sludge or other contaminated soils. Providing that isotopically exchangeable Cd is similar to phytoavailable Cd, the portion of phytoavailable Cd in soils that is absorbed by plants can be calculated.

CONCLUSION

The study shows that IEK methods allowed Cd phytoavailability in high-pH-contaminated soils to be assessed. The optimal choice of IEK operational conditions, including soil-to-solution ratio and soil–solution equilibration procedure, is highly dependent on the physical and chemical properties of the studied soil. Isotopic methods showed that the isotopically exchangeable pool was closely related to that accessed by plants. These methods allow the available pools to be described as a function of time and are pertinent tools when assessing the risk of metal transfer from soil to plants. Work is underway to determine whether these findings can be generalized for other soil conditions.

ACKNOWLEDGMENTS

This project was funded by the Nord Pas de Calais Regional Council (Conseil Régional, France), the Ministère de l'Education Nationale, de la Recherche et de la Technologie (France), and the National Institute for Agronomic Research (INRA, France). The authors wish to thank J.C. Fardeau (INRA) and Dr. C. Schwartz (LSE-INRA) for their helpful comments and advice.

REFERENCES

1. Ross, S.M., Retention, transformation and mobility of toxic metals in soils, in *Toxic Metals in Soil-Plant Systems,* Ross, S.M., Ed., John Wiley and Sons, New York, 27, 1994.
2. Alloway, B.J., Cadmium, in *Heavy Metals in Soils,* Alloway, B.J., Ed., Blackie Academic and Professional, London, 123, 1995.
3. Cabrera, D., Young, S.D., and Rowell, D.L., The toxicity of cadmium to barley plants as affected by complex formation with humic acid, *Plant and Soil,* 105, 195, 1988.
4. Smolders, E. and McLaughlin, M.J., Chloride increases cadmium uptake in Swiss chard in a resin-buffered nutrient solution, *Soil Sci. Soc. Am. J.,* 60, 1443, 1996.
5. Bitton, G., Jung, K., and Koopman, B., Evaluation of a microplate assay specific for heavy metal toxicity, *Arch. Environ. Contam. Toxicol.,* 27, 1, 25, 1994.
6. Morel, J.L., Hosy, C., and Bitton, G., Assessment of bioavailability of metals to plants with MetPLATE, a microbiological test, 1995, in *Contaminated Soils 95,* Van den Brink, W.J., Ed., Kluwer Acad. Publ., Dordrecht, The Netherlands, 527, 1995.
7. Roca, J. and Pomares, F., Prediction of available heavy metals by six chemical extractants in a sewage sludge-amended soil, *Commun. Soil Sci. Plant Anal.,* 22, 19, 1991.
8. Brown, P.H., Dunemann, L., Schulz, R., and Marshner, H., Influence of redox potential and plant species on the uptake of nickel and cadmium from soils, *Z. Pflanznernähr Bodenk.,* 152, 85, 1989.
9. Hooda, P.S. and Alloway, B.J., Effects of time and temperature on the bioavailability of Cd and Pb from sludge-amended soils, *J. Soil Sci.,* 44, 1, 97, 1993.
10. Lindsay, W.L. and Norwell, W.A., Development of a DTPA soil test for zinc, iron, manganese, copper, *Soil Sci. Soc. Am. J.,* 42, 421, 1978.

11. Lebourg, A., Sterckeman, T., Ciesielski, H., and Proix, N., Intérêt de différents réactifs d'extraction chimique pour l'évaluation de la biodisponibilité des métaux en traces du sol, *Agronomie*, 16, 201, 1996 (in French).

12. Tiller, K.G. and Wassermann, P., Radioisotopic techniques and zinc availability in soil, in *Proceedings of the Symposium on the Use of Isotopes and Radiation in Research on Soil–Plant Relationships Including Applications in Forestry,* Proceedings series, International Atomic Energy Agency, Vienna, 13–17 December 1971, 517.

13. Tiller, K.G., Honeysett, J.L., and De Vries, M.P.C., Soil zinc and its uptake by plants, I. Isotopic exchange equilibria and the application of tracer techniques, *Aust. J. Soil Res.,* 10, 151, 1972.

14. Fardeau, J.C., Hétier, J.M., and Jappé, J., Potassium assimilable du sol: Identification du compartiment des ions isotopiquement diluables, *C. R. Acad. Sci.,* (Paris) 288, 1039. 1979 (in French).

15. Echevarria, G., Morel, J.L, Fardeau, J.C., and Leclerc–Cessac, E., Assessment of phytoavailability of nickel in soil, *J. Environ. Qual.,* 27, 1064, 1998.

16. Gérard, E., Echevarria, G., Sterckeman, T., and Morel, J.L., Cadmium availability to three plant species varying in Cd accumulation pattern, *J. Environ. Qual.,* 29, 1117–1123, 2000.

17. Association Française de Normalisation (AFNOR), NF X 31-107 standard, Qualité des sols, Analyse granulométrique par sédimentation, Méthode de la pipette, 1983 (in French).

18. AFNOR, NF X 31-109 standard: Détermination du carbone organique par oxydation sulfochromique; NF X 31-161 standard: Détermination du phosphore soluble dans une solution d'oxalate d'ammonium; NF X 31-130 standard: Détermination de la capacité d'échange cationique (CEC) et des cations extractibles, Qualité des sols; 1993 (in French).

19. AFNOR, NF ISO 13878 standard, Qualité du sol, Détermination de la teneur totale en azote par combustion sèche (analyse élémentaire), 1998 (in French).

20. AFNOR, NF ISO 10390 standard, Qualité du sol, Détermination du pH, 1994 (in French).

21. AFNOR, NF ISO 10693 standard, Qualité du sol, Détermination de la teneur en carbonate. Méthode volumétrique, 1995 (in French).

22. Fardeau, J.C., Cinétiques de dilution isotopique et phosphore assimilable des sols, doctoral dissertation, État de Paris VI, 1981, 198 (in French).

23. Fardeau, J.C., Morel, C., and Boniface, R., Cinétiques de transfert des ions phosphate vers la solution du sol: Paramètres caractéristiques, *Agronomie* (Paris), 11, 787, 1991 (in French).

24. Frossard, E., Fardeau, J.C., Brossard, M., and Morel, J.L., Soil isotopically exchangeable phosphorus: A comparison between E and L values, *Soil Sci. Soc. Am. J.,* 58, 846, 1994.

25. Sinaj, S., Frossard, E., and Fardeau, J.C., Isotopically exchangeable phosphate in size fractionated and unfractionated soils, *Soil Sci. Soc. Am. J.,* 61, 1413, 1997.

26. Probert, M.E. and Larsen, S., The kinetics of heterogeneous isotopic exchange, *J. Soil Sci.,* 23, 76, 1972.

27. Sposito, G. and Bingham, F.T., Computer modeling of trace metal speciation in soil solutions: Correlation with trace metal uptake by higher plants, *J. Plant Nutr.,* 3, 35, 1981.

28. Beaux, M.F., Gouet, H., Gouet, J.P., Morleghem, P., Philippeau, G., Tranchefort, J., and Verneau, M., STAT–ITCF, Manuel d'utilisation, Institut Technique des Céréales et des Fourrages (ITCF) Editor, Paris, 1991 (in French).

29. Fardeau, J.C., Le phosphore assimilable des sols: Sa représentation par un modèle fonctionnel à plusieurs compartiments, *Agronomie* (Paris), 13, 1, 1993 (in French).

30. Shallari, S., Disponibilité du nickel du sol pour l'hyperaccumulateur *Alyssum murale,* (abstract in English), doctoral dissertation, Inst. National Polytechnique de Lorraine, Nancy, France, 1997 (in French).

31. Street, J.J., Lindsay, W.L., and Sabey, B.R., Solubility and plant uptake of cadmium in soils amended with cadmium and sewage sludge, *J. Environ. Qual.,* 6, 1, 72, 1977.
32. Street, J.J., Sabey, B.R., and Lindsay, W.L., Influence of pH, phosphorus, cadmium, sewage sludge, and incubation time on the solubility and plant uptake of cadmium, *J. Environ. Qual.,* 7, 2, 286, 1978.
33. Echevarria, G, Contribution à la prévision du transfert sol-plante des radionucléides, doctoral dissertation, Inst. National Polytechnique de Lorraine, Nancy, France, 1996.
34. Jopony, M. and Young, S.D., The solid-solution equilibria of lead and cadmium in polluted soils, *Euro. J. Soil Sci.,* 45, 59, 1994 (in French).
35. Candelaria, L.M. and Chang, A.C., Cadmium activities, solution speciation, and solid phase distribution of Cd in cadmium nitrate and sewage sludge-treated soil systems, *Soil Science,* 162, 10, 722, 1997.
36. Lamy, L., Cambier, P., and Bourgeois, S., Pb and Cd complexation with soluble organic carbon and speciation in alkaline soil leachates, in *Biogeochemistry of Trace Elements,* Adriano, D.C., Chen. Z.S., and Yang, S.S., Eds., Northwood (Great Britain), 1, 1994.
37. Hirsch, D. and Banin, A., Cadmium speciation in soil solutions, *J. Environ. Qual.,* 19, 366, 1990.

8 Accumulation, Redistribution, Transport and Bioavailability of Heavy Metals in Waste-Amended Soils

F.X. Han, W.L. Kingery, and H.M. Selim

ABSTRACT

Heavy metals accumulate in agricultural soils amended with various agricultural and industrial wastes. There is some evidence of metal transport in long-term waste-amended soils, but most data show limited mobility of heavy metals in waste-amended soil profiles. The bioavailability of heavy metals and their mobility in soils are largely determined by their distribution among various solid-phase components. Heavy metals in soils amended with various wastes are redistributed and transferred with time from the labile forms to the more stable forms, and the redistribution processes are dependent upon the source and process of waste, level of waste input, nature of metal, time scale, and soil properties such as pH, Eh, texture, and moisture regime.

Long-term studies on the kinetics of transformation and redistribution of heavy metals in various waste-amended soils, both under laboratory-controlled and field conditions, are needed to determine the dosage limits of various wastes in different soils and under different agricultural practices, and to select crop rotation systems and optimize management protocols to achieve minimum metal availability in soils. This will achieve the lowest toxicity to plants and minimum potential contamination of groundwater from the application of various wastes to soils. This understanding will enable one to assess the merits and disadvantages of irrigation with reclaimed sewage water, the prolonged use of animal waste and sewage sludge, and the use of municipal compost in agricultural land disposal.

INTRODUCTION

Soil is the indispensable land resource that produces food and other raw materials for humans. Yet, soil has become and continues to be the sink for wastes, including various animal wastes and metals released from ores in the earth through increasing human activities, in particular industrialization. Mining, smelting, and the combustion of fossil fuels are the greatest sources of contamination. Vast amounts of wastewater, sewage sludge, and city garbage compost are produced every year. Recently, a rapid growth in animal production has resulted in large amounts of animal waste. Other sources of heavy metals in soils include atmospheric fallout resulting from gaseous emissions from fuel and coal-burning power plants, auto emissions, and industrial emissions. Some fertilizers also contain heavy metals, such as Cd in certain P-fertilizers.

Among the heavy metals of greatest concern produced from sludge and sewage water are Cd, Zn, Pb, Ni, Cu, and Cr. In animal waste, Cu, Zn, Mn, As, and Se are the most important elements. Of these, Cd is of particular concern because of its high proportion in available (exchangeable) form in the soils[1-3] and its possible accumulation to potentially harmful levels in the food chain. In addition, Cd is fairly mobile, which can result not only in high availability to plants, but potential contamination of surface- and groundwater as well, especially in sandy soils. Adequate assessment of the potential hazards of heavy metal accumulation must take into account not only the total concentration, but also the solid-phase chemical forms.

The risk from heavy metal input into a given soil, as related to its introduction into the food chain and/or migration to groundwater, is determined by the partition of the added heavy metals between the solution and the solid phase, and their partition among the various components of the solid phase, including clay and organic matter surfaces (EXC); carbonates (CARB); easily reducible oxides, e.g., manganese oxides (ERO); organic matter (OM); reducible oxides, e.g., iron oxides (RO); and a residual phase mostly related to alumosilicate minerals (RES).[4-7]

The sequential selective dissolution method (SSD) has been developed to study the forms, availability, mobility, and transformation of heavy metals in sludge, manure, soils, sludge-amended soils, and sediments.[4-24] This method is based on both the solubility of individual solid-phase components and the selectivity and specificity of chemical reagents. The procedure should provide a gradient for the physicochemical association strength between trace elements and solid particles rather than actual speciation,[25] thus giving a semiquantitative indication of their relative availability to plants or to further migration to groundwater. A number of such procedures were developed for specific elements, matrices, regional soils, and specific purposes. In addition, various extractants were used by different investigators for the same targeted solid-phase component. The latest review on the extractants of individual fractions of metals in soil was made by Shuman.[26] However, in most protocols heavy metals in native and waste-amended soils were divided into the following physicochemical forms:

1. As simple or complexed ions in solution, in equilibrium with exchangeable and other phases.
2. As exchangeable ions (EXC), sometimes including ions nonspecifically adsorbed and specifically absorbed on the surface of various soil components,

such as carbonate, organic matter, Fe, Mn, Si, and Al oxides and clay miner-als. This part is controlled by adsorption-desorption processes.

3. Organically bound (OM). Heavy metals may be bound in living organisms, detritus, and the organic matter of the soil. This part is affected by the production and decomposition of organic matter.

4. Occluded or coprecipitated by carbonate (CARB). This fraction would be susceptible to changes in soil pH.

5. Bound to iron and manganese oxides, including amorphous and crystalline oxides, which appear as nodules, concretions, and coatings on particles. This fraction is reduced or oxidized by change of Eh, including three divisions: metals bound to Mn oxides or easily reducible oxides (ERO), bound to amorphous Fe oxides (AmoFe), and to crystalline Fe oxides (CryFe) (the last two divisions are collectively called the reducible oxides fraction (RO)).

6. Residual. This fraction mainly contains primary and secondary minerals, which hold elements within their crystal structure. These metals are not expected to be released to soil solution over a reasonable time span under conditions normally encountered in nature.

In the selective sequential dissolution procedures, the terms of all fractions are more likely to be defined operationally rather than chemically.[27] However, each extractant in the procedures effectively targets one major solid-phase component. It is recognized that in no case can an extractant remove all of a targeted solid-phase component without any attack on other components. No selective dissolution scheme can be considered completely accurate in distinguishing between different forms of an element, i.e., various organic–inorganic solid-phase components. In addition, there may be redistribution and readsorption during sequential dissolution extraction.[27,28] However, readsorption observed during sequential extraction may possibly be due to the use of either large spikes or simple model materials.[29] It was reported that readsorption of metals (Cd, Cu, Pb, and Zn) during sequential extraction was minimal.[29,30] Despite these shortcomings common to any chemical extraction procedure, sequential dissolution techniques still furnish more useful information on metal binding, mobility, and availability than can be obtained with a single extractant. Moreover, various extractants with different selectivities and specificities were used in different sequential procedures. These made comparisons of distribution of metals in soils or waste-amended soils obtained by different sequential procedures more difficult. In this chapter, we emphasize the common characterization of the distribution of heavy metals among solid-phase fractions in soils and waste-amended soils rather than their detailed differences in targeted solid-phase fractions obtained by different sequential procedures. Consequently, we will name the metal fractions mostly in terms of targeted solid-phase components instead of extractants.

The accumulation of heavy metals, their distribution among various solid-phase components, and their transport through soil profiles in animal waste-, sewage sludge-, inorganic salt-, and mining tailing-amended soils are reviewed.

ANIMAL WASTE-AMENDED SOILS

Even though cattle wastes are far greater than poultry and swine in terms of quantity, poultry and swine wastes are two of the most important from the point of view of possible metal pollution. They have recently received intensive study. During the last 30 years, the U.S. poultry industry has increased rapidly. By 1994 there were 1.4 billion birds.[31] The gross farm income in 1990 was $15.5 billion.[31] Approximately 13 million Mg (14 million tons) of poultry litter and manure was produced on U.S. poultry farms in 1990, most of which (68%) was broiler litter.[32] There is a clear trend of increase in annual waste production. In 1993, approximately 54 million tons of fresh poultry manure was produced.[31] In addition, the major portion (90%) of poultry litter produced was applied to agricultural land.[32]

Several heavy metals and metalloids, including As, Co, Cu, Fe, Mn, Se, and Zn, are added to poultry diets for various purposes, resulting in fairly high concentrations of heavy metals in the waste.[33] High concentrations of heavy metals in poultry waste-amended soils may result in environmental concerns due to the potential contamination of surface and groundwater via runoff and leaching.[34] In addition, phytotoxicity may result where there has been long-term application of poultry waste.[33,35] A number of different soils receiving repeated applications of poultry litter for several years have exhibited high concentrations of extractable Cu and Zn,[34,36–38] and Moore et al.[39] found that heavy metal concentrations in runoff from poultry waste-amended soils increased as the litter application rate increased.

During the 1990s the pork industry both consolidated and grew. In 1994 the total hogs and pigs in the U.S. reached 57.5 billion.[31] Accordingly, the U.S. hog and pig industry annually generates approximately 100 million tons of fresh manure (12.3 million tons dry solids).[31] The use of high levels of Cu in swine production to stimulate growth and improve feed efficiency results in elevated Cu concentrations in the manure. After an extended period of application of elevated-Cu manure, it was found that Cu had accumulated in soils, and in plants such as grass and corns.[40–43]

ACCUMULATION OF HEAVY METALS IN ANIMAL WASTE-AMENDED SOILS

Elevated levels of heavy metals were found in long-term poultry and pig manure waste-amended soils.[34,36,38,40–43] The metal accumulation in the soils did not usually reach levels toxic to plants.[34] Kornegay et al.[40] reported that Cu levels in the plow layer of the soil increased after three years of the application of pig manure. Sutton et al.[41] found that Cu-enriched swine manure increased the 1-N HCl-extractable Cu in the upper portion of the soil profile (0–31 cm) after five years of application, but not at lower depths. Mullins et al.[42] indicated that DTPA-extractable Cu in the surface soil increased after three years of application, and Cu was found to move downward in a sandy soil. Payne et al.[43] further pointed out that after eight years of application of manure, DTPA-extractable Cu was linearly related to the amount of Cu applied.

Total concentrations of selected heavy metals (Cu, Zn, Pb, Ni, Cr, and Mn) in a Mississippi surface soil receiving 25 years of poultry waste are presented in

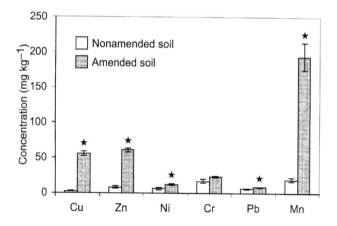

FIGURE 8.1   Total concentrations (averages and standard errors) of Cu, Cr, Mn, Ni, Pb, and Zn as extracted with 4-M HNO_3 in nonamended and poultry waste-amended soils (metal concentrations with * are significantly different at $p < 0.05$ level).

Figure 8.1. Statistically, concentrations of Cu, Zn, Mn, Ni, and Pb in nonamended and amended soils were significantly different (Figure 8.1). Copper, Zn, and Mn accumulated considerably in the poultry waste-amended soil over 25 years. Other metals such as Ni, Cr, and Pb increased slightly in the amended as compared to nonamended soil. Based on total concentrations, Cu and Zn in the amended soil increased at an average rate of approximately 2 mg $kg^{-1} yr^{-1}$. The difference in Mn represents an average rate of accumulation of 6.5 mg $kg^{-1} yr^{-1}$.

DISTRIBUTION OF HEAVY METALS AMONG SOLID-PHASE COMPONENTS IN ANIMAL WASTE-AMENDED SOILS

Copper in the Coastal Plain (Aquic Hapludults), Ridge and Valley (Aquic Hapludalfs), and Piedmont regions (Typic Rhodudults) of Virginia following three and eight years of application of Cu-enriched manure was mostly bound in the organic matter fraction, followed by the oxide-occluded fraction.[42–43] These manure treatments also resulted in higher Cu contents in the organically-bound fraction than did sulfate salt application.[43]

A short-term laboratory incubation experiment showed that in a Hawaii soil amended with poultry waste, organically-bound Zn decreased with time due to the decomposition of organic matter, whereas, depending upon soil pH, the carbonate bound, residual, and exchangeable Zn fractions increased.[44]

As illustrated in Figure 8.1, long-term application of poultry waste resulted in significant accumulations of Cu, Zn, and Mn in soil. The distribution of these heavy metals among the solid-phase components in both poultry waste-amended and nonamended Mississippi soils is presented in Figure 8.2. Copper in the amended soil was mostly present in the organically-bound fraction (46 ± 7.1%), and Cu in the nonamended soil was mainly in the residual fraction (52 ± 6%) (see Figure 8.2). In both

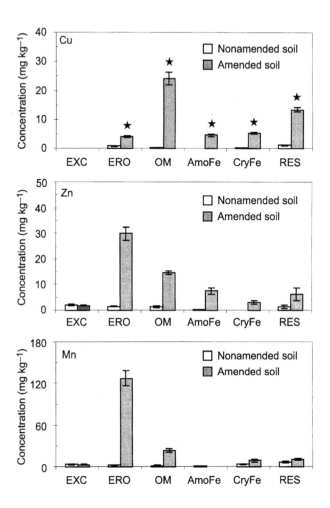

FIGURE 8.2 Concentrations (averages and standard errors) of Cu, Zn, and Mn in solid-phase components of nonamended and poultry waste-amended soils (EXC, ERO, OM, AmoFe, CryFe, and RES represent exchangeable, easily reducible oxide-bound, organic matter-bound, amorphous Fe oxide-bound, crystalline Fe oxide-bound, and residual fractions) (metal concentrations with * are significantly different at p <0.05 level).

soils, Cu concentrations in the exchangeable fraction were low. In contrast, Cu in the organically-bound fraction in amended soil was approximately 127 times that in non-amended soil. High Cu concentrations in the organically-bound fraction may sometimes pose potential risks after Cu is released upon breakdown of soil organic matter.

Compared with Cu, Zn concentrations in the exchangeable fraction for both Mississippi soils were quite high. This suggests a likelihood that Zn had higher solubility than Cu, and thus may be bioavailable. Zinc in the amended soil was predominantly in the easily reducible oxide (48 ± 10%) and organically-bound fractions (23 ± 2.8%), and, to a lesser degree, in the amorphous Fe oxide fraction (12 ± 6%) (Figure 8.2). Zinc in the easily reducible oxide fraction in amended soil was 20 times that in nonamended soil. It may then be suggested that adsorption–desorption (on the exchangeable fraction) and oxidation–reduction (on the easily reducible oxide fraction) processes may govern Zn availability, phytotoxicity, and mobility in this soil.

Our results indicate distinctly different distributions of Cu and Zn among the solid-phase components in these soils (Figure 8.2). The distribution patterns of these two metals can be attributed to their chemical properties. Copper is known to bind strongly as an inner-sphere complex with organic matter,[45] and Zn may be bound by outer-sphere complexation, with the metal retaining its inner hydration sphere. Tan et al.[46] reported that organic matter from poultry litter chelated Cu and Zn, and the stability of the Cu complexes was larger than those of Zn. Due to its similarity with Fe and Mg ionic radii, Cu may be capable of isomorphous substitution of Fe^{2+} and Mg^{2+} in layer silicates,[6,15] which explains the concentrations of Cu observed in the residual fraction in nonamended soils. Copper incorporation into alumino-silicate minerals in the soil, however, was perhaps a slow process. The significant proportion of metal (Cu and Zn) concentrations in the residual fraction in amended soil was, in part, due to metal incorporation into alumino-silicates, and binding to humin, which was not completely removed in the organically-bound fraction step. In addition, Cu and Zn may also coprecipitate with and form solid solutions of iron oxides.[47]

Manganese in the nonamended Mississippi soil was mainly in the residual fraction (35%), followed by the crystalline Fe oxide (20%), exchangeable (20%), and easily reducible oxide (12%) fractions (see Figure 8.2).[34] For the amended soil, Mn was predominantly in the easily reducible oxide fraction (73%), followed by the organically bound fraction (14%). This is indicative of the presence of secondary minerals such as Mn oxides in the amended soil.[48] As a result, in the amended soil, greater lability and mobility of Mn in comparison to the nonamended soil might be expected. McKenzie[49] showed that the affinity of synthetic Mn oxides to adsorb Zn^{2+} and Cu^{2+} ions is stronger than that of Fe or Al oxides. Since organic matter competitively binds Cu, the role of Mn oxides in binding Cu is not very significant for the amended soil. On the other hand, Mn oxides, as extracted in the ERO fraction, exhibited strong solid-phase affinity for added Zn in the poultry waste-amended soils.

MOBILITY IN AMENDED SOIL PROFILE

Since less attention has been paid to heavy-metal accumulations in poultry waste-amended soils, reports on the mobility of metals in amended-soil profiles are scarce. In the Sand Mountain region of Alabama, double-acid-extractable Cu and Zn were found to accumulate in long-term poultry waste-amended surface soils, but there was no evidence of metal transport in the soil profile using soil testing procedures.[36] In soils receiving a long-term application of swine manure, DTPA-extractable Cu in the B horizons of the soils was found to increase, indicating some extent of downward movement of the applied Cu in these soils; however, larger increases in extractable Cu in one subsoil were attributed to the downward movement caused by plowing.[42,43] Han et al.[34] clearly demonstrated that long-term application of poultry waste resulted in the slight movement of Zn to the deep soil profiles, and movement of Cu to the 40-cm depth.

Most heavy metals applied in poultry waste to a Mississippi soil accumulated in the upper 0–20 cm and, to a lesser extent, in the 20–40 cm depth.[34] In contrast, for the nonamended soil, the total concentrations of Cu and Zn showed a slight decrease close to the soil surface (0–20 cm) (see Figure 8.3). This decrease may be due to plant uptake of these metals from the surface zone. For Cu, the total concentration was almost uniform with depth below 40 cm in both profiles (Figure 8.3), whereas total Zn in the amended soil exhibited a noticeable increase with soil depth. Specifically, concentrations of Zn in amended soil were nearly 1.5 to 2 times those of the nonamended soil. These results reflect the potential downward movement of Zn and, to some extent, Cu, following long-term poultry waste additions.

The distributions of Cu and Zn in the various solid-phase fractions with depth in the soil profile for both poultry waste-amended and nonamended soils are shown in Figures 8.4 and 8.5. Copper concentrations in all solid-phase fractions at depths from 0 to 60 cm in amended soil were much higher than those in nonamended soils, indicating some movement of Cu to the subsurface. Further, Cu in the crystalline Fe oxide fraction in amended soil profiles was higher than in the nonamended soil (Figure 8.4). For all soil depths of the amended soil, Zn concentrations in the crystalline Fe oxide, residual, exchangeable, organic matter, and, to some extent, amorphous Fe oxide fractions were higher than those in the nonamended soil, even though they were not statistically significant (Figure 8.5). At depths below 60 cm in the amended soil, Zn in the iron oxide (CryFe) and residual fractions was roughly twice that measured in the nonamended soil. According to Shuman,[48] Zn and Cu in the residual fraction resided primarily in the clay- and silt-sized subfractions of fine-textured soils. Such findings indicate that metals in the amended soil profile may be mainly present in colloidal forms, such as iron oxides and fine clay particles.

The waste-amended surface soil exhibited high organic matter content due to poultry waste applications. Organic matter content was highest near the surface and decreased with soil depth in both amended and nonamended soils (data not shown).[50] This is consistent with the accumulation of metals (Cu and Zn) in the organically

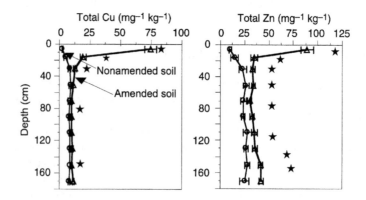

FIGURE 8.3 Total concentrations (averages and standard errors) of Cu and Zn as extracted with 4-M HNO_3 in the profiles of nonamended and poultry waste-amended soils (n = 8 and 6 for amended soil and nonamended soils, respectively) (metal concentrations with * are significantly different at p <0.05 level).

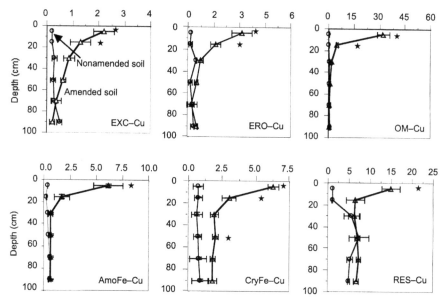

FIGURE 8.4   Distribution of Cu in solid-phase components in profiles of nonamended and poultry waste-amended soils (in mg kg^{-1}, averages and standard errors and n = 2 and 5 for nonamended and amended soils, respectively) (metal concentrations with * are significantly different at p <0.05 level).

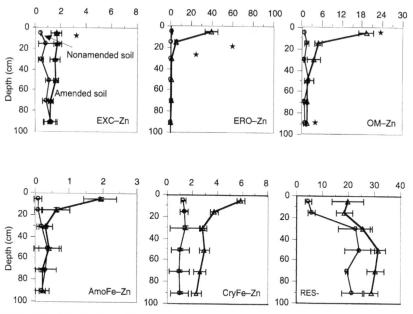

FIGURE 8.5   Distribution of Zn in solid-phase components in profiles of nonamended and poultry waste-amended soils (in mg kg^{-1}, averages and standard errors and n = 2 and 5 for nonamended and amended soils, respectively) (metal concentrations with * are significantly different at p <0.05 level).

bound fraction in soil profiles. Both waste-amended and nonamended surface soils were dominated by sand- and silt-sized fractions, but the clay-sized fraction increased with depth (data not shown).[50]

In contrast to Cu, Zn exhibited higher mobility in the poultry waste-amended soil. As shown in Figure 8.2, the amended soil had a higher portion of Mn (possibly Mn oxides) in the easily reducible oxide fraction. Heavy metals bound in the easily reducible oxide fraction may be of higher mobility than those in the organically-bound fraction.[9] It is conceivable that during periods of soil-water saturation and/or water ponding on the soil surface, the soil was periodically in a reduced state. When the soil was saturated, easily reducible oxides (e.g., Mn oxides) were reduced and possibly released metals into the exchangeable fraction.[14,15] Following exchange by other cations, these forms were susceptible to transport in the soil profile. Similarly, Zn in the organically-bound fraction was also subject, although to a lesser extent, to transport in the soil profile. In contrast, Cu in the organically-bound as well as in the easily reducible oxide fractions appeared to be strongly stable and exhibited limited mobility in the soil profile below 20- to 30-cm depths (Figures 8.4 and 8.5).

Zinc concentrations in the residual and crystalline Fe oxide fractions in the poultry waste-amended soil profile were higher than in the nonamended soil (Figure 8.5). It is recognized that the mobility of heavy metals may be enhanced by the transport of colloids such as oxides and fine clay minerals.[51] Our results may support the possibility of colloidal-enhanced transport processes of Zn in the amended soil. However, it is also possible that metals were transformed into the more labile form and transported to the lower parts of profiles, as discussed above; they were then redistributed into the present solid-phase components. Further studies are needed to quantify possible mechanisms that govern the mobility of heavy metals, including colloidal-enhanced transport in long-term waste-amended soil.

BIOAVAILABILITY OF METALS IN WASTE-AMENDED SOILS

The bioavailability of heavy metals in poultry waste-amended soils is strongly affected by soil pH. Van der Watt et al.[37] found that plant uptake was positively correlated with the DTPA- and 0.1-M $NaNO_3$-extractable metal contents, and negatively correlated with soil pH. They also found that 3 to 5% of the Zn and 0.5% of the Cu in the added poultry waste were taken up by sudax [*Sorghum bicolor (L.) Moench*] in greenhouse experiments. Additions of poultry litter significantly decreased both plant tissue Zn concentration and KNO_3-extractable Zn levels in the amended soil,[52] and decreased Cd extracted in the exchangeable, complexed, and HCl-extractable fractions.[35] However, tissue Cu levels from poultry waste-amended fescue showed a slight trend toward higher levels than those of nonamended fescue, and Zn levels from both treatments were similar.[53]

SEWAGE SLUDGE-AMENDED SOIL

Heavy metals may enter soils through irrigation with reclaimed sewage water. This is especially important in arid and semi-arid areas where water supplies are limited and fresh water is very valuable. Irrigation with reclaimed sewage water is the most read-ily available and economically feasible way to supplement fresh water in these areas. In Israel, sewage irrigation on a large scale was initiated in 1972. About 66% of efflu-ents are reused for irrigation or recharging the aquifer, as compared with about 2.4% of the reclaimed municipal wastewater in the U. S., and about 11% of the total annual flow in Australia.[54,55] In fact, Israel has gained experience from irrigation with recy-cled sewage water for the past 30–40 years.[54,56] Banin et al.[56] found that prolonged irrigation with treated sewage effluents from rural sources led to the accumulation of heavy metals (Cd, Cu, Ni, and Pb) in the top 10–15-cm layer of the coastal plain soils of Israel after 28 years of irrigation. This led to increased uptake of certain metals to crops and their possible introduction to animals and humans through the food chain.[56] However, at the present time in the U.S., most reclaimed municipal wastewater meets current irrigation water-quality criteria.

Sewage sludge is another potential source of heavy metals to soils. Sludge pro-duction in the U.S. reached 4.5×10^6 tons in the early 1980s, and the European Economic Community exceeded 6×10^6 tons of dry solids annually during the same period.[57] An estimated 5.3×10^6 tons per year of dry weight of sludge are currently produced in the U.S. from publicly owned treatment plants,[58] which has no doubt increased substantially with population growth. Sewage sludge may be a valuable source of essential nutritional elements (N, P, K, and some micronutrients) for agri-cultural crops, especially in soils deficient in major, minor, and micronutrients. In addition, it may improve the physical properties of soils by the addition of organic matter. The application of sewage sludge to agricultural land is considered a suitable method for its disposal. Agricultural use of sewage sludge consumed approximately 44%, 20–25%, and more than 31% of sludge production in the U. K., Germany, and the U. S., respectively.[54,57,59] However, sewage sludges have elevated levels of poten-tially toxic trace elements (Zn, Cd, Cu, Pb, Cr, and Ni)[2,57,60] that may exceed natural soil concentrations by two orders of magnitude or more.[57] The application of sewage sludge and irrigation with recycled sewage water over long periods of time have been shown to result in a substantial accumulation of heavy metals in soils and the crops grown on them.[2,9,11,56,57,60–63] Groundwater could also be in danger of potential conta-mination in the long run.

The distribution of heavy metals among the solid-phase fractions of soils is affected by climate, soil formation processes, elemental make-up of the parent mate-rial, soil properties such as pH and Eh, texture, mineralogical composition, organic matter, and carbonate contents, as well as the chemical properties of the elements.

DISTRIBUTION OF HEAVY METALS IN SEWAGE SLUDGE

In sludge and manure, metals are predominantly in the carbonate-bound, residual, and organically-bound fractions, except Cu, which is mainly in the organically-bound frac-tion, and Cd, which is in the exchangeable form. However, the major forms of metals in sludge are dependent upon the source of the wastewater and the wastewater treat-ment process. In addition, sewage sludge may contain a substantial amount of soil, giving rise to aluminosilicates, hydrous oxides of iron, manganese, and aluminum.

Exchangeable, carbonate, and easily reducible oxide-bound fractions are the main forms for Cd in sludge.[6] Sposito et al.[65] found that most Cd in sludge was in the carbonate fraction, followed by the organically-bound and residual fractions. Emmerich et al.[5] found that Cd was present mainly in the carbonate-bound fraction (56%), and to a lesser extent in the residual (36%) and organically-bound fractions (22%). McGrath and Cegarra[11] reported that Cd in sludge is mostly present in the carbonate, residual, and exchangeable fractions.

In one study, Cr in sludge occurred mostly in the residual and carbonate fraction of the sludge.[11]

Copper in sludge prevailed in the organically bound fraction.[6] Sposito et al.[65] and Emmerich et al.[5] also reported that Cu mainly occurred in the organic fraction, followed by the carbonate and residue fractions. However, McGrath and Cegarra[11] reported that Cu occurred mostly in the residual, organically-bound, and carbonate fractions.

Iron in sludge was mainly (90%) present in the residual fraction, according to the findings of Knudtsen and O'Connor.[64]

Nickel has been found to be roughly equally present in the carbonate (32%), residual (26%), and organically-bound fractions (24%) in sludge.[5] However, McGrath and Cegarra[11] reported that Ni in the sludge was mostly in the residual, organically-bound, and carbonate fractions.

Banin et al.[6] reported that most of the Pb was bound in the reducible oxide fraction, followed by the readily reducible oxides and carbonate fractions. McGrath and Cegarra[11] found that Pb mainly existed in the residual and carbonate fractions.

Zinc in sludge has been reported to occur in the carbonate, easily reducible oxide, and organically-bound fractions.[6] Emmerich et al.[5] reported that Zn in sludge mainly existed in carbonate-bound (57%) and organically-bound fractions (28%). Knudtsen and O'Connor[64] found that 53% of the Zn in the sludge occurred in the carbonate fraction. McGrath and Cegarra[11] reported that Zn in sewage sludge mainly resided in the carbonate, organically-bound, and residual fractions.

DISTRIBUTION AND REDISTRIBUTION OF HEAVY METALS IN AMENDED SOILS

The addition of heavy metals to soils, in any form of addition, is known to change the original distribution patterns of heavy metals in soils. There is a large amount of data on the distribution of Zn and Cu in native soils and soils amended with sewage sludge. Much less information has been obtained for Cd, Cr, Ni, and Pb. For example, one study revealed that the application of sludge tended to reduce the relative amounts of Cu, Cd, and Pb in the residual fraction due to dilution from added material, and that it increased organically-bound Cu and carbonate-bound Cd, Pb, and, to some extent, Zn and Ni after four years of sludge application.[9] After being treated with sludge for seven years, the carbonate- and organically-bound fractions became the most prevalent solid phase for Cu, Ni, and Zn, but the distribution patterns of Cd, Cr, and Pb were not significantly affected by the amounts of sludge added.[24] However, Banin et al.[56] found that the distribution of some heavy metals in calcareous soils in Israel after prolonged (up to 28 years) irrigation with treated sewage effluents was characterized by less than 0.2% in the water-soluble fraction, and 9–20%, 5–11%, 16–23%, 10–14%, 1–3.9% in the fraction extracted by hydroxylamine hydrochloride in 25% acetic acid for Cd, Cr, Cu, Ni, and Pb, respectively.

Cadmium

In waste-amended soils, most of the Cd was in the EXC and CARB fractions. In a Typic Udipsamment from England amended with sludge, the CARB fraction is a predominant fraction for Cd, followed by the residual and exchangeable fractions.[11] In California soils, Emmerich et al.[5] found that most of the Cd was in the carbonate-bound (46%) and residual fractions (42%). In acidic soils amended with composted sewage sludge from Delaware, Cd was mainly in the RES (40–56%), the EDTA-extractable (23–28%), and EXC (20–27 %) fractions.[66] Cadmium in the carbonate and exchangeable fractions increased, and the percentage of residual fraction decreased with time during the first ten years of annual application of sludge.[11] Sposito et al.[65] and Chang et al.[24] also found that the carbonate Cd increased with time in arid-zone soils amended with sludge. After the sludge was no longer applied, Cd still remained highly extractable by DTPA and $Ca(NO_3)_2$ (EXC) (50% of total Cd), and a larger percentage of the total soil Cd was extracted with the paracrystalline iron oxide fraction of the sludge-amended soil (35%) than in the salt-amended soil (23%).[67] Under the saturated soil–sludge–water system, during 28 days of incubation, DTPA-Cu increased and HNO_3-Cd (RES) decreased with time.[68] Cd uptake by corn decreased over time at all sludge rates and was inversely related (p > 0.01) to time in years after the last application.[69] Khalid and his colleagues[70] studied the transformation of Cd among solid-phase forms in sediments under different pH and Eh conditions. They found that increasing oxidation intensity increased the DTPA-extractable fraction and the exchangeable fraction, and decreased the insoluble organic-bound fraction at pH 8 and 6.5. Furthermore, the larger increase of the exchangeable and water-soluble forms was found at pH 5. This implied that oxidation conditions could cause the transformation of Cd from insoluble organic-bound into more available fractions. Khalid et al.[70] indicated that Cd–organic complexes are stable under strongly reducing conditions and may be responsible for the low solubility of Cd, particularly in the absence of sulfides; however, in acid-oxidizing conditions, Cd chelation was not an important process.

Chromium

McGrath and Cegarra[11] reported that more than 80% of Cr was present in the residual fraction (included iron oxides bound) in a sludge-treated Typic Udipsamment from England (because the procedure did not separate iron oxides bound from the residual fraction). The percentages of carbonate fraction increased and the residual fraction decreased with time during long-term application of sludge. In the following years, Cr distribution seemed constant.[11] In the acidic soils amended with composted sewage sludge from Delaware, Cr was mainly in the RES fraction, but application of sludge decreased the percentage of the RES fraction and increased the percentages of the EDTA-extractable and organic-bound fractions.[66]

Copper

Copper in a sludge-treated Typic Udipsamment from England and in Israeli arid-zone soils predominated in the organically-bound fraction, followed by the carbonate fraction.[6,11] Over time, Cu was redistributed into the RO fractions. The addition of sludge increased the organically-bound and, to some extent, the carbonate-bound fractions. Emmerich et al.[5] pointed out that in sludge-amended California soils, organically-bound Cu accounted for 54%, and the carbonate and residual fractions for 25% and 22%, respectively. In acidic soils amended with composted sewage sludge from Delaware, Cu was mainly in the OM fraction.[66] Copper in soils shifts from the residual fraction in nonamended soil to the organically-bound fraction after the application of sludge.[64] However, McGrath and Cegarra[11] found that Cu in the carbonate fraction increased and the percentage of the residual fraction decreased with time during the first 10 years of sludge amending, and all fractions remained quite stable during the following 20 years after cessation of sludge application. In a water-saturated soil–sludge system, DTPA-Cu increased and HNO_3-Cu (RES) decreased with time during 28 days of incubation.[68]

Nickel

Nickel in a sludge-treated Typic Udipsamment from England has been shown to occur mainly in the residual fraction, but the percentage of the residual fraction decreased and the carbonate and organically-bound fractions increased with the annual application of sludge.[11] In the acidic soils amended with composted sewage sludge from Delaware, Ni was mainly in the RES fraction, but application of sludge decreased the percentage of the RES fraction and increased percentages of the organically-bound fractions.[66] Another study also showed that Ni in California soils amended with sludge for seven years was mostly present in the residual (64%) and less in the organically-bound (12%) and carbonate fractions (18%).[24] Nickel in the carbonate fraction was found to increase with time in arid-zone soils amended with sludge.[64] During the first 10 years of annual application of sludge, Ni in the carbonate, organically-bound, and exchangeable fractions increased, and the percentage of the residual fraction decreased with time. During the following 20 years after cessation of application of sludge, however, the exchangeable, organically-bound, and carbonate fractions declined slightly, and the residual fraction increased with time.[11]

Lead

Most of the Pb was found to be in the carbonate fraction in a sludge-treated Typic Udipsamment from England during 20 years and after 25 years of applications of sewage sludge, according to the findings of McGrath and Cegarra.[11] In addition, Sposito et al.[9] reported that Pb in California soils was mostly present in the carbonate fraction (80–90%) and less so in the residual fraction (18%). Lead in the carbonate fraction in sludge-amended soils increased and the percentage of residual fraction decreased with time.[11,64] After cessation of application of sludge, Pb in all fractions seems to remain unchanged.[11] In the acidic soils from Delaware amended with composted sewage sludge, Pb was mainly in the EDTA-extractable fraction.[66] Silviera and Sommers[68] also found that DTPA-extractable Pb remained constant with time in a water-saturated soil–sludge system during 28 days of incubation. Similarly, Misra et al.[71] reported that the transformation of Pb into DTPA-nonextractable forms was relatively slow during 30–60 days of flooding-incubation.

Zinc

Zinc in a Typic Udipsamment amended with sewage sludge has been shown to occur mainly in the carbonate-bound fraction, followed by the residual fraction.[11] In acidic soils from Delaware amended with composted sewage sludge, Zn was mainly in the RES and EDTA-extractable fractions.[66] Application of sludge increased the percentage of Zn in the carbonate fraction. Emmerich et al.[5] indicated that Zn in California soils existed mainly in the carbonate-bound (46%) and residual fractions (36%), and less in the organically-bound fraction (19%). After eight to nine years of cessation of application of sewage sludge, Zn in a Maryland fine sandy loam soil was dominantly in the RES fraction, followed by the Fe oxide-bound fraction.[67] In Israeli soils, Zn was mostly in the OM, CARB, and ERO fractions. With time, more Zn was transferred into the CARB and ERO fractions.[6]

The prolonged application of sludge in soils increased Zn in the carbonate,[11,64] organically-bound, and exchangeable fractions, and decreased the percentage in the residual fraction.[11] After cessation of the application of sludge, the organically-bound and exchangeable Zn fractions seemed to decrease with time, and Zn was transformed to the residual fraction. Seventeen years of application of sludge increased mostly oxide-bound and HOAc-extractable Zn fractions.[43] Silviera and Sommers[68] reported that DTPA-extractable Zn in a water-saturated-sludge–soil system increased with time during 28 days of incubation, whereas HNO_3-extractable Zn (RES) decreased with time in this system. This indicated that Zn was transformed from less labile into more labile forms under a saturated water regime.

The distribution of heavy metals in amended soils in various forms of addition, mainly sludge, was also influenced by soil properties (organic matter, oxide, carbonate contents, soil texture, etc.), soil condition (soil pH and Eh), and the loading levels of sewage sludge.[6] Banin et al.[6] reported that a sandy soil amended with sludge contained more Cu in the carbonate and easily reducible oxide fractions than a clay soil did, but the clay soil had higher Zn in the reducible oxide and less in the carbonate fraction than the sandy soil. They also found that Cd in soils at low sludge rate (1%) was mainly in the reducible oxide (35–45%) and easily reducible oxide (25–30%) fractions, whereas at a higher sludge rate (10%), most of the Cd was in the carbonate (20–30%) and easily reducible oxide (20–30%) fractions. The effects of levels of sewage sludge on metal redistribution in soil may be, in part, related to dilution. Sposito et al.[9] found that in sludge-treated California soils at a high rate of sludge application (90 tons ha^{-1} yr^{-1}), the predominant forms of the metals were the organic fraction for Cu, the residual fraction for Ni, and the carbonate fraction for Zn, Cd, and Pb.

TRANSPORT OF HEAVY METAL IN SEWAGE SLUDGE-AMENDED SOIL PROFILES

Limited transport of heavy metals in the profile of sludge-amended soil has been observed by several researchers.[72,73] Sludge-borne Cd, Cr, Cu, and Zn in a silt loam loessial soil were reported to move into the 16- to 30-cm soil layer, directly below the zone of incorporation of sludge, after 10 years of application.[72] After 17 years of sludge use, some further movement of Cr and Cu into the 45- to 60-cm layer was identified.[72] Barbarick et al.[73] also observed that Zn in sewage sludge-amended loam soils significantly and consistently increased in extractable levels (DTPA-Zn) below the plow layer.

BIOAVAILABILITY OF HEAVY METALS IN AMENDED SOILS

LeClaire et al.[74] reported that soluble and exchangeable Zn, as extracted by KNO_3 and H_2O, were highly labile, the organic matter fraction extracted by NaOH was labile, and the carbonate fraction extracted by EDTA represented a reservoir of potentially bioavailable Zn to plants. The soluble and exchangeable Zn as extracted by KNO_3, H_2O, and $CaCl_2$ was significantly correlated with plant Zn in wheat, soybean, and maize tissues,[66,75] and the organically-bound Zn (H_2O_2) and complexed Zn (EDTA) were also closely related to the bioavailability of Zn to maize.[75] Murthy[76] found that the exchangeable and soluble organically complexed Zn extracted by $Cu(OAc)_2$ were more available to rice in wetland soils, and amorphous iron and aluminum oxides by oxalate reagents also contributed to availability. Organically complexed Zn played the most important role in Zn nutrition of rice in paddy soils.[19] The exchangeable and chelated or soluble organically-bound fractions were more available sources to plants.[18,21,23,77] In calcareous soils, concentrations of Zn and Cd in the plants also had significant regression relationships, with Zn concentration in the CARB fraction and Cu concentration in the OM fraction.[6,18,21,23] Metal concentration as measured by various extractants is an index of their availability. Actually, the bioavailability of metals in soils is largely dependent upon soil pH.

Cadmium uptake by plants grown on sludge-amended soils was suppressed by the application of phosphorus. Gonzalez et al.[78] reported that application of waste phosphoric clay decreased Cd uptake by alfalfa grown on sludge-amended soils, and, after application of phosphoric clay, the extractable level of Cd from soil decreased with time (within 155 days).

This brief review shows that the metal distribution in sludge-amended soils depends on the element, the soil properties, and soil conditions such as pH, Eh, etc. Furthermore, it has been observed that the distribution changes with time. Metals are transferred from labile forms to more stable forms in the soils.[6]

SALT-AMENDED SOILS AND MINE-TAILING SOILS

Metals in salt-amended soils are characterized by two processes of redistribution and transformation: the initial fast retention process, followed by the slow, long-term process.[15,16,79,80] Both processes are the transfer of heavy metals from the more labile fractions into the more stable fractions with time. Within the first hour and/or the first step of sequential selective dissolution, Cd, Pb, Cu, Ni, and Zn added to arid-zone soils were transferred mainly from the soluble and exchangeable fractions into the carbonate fraction; added Cr transferred mainly into the organic matter fraction.[15,16,79,80] However, during the slow, long-term process, incubation under the saturated and wetting–drying cycle conditions caused more transfer of heavy metals into the more stable fractions, such as the reducible oxides, e.g., Fe oxides and the organically-bound fraction.[15,16,80] In other work, incubation at the field capacity regime resulted in more transfer of metal into the easily reducible oxides, e.g., the Mn oxides fraction and, to some extent, the organically-bound fraction.[16,80]

CADMIUM

Cadmium distribution in salt-amended soils apparently was affected by soil pH and not affected by soil moisture. Cadmium in salt-applied arid-zone soils was mostly redistributed in the carbonate fraction during one year of incubation at the saturated-paste, field-capacity, and wetting–drying cycle regimes.[15,16,79,80] Moisture regime did not have much effect on Cd redistribution pathways. In acid and neutral soils of China, added soluble Cd was evenly distributed in the exchangeable, complexed organic matter; insoluble organic matter; and residual fractions. In calcareous soil, Cd in the exchangeable fraction was still an important fraction.[3] In German soils receiving Cd salt, Cd was mostly (90%) in the exchangeable fraction in very acid soil (pH 3.3–4.5), and Cd in the exchangeable fraction deceased with increasing soil pH.[1] At acid soils (pH 5–5.9), Cd in the exchangeable fraction decreased to 50% and Cd in the sodium acetate (at pH 6) extractable, the Mn oxide-bound, and organically-bound fractions increased up to 30–40% in soils with pH 6.0–7.5.[1] In western Australian soils incubated at field capacity regime, added Cd was redistributed from soluble to less soluble fractions with time, depending upon the type of soil, soil pH, and rate of application: in sandy soil, Cd from the soluble to exchangeable fraction; in lateritic podzolic soils from the exchangeable to organic matter fraction, especially at higher pH; and in soils dominated by goethite from exchangeable to oxide and residual fractions with time.[81] Eight to nine years after cessation of the application of salts, Cd in a Maryland sandy loam soil remained highly extractable by DTPA and $Ca(NO_3)_2$ (EXC) (65% of total Cd), and both reagents extracted much higher Cd from the salt-amended soil (65%) than from the sludge-amended soil (50%).[67] However, Cd in soils contaminated with $PbHAsO_4$ and in battery-breaking sites was mostly in the residual fraction (>90%), but at a smelter site Cd was mostly in the exchangeable, carbonate, and residual fractions.[82]

Chromium

In arid-zone soil, added Cr was initially in the organically-bound fraction, but with time it was, to some extent, continually transferred from the carbonate fraction to the organically-bound fraction (Figure 8.6). Statistically, Cr concentrations in the

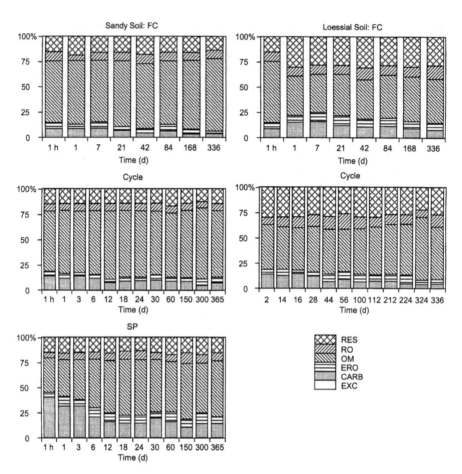

FIGURE 8.6 Redistribution of Cr in arid-zone soils receiving nitrate salt (3T, T = Cr concentration in native soil) during incubation at saturated-paste (SP), field-capacity (FC), and wetting–drying cycle (Cycle) regimes. EXC, CARB, ERO, OM, RO, and RES represent exchangeable, carbonate-bound, easily reducible oxide-bound, organic matter-bound, reducible oxide-bound, and residual fractions.

carbonate and organic-matter-bound fractions in soils at three moisture regimes, and Cr in the easily reducible oxide bound fraction in soils under field-capacity regime, were significantly different at $p = 0.05$ level between an initial and a year of incubation (Table 8.1). Moisture does not considerably affect Cr redistribution in soil.[15,16,80] However, under saturated conditions, Cr in the sandy soil had a higher proportion in the carbonate fraction than field capacity and wetting–drying cycle regimes, partly due to increased pCO_2 under saturated condition (Figure 8.6). In German soils, Cr was primarily in the poorly crystalline Fe oxide and well crystalline Fe oxide and residual fractions after one year of incubation.[1] Soil pH does not considerably affect its redistribution.[1]

TABLE 8.1

Statistical analysis (t-test)[a] of concentrations (in mg kg^{-1}) of metals in individual fractions in two arid-zone soils receiving soluble salts at 3T level of addition after an initial period[b] and a year of incubation under various moisture regimes

Moisture	Soil	Metal	EXC[c]	CARB	ERO	OM	RO	RES
Saturated paste	Sandy	Cr	*[d]	*	—	*	*	—
		Cu	ND[e]	ND	ND	ND	ND	ND
		Ni	*	*	—	*	—	—
		Zn	*	*	—	*	*	*
	Loessial	Cr	ND	ND	ND	ND	ND	ND
		Cu	*	*	*	—	*	*
		Ni	*	*	—	—	*	*
		Zn	*	*	*	*	*	*
Field capacity	Sandy	Cr	BD[f]	*	*	*	—	*
		Cu	*	*	*	*	—	*
		Ni	*	*	*	*	—	*
		Zn	*	*	*	*	*	—
	Loessial	Cr	BD	*	*	*	*	—
		Cu	*	—	*	*	—	*
		Ni	*	*	*	*	*	*
		Zn	*	*	*	*	*	*
Wetting-drying cycle	Sandy	Cr	BD	*	*	*	*	—
		Cu	—	*	*	*	—	—
		Ni	*	*	*	*	*	*
		Zn	*	*	*	*	*	*
	Loessial	Cr	BD	*	—	*	*	—
		Cu	*	*	—	*	*	*
		Ni	*	*	*	*	*	*
		Zn	*	*	*	*	*	*

[a]Statistical analysis for each metal was done in soils at 3T level of addition.

[b]Initial period—1 hr incubation for saturated and field capacity regimes and 2 d incubation for wetting–drying cycle regime.

[c]EXC, CARB, ERO, OM, RO, and RES represent exchangeable, carbonate-bound, easily reducible oxide-bound, organic matter-bound, reducible oxide-bound, and residual fractions, respectively.

[d]* Indicates metal concentrations in individual fractions after initial period and year of incubation were significantly different (p $<$ 0.05), and—indicates not significant.

[e]ND—No addition of metal at this level.

[f]BD—Metal concentration below detection limits of ICP-AES.

Copper

Copper distribution in salt-amended soils was affected by moisture and soil properties (Figure 8.7). In arid-zone soils, Cu was initially in the carbonate, organically-bound, and exchangeable fractions (Figure 8.7). After one year of incubation, Cu was

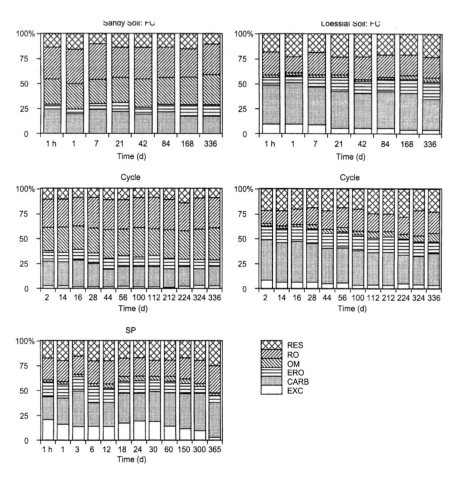

FIGURE 8.7 Redistribution of Cu in arid-zone soils receiving nitrate salt (3T, T = Cu concentration in native soil) during incubation at saturated-paste (SP), field-capacity (FC), and wetting–drying cycle (Cycle) regimes. EXC, CARB, ERO, OM, RO, and RES represent exchangeable, carbonate-bound, easily reducible oxide-bound, organic matter-bound, reducible oxide-bound, and residual fractions.

mainly in the carbonate, reducible oxide, easily reducible oxide, and residual fractions in the loessial soil at various moistures, and in the organically-bound, reducible oxide, and carbonate fractions in the sandy soils at a field capacity and wetting-drying cycle regime (Figure 8.7). Copper in the loessial soil under the saturated regime was mainly in the carbonate, reducible oxide, and residual fractions.[15] The slow redistribution processes were considerably affected by moisture regimes.[15,16,80] In the field-capacity regime, added Cu was transferred from the exchangeable and carbonate fractions into the easily reducible oxide fraction and, to some extent, organically-bound fraction (Figure 8.7, Table 8.1). In the saturated paste regime, Cu was mainly changed from the exchangeable and easily reducible oxide bound fractions into the carbonate and reducible oxide fractions, but Cu concentration in the exchangeable fraction was initially much higher compared with other moisture regimes (Figure 8.7, Table 8.1). In the wetting–drying cycle regime, Cu was mainly transferred into the reducible oxide, residual, and organically-bound fractions with time, depending upon soil properties (Figure 8.7). Payne et al.[43] also found that, after eight annual applications, Cu applied to soils as sulfate salt was mostly distributed in the organically-bound and oxide-occluded fractions. McLaren and Ritchie[12] reported that a high proportion of the applied Cu, as copper sulfate in a lateritic sandy soil in Western Australia, was initially associated with the soil organic matter, and during the course of the trial (20 years), a substantial proportion of Cu was redistributed to the residual fraction. In smelter sites contaminated with Cu, Cu was mainly in the organically-bound fraction, followed by the residual, Fe–Mn oxide, and carbonate fractions.[82] Copper in Colorado soils contaminated with mine tailings was predominantly associated with the organic fraction.[83] In very acidic salt-amended German soils (pH 3.3–4.5), Cu was mainly in the exchangeable fraction (70%), which decreased with time, but in the acid soils (pH 5.0–5.9) Cu was mostly in the sodium acetate (at pH 6) extractable fraction, the Mn oxide-bound, and organically-bound fractions; and in soils with high pH (>6.7), Cu was transferred from the sodium acetate (at pH 6) extractable, the Mn oxide-bound, and organically-bound fractions into the well-crystalline Fe oxide and residual fractions.[1] Han et al.[84] reported that in catfish pond soils of Mississippi that received a long term of application of $CuSO_4$, total Cu concentration in the amended soils was 4 to 8 times higher than in nonamended sediments, and Cu was mainly in the organic matter (30.7%), carbonate (31.8%), and amorphous iron oxide (22.1%)-bound fractions, with a considerably higher amount of Cu in the soluble and exchangeable (3–8 mg kg^{-1}). This suggests that Cu had accumulated considerably in the catfish pond soils and posed a high lability.[84]

Nickel

Nickel distribution was affected by moisture regime.[15,16,79,80] Added soluble Ni in arid-zone soils was bound during the first stage of rapid retention to the carbonate, exchangeable, and easily reducible oxide fractions (Figure 8.8). After one year of incubation, the carbonate fraction was still an abundant fraction (Figure 8.8). However, during prolonged incubation at field capacity moisture regime, Ni was slowly transferred from the exchangeable and carbonate fractions into the easily reducible oxide and organically-bound fractions in soils (Figure 8.8, Table 8.1). Under saturated conditions, added Ni was transferred from the exchangeable and carbonate fractions mainly into the reducible oxide fraction in the loessial soil, and into the organically-bound and easily reducible oxide fractions in the sandy soils (Figure 8.8, Table 8.1). In the wetting–drying cycle regime, added Ni in soil was transferred mainly into the organically bound fraction, and Ni in all stable fractions significantly increased after a year of incubation (Figure 8.8, Table 8.1). Incubation under the wetting–drying cycle regime resulted in more transfer of added Ni into the stable fractions than at the field-capacity and saturated regimes,[15,16,80] as indicated by the

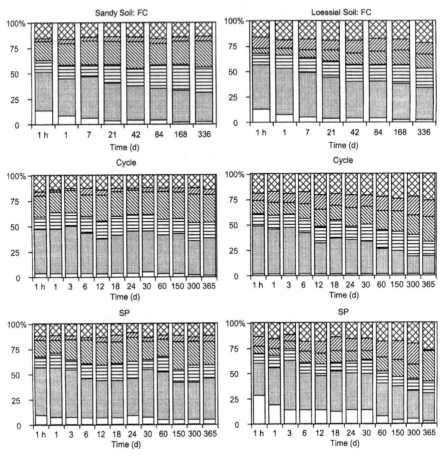

FIGURE 8.8   Redistribution of Ni in arid-zone soils receiving nitrate salt (3T, T = Ni concentration in native soil) during incubation at saturated-paste (SP), field-capacity (FC), and
wetting–drying cycle (Cycle) regimes. EXC, CARB, ERO, OM, RO, and RES represent
exchangeable, carbonate-bound, easily reducible oxide-bound, organic matter-bound,
reducible oxide-bound, and residual fractions.

lowest concentration of Ni in the exchangeable fraction both initially and after one
year of incubation in the wetting–drying cycle regime (Figure 8.8, Table 8.1).

Lead

Most of the added Pb in arid-zone soils remained bound to the carbonate fraction.[15,79]
During incubation at field-capacity moisture, small amounts of Pb were transferred
from the carbonate fraction into the easily reducible oxide fraction, but generally the
distribution pattern did not change.[16,79] In salt-amended German very acid soil (pH
3.7), Pb was mainly in the exchangeable fraction, but in the neutral soil Pb was

mostly in the sodium acetate-extractable, Mn oxide-bound, and organically-bound fractions. Redistribution of Pb took place very slowly compared with other metals.[1] Levy et al.[83] found that Pb in mine-contaminated soils was present mostly in the Fe and Mn oxide fractions.

Zinc

Zinc distribution was affected by soil pH, soil properties, time, and soil mois-ture.[15,16,18,79,80] Han et al.[18,20,21] reported that after seven months of application of Zn as sulfate in acid, neutral, and calcareous soils in China, Zn was redistributed in the residual (20–30% of the added soluble Zn), soluble organically-bound (30%), iron oxide-bound (13–20%), and carbonate-bound (only 5–13%) fractions. They pointed out that metal in the carbonate fraction seems higher in the amended soils than in the native soils. Added Zn in arid-zone soils in the field capacity and wetting–drying cycle regimes was initially attached mostly to the carbonate and exchangeable frac-tions, and after one year the carbonate fraction was still the most abundant fraction. In loessial soil, the easily reducible oxide and residual fractions were also important fractions (Figure 8.9). During incubation, redistribution took place from the exchangeable and carbonate fractions mainly into the easily reducible oxide fraction, and a small part of it into the organically-bound and reducible oxide fractions in the sandy soil.[15,16,79,82] In comparison, when soils were incubated in the saturated paste regime, added Zn was initially in the carbonate fraction, and after one year Zn was mainly in the carbonate fraction in the sandy soil and in the reducible oxide, residual, and carbonate fractions in the loessial soil (Figure 8.9). With time, however, Zn in soils in the saturated regime was transferred from the exchangeable and carbonate fractions and the easily reducible oxide fraction, mainly into the reducible oxide fraction in the loessial soil, and from the exchangeable and carbonate fractions mainly into the organically-bound and reducible oxide fractions in the sandy soil (Figure 8.9). In arid-zone soils, redistribution of Zn took place in almost all fractions (Table 8.1). Zinc added to the calcareous paddy soils after three rice crops was transported into the amorphous-iron oxide > crystalline iron oxide > complexed Zn > residual > exchangeable fractions.[77] The major part of total Zn occurred in the residual fraction in acid paddy soils, and waterlogging increased the organic com-plexed[85] and amorphous sesquioxide Zn fractions, and thus decreased the crystalline sesquioxide-bound fraction.[86] Preflooding for 15 days before application of Zn in a lateritic paddy soil increased the transfer of Zn into exchangeable, organically-bound, and amorphous iron oxide-bound fractions, as compared with the transformation when Zn was applied immediately after flooding.[87] Han et al.[18] found, however, that Zn decreased in the amorphous iron oxide fraction and Zn in the crystalline iron oxide fraction increased with time in upland soils, especially in calcareous soils, and the soluble organically-bound fraction was transformed into the insoluble organic matter fraction with time in upland soil. In salt-amended German acid and neutral soils, Zn was mostly in the exchangeable fraction (70–80% and 40–60%, respec-tively). The lower the soil pH, the higher the Zn in the exchangeable fraction.

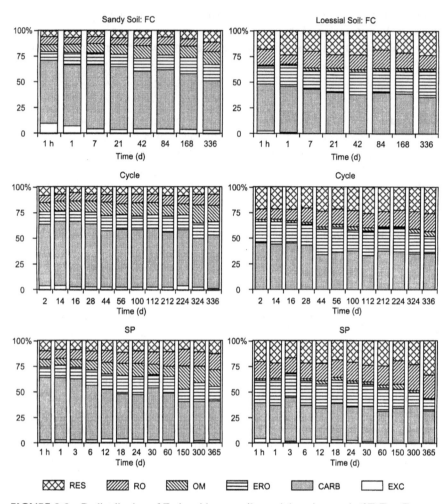

FIGURE 8.9 Redistribution of Zn in arid-zone soils receiving nitrate salt (3T, T = Zn concentration in native soil) during incubation at saturated-paste (SP), field-capacity (FC), and wetting–drying cycle (Cycle) regimes. EXC, CARB, ERO, OM, RO, and RES represent exchangeable, carbonate-bound, easily reducible oxide-bound, organic matter-bound, reducible oxide-bound, and residual fractions.

Furthermore, at low pH, Zn was very seldom redistributed with time; at high pH, Zn was redistributed with time into the more stable fractions. In the soil with pH > 6.7–7, Zn was initially in the sodium acetate (at pH 6) extractable, the Mn oxide-bound, and organically-bound fractions. With time Zn was transferred into the well-crystalline Fe oxide and residual fractions.[1] In soils contaminated with mine tailings, Zn has been shown to occur mainly in the Fe and Mn oxide fractions.[83] At smelter sites and battery-breaking sites, Zn was mainly in the residual fraction, followed by the Fe–Mn oxide and carbonate fractions.[82]

In addition to the heavy metals discussed above, distribution of other trace elements such as Co and V were studied.[88,89] The bulk of Co mainly resides in the primary and secondary mineral-bound (residual) fraction, probably as isomorphous substitution in the secondary and primary minerals[88,90] and the easily reducible oxide (Mn oxide) fraction.[88] Vanadium in arid-zone soils was mainly in the residual fraction and organic matter fraction.[89] A significant linear correlation was found between the total concentrations of Co and Mn and their fractionation patterns in the arid-zone soils.[88,91,92] During saturated incubation, Co in two arid-zone soils was redistributed mainly from the easily reducible oxide (Mn oxide-bound) fraction and, to some extent, the reducible oxide (Fe oxide-bound) and organic matter-bound fractions into the carbonate fraction.[88,89] V was redistributed from the reducible oxide and organic matter fractions into the carbonate fraction and, to some extent, the exchangeable and easily reducible oxide fractions.[88,89] Changes in Co concentrations in the easily reducible oxide, reducible oxide, and carbonate fractions were highly correlated with changes in Mn content.[88]

ACKNOWLEDGMENTS

The authors are grateful to Dr. A. Banin for his revision of most parts of this paper when the senior author was a research associate at the Hebrew University of Jerusalem, Israel. We also thank anonymous reviewers for their helpful comments and suggestions.

REFERENCES

1. Banin, A., Nir, S., Brummer, G.W., Han, F.X., Serban, C., and Krumnohler, J., Cd pollution in soils: Long-term processes in the solid phase, their characterization and models for their prediction, Report ISC-8911-ISR (ENV), Joint Israel–Commission of the European Communities research projects, 1995.
2. Jung, J. and Logan, T.J., Effects of sewage sludge cadmium concentration on chemical extractability and plant uptake, *J. Environ. Qual.*, 21, 73, 1992.
3. Han, F.X., Hu, A.T., and Qin, H.Y., Fractionation and availability of added cadmium in soil environment, *Environ. Chem.*, 9, 49, 1990 (in Chinese).
4. Tessier, A., Campell, P.G.C., and Bisson, M., Sequential extraction procedure for the speciation of particulate trace metals, *Anal. Chem.*, 51, 844, 1979.
5. Emmerich, W.E., Lund, L.J., Page, A.L., and Chang, A.C., Solid phase forms of heavy metals in sewage sludge-treated soils, *J. Environ. Qual.*, 11, 178, 1982.
6. Banin, A., Gerstl, Z., Fine, P., Metzger, Z., and Newrzella, D., Minimizing soil contamination through control of sludge transformations in soil, Report Wt 8678/458, Joint German–Israel research projects, 1990.
7. Banin, A., Han, F.X., Serban, C., Ben-Dor, E., and Schachar, Y., The dynamics of heavy metals partitioning and transformations in arid-zone soils, in *Proc. 4th Intl. Conf. on Biogeochemistry of Trace Elements,* Berkeley, CA, 1997.
8. Shuman, L.M., Fractionation method for soil microelements, *Soil Sci.*, 140, 11, 1985.

9. Sposito, G., Lund, L.J., and Chang, A.C., Trace metal chemistry in arid-zone field soils amended with sewage sludge: I. Fractionation of Ni, Cu, Zn, Cd, and Pb in solid phases, *Soil Sci. Soc. Am. J.,* 46, 260, 1982.
10. Lindau, C.W. and Hossner, L.R., Sediment fractionation of Cu, Ni, Zn, Cr, Mn, and Fe in one experimental and three natural marshes, *J. Environ. Qual.,* 11, 540, 1982.
11. McGrath, S.P. and Cegarra, J., Chemical extractability of heavy metals during and after long-term applications of sewage sludge to soil, *J. Soil Sci.,* 43, 313, 1992.
12. McLaren, R.G. and Ritchie, G.S.P., The long-term fate of copper fertilizer applied to a lateritic sandy soil in Western Australia, *Aust. J. Soil Res.,* 93, 39, 1993.
13. Soon, Y.K., Changes in forms of soil zinc after 23 years of cropping following clearing of a boreal forest, *Can. J. Soil Sci.,* 74, 179, 1994.
14. Han, F.X. and Banin, A., Solid-phase manganese fractionation changes in saturated arid-zone soils: Pathways and kinetics, *Soil Sci. Soc. Am. J.,* 60, 1072, 1996.
15. Han, F.X. and Banin, A., Long-term transformations and redistribution of potentially toxic heavy metals in arid-zone soils. I: Incubation under saturated conditions, *Water Air Soil Poll.,* 95, 399, 1997.
16. Han, F.X. and Banin, A., Long-term transformations and redistribution of potentially toxic heavy metals in arid-zone soils. II: Incubation under field capacity conditions, *Water Air Soil Poll.,* 114, 221, 1999.
17. Han, F.X. and Banin, A., Selective sequential dissolution techniques for trace metals in arid-zone soils: The carbonate dissolution step, *Commun. Soil Sci. Plant Anal.,* 26, 553, 1995.
18. Han, F.X., Hu, A.T., and Qin, H.Y., Transformation and distribution of forms of zinc in acid, neutral and calcareous soils of China, *Geoderma,* 66, 121, 1995.
19. Han, F.X. and Zhu, Q.X., Fractionation and availability of zinc in paddy soils of China, *Pedosphere,* 2, 283, 1992.
20. Han, F.X., Hu, A.T., Qin, H.Y., and Shi, R.H., Enrichment capability of native zinc by some components of soils of China, *Acta Pedologica Sinica,* 28, 327, 1991.
21. Han, F.X., Hu, A.T., Qin, H.Y., and Shi, R.H., Fractionation and availability of added soluble zinc in various soil environments, *China Environ. Sci.,* 12, 108, 1992.
22. Han, F.X., Hu, A.T., Qin, H.Y., and Shi, R.H., Study on mechanism of zinc deficiency in calcareous soils, *Environ. Chem.,* 12, 36, 1993 (in Chinese).
23. Han, F.X., Hu, A.T., and Qin, H.Y., Fractionation of zinc bound to organic matter in soil, *J. Nanjing Agri. Univ.,* 13, 68, 1990.
24. Chang, A.C., Page, A.L., Warneke, J.E., and Grgurevic, E., Sequential extraction of soil heavy metals following sludge applications, *J. Environ. Qual.,* 13, 33, 1984.
25. Martin, J.M., Nirel, P., and Thomas, A.J., Sequential extraction techniques: Promises and problems, *Mar. Chem.,* 22, 313, 1987.
26. Shuman, L.M., Chemical forms of micronutrients in soils, in *Micronutrients in Agriculture* (2nd Edn.), Mortvedt, J.J., Cox, F.R., Shuman, L.M. and Welch, R.M., Eds., Soil Sci. Soc. Am. Inc., Madison, WI, 1991.
27. Kheboian, C. and Bauer, C., Accuracy of selective extraction procedures for metal speciation in model aquatic sediments, *Anal. Chem.,* 59, 1417, 1987.
28. Rendell, P.S., Batley, G.E., and Cameron, A.J., Adsorption as a control of metal concentrations in sediment extracts, *Environ. Sci. Technol.,* 14, 314, 1980.
29. Belzile, N., Lecomte, P., and Tessier, A., Testing readsorption of trace elements during partial chemical extractions of bottom sediments, *Environ. Sci. Technol.,* 23, 1015, 1989.
30. Kim, N.D. and Fergusson, J.E., Effectiveness of a commonly used sequential extraction techniques in determining the speciation of cadmium in soils, *Sci. Total Environ.,* 105, 191, 1991.

31. Council for Agricultural Science and Technology, *Waste Management and Utilization in Food Production and Processing,* Council for Agricultural Science and Technology, Ames, IA, 1995.
32. Moore, P.A, Jr., Daniel, T.C., Sharpley, A.N., and Wood, C.W., Poultry manure management: Environmentally sound options, *J. Soil Water Conserv.,* 50, 321, 1995.
33. Sims, J.T. and Wolf, D.C., Poultry manure management: Agricultural and environmental issues, *Adv. Agron.,* 52, 1, 1994.
34. Han, F.X., Kingery, W.L., Selim, H.M., and Gerald, P., Accumulation of heavy metals in a long-term poultry waste-amended soil, *Soil Sci.,* 165, 260, 2000.
35. Warman, P.R. and Thomas, R.L., Effect of poultry manure additions on the extractability of an added metal, *Commun. Soil Sci. Plant Anal.,* 7, 405, 1976.
36. Kingery, W.L., Wood, C.W., Delaney, D.P., Williams, J.C., and Mullins, G.L., Impact of long-term land application of broiler litter on environmentally related soil properties, *J. Environ. Qual.,* 23, 139, 1994.
37. Van der Watt, H.v.H., Sumner, M.E., and Cabrera, M.L., Bioavailability of copper, manganese, and zinc in poultry litter, *J. Environ. Qual.,* 23, 43, 1994.
38. Mitchell, C.C., Windham, S.T., Nelson, D.B., and Baltikausiki, M.N., Effects of long-term broiler litter application on coaster plain soils, in *Proc. Natl. Poultry Waste Mgmt. Symp.,* Blake, J.P., Donald, J.O., Patterson, P.H., Eds., Auburn University, Alabama, 1992.
39. Moore, P.A., Jr., Daniel, T.C., Gilmour, J.T., Shreve, B.R., Edwards, D.R., and Wood, B.H., Decreasing metal runoff from poultry litter with aluminum sulfate, *J. Environ. Qual.,* 27, 92, 1998.
40. Kornegay, E.T., Hedges, J.D., Martens, D.C., and Kramer, C.Y., Effect on soil mineral storage and plant leaves following application of manures of different copper contents, *Plant Soil,* 45, 151, 1976.
41. Sutton, A.L., Nelson, D.W., Mayrose, V.B., and Kelly, D.T., Effect of copper levels in swine manure on corn and soil, *J. Environ. Qual.,* 12, 198, 1983.
42. Mullins, C.L., Martens, D.C., Miller, W.P., Kornegay, E.T., and Hallock, D.L., Copper availability, form and mobility in soils from three annual copper-enriched hog manure applications, *J. Environ. Qual.,* 11, 316, 1982.
43. Payne, G.G., Martens, D.C., Winarko, C., and Perera, N.F., Availability and form of copper in three soils following eight annual applications of copper-enriched swine manure, *J. Environ. Qual.,* 17, 740, 1988.
44. Li, M., Hue, N.V., and Hussain, S.K.G., Changes of metal forms by organic amendments to Hawaii soils, *Commun. Soil Sci. Plant Anal.,* 28, 381, 1997.
45. McBride, M.B., Forms and distribution of copper in solid and solution phases of soils, in *Copper in Soils and Plants,* Loneragan, J.F., Robson, A.D. and Graham, R.D., Eds., Academic Press, Sydney, 1981.
46. Tan, K.H., Leonard, R.A., Bertrand, A.R., and Wilkinson, S.R., The metal complexing capacity and the nature of the chelating ligands of water extract of poultry litter, *Soil Sci. Soc. Am. Proc.,* 35, 265, 1971.
47. Lindsay, W.L., *Chemical Equilibria in Soils,* John Wiley & Sons, New York, 1979.
48. Shuman, L.M., Separating soil iron- and manganese-oxide fractions for microelement analysis, *Soil Sci. Soc. Am. J.,* 46, 1099, 1982.
49. McKenzie, R.M., Manganese oxides and hydroxids in minerals in soil environments, in *Minerals in Soil Environments* (2nd Edn.), Dixon, J.B. and Weed, S.B., Eds., Soil Sci. Soc. Am, Madison, WI, 1989.
50. Curtis, J.L., The distribution and dynamics of phosphorus in an agricultural watershed with a long-term history of poultry waste application, M.S. thesis, Mississippi State University, Mississippi State, Mississippi, 1998.

51. Barton, C.D. and Karathanasis, A.D, Colloid-facilitated transport of atrazine and zinc through soil monoliths, *Agronomy Abstracts,* Annual meeting of Soil Sci. Soc. Am., Baltimore, MD, 1998.

52. Pierzynski, G.M. and Schwab, A.P., Bioavailability of zinc, cadmium, and lead in a metal-contaminated alluvial soil, *J. Environ. Qual.,* 22, 247, 1993.

53. Kingery, W.L., Wood, C.W., Delaney, D.P., Williams, J.C., Mullins, G.L., and van Santen, E., Implications of long-term land application of poultry litter on tall fescue pastures, *J. Prod. Agric.,* 6, 390, 1993.

54. Feigin, A., Ravina, I., and Shalhevet, J., *Irrigation with Treated Sewage Effluent,* Springer-Verlag, New York 1991, 99.

55. Avnimelech, Y., Irrigation with sewage effluents: The Israeli experience, *Environ. Sci. Technol.,* 27, 1278, 1993.

56. Banin, A., Navrot, J., Noi, Y., and Yoles, D., Accumulation of heavy metals in arid-zone soils irrigated with treated sewage effluents and their uptake by Rhodes grass, *J. Environ. Qual.,* 10, 536, 1981.

57. Lake, D.L., Kirk, P.W.W., and Lester, J.N., Fractionation, characterization, and speciation of heavy metals in sewage sludge and sludge-amended soils: A review, *J. Environ. Qual.,* 13, 175, 1984,

58. National Research Council, *Use of Reclaimed Water and Sludge in Food Crop Production,* National Academy Press, Washington, D.C., 1996.

59. Bilitewski, B., Kardtle, G., Marek, K., Weissbach, A., and Boeddicker, H., *Waste Management,* Springer, New York, 1994.

60. Essington, M.E. and Mattigod, S.V., Trace element solid-phase associations in sewage sludge and sludge-amended soil, *Soil Sci. Soc. Am. J.,* 55, 350, 1991.

61. Beckett, P.H.T., The use of extractants in studies on trace metals in soils, sewage sludges and sludge-treated soils, *Adv. Soil Sci.,* 9, 143, 1989.

62. Villarroel, J.R. de, Chang, A.C., and Amrhein, C., Cd and Zn phytoavailability of a field-stabilized sludge-treated soil, *Soil Sci.,* 155, 197, 1993.

63. Shahar, Y. and Banin, A., The effect of long-term irrigation with treated sewage effluents on the concentration and partitioning of trace elements in arid-zone soils, in *Proc. 4th Intl. Conf. on Biogeochemistry of Trace Elements,* Berkeley, CA, 1997.

64. Knudtsen, K. and O'Connor, G.A., Characterization of iron and zinc in Albuquerque sewage sludge, *J. Environ. Qual.,* 16, 85, 1987.

65. Sposito, G., LeVesque, C.S., LeClaire, J.P., and Chang, A.C., Trace elements chemistry in arid-zone field soils amended with sewage sludge: III. Effect of the time on the extraction of trace metals, *Soil Sci. Soc. Am. J.,* 47, 898, 1983.

66. Sims, J.T. and Kline, J.S., Chemical fractionation and plant uptake of heavy metals in soils amended with composted sewage sludge, *J. Environ. Qual.,* 20, 387, 1991.

67. Bell, F.B., James, B.R., and Chaney, R.L., Heavy metal extractability in long-term sewage sludge and metal salt-amended soils, *J. Environ. Qual.,* 20, 481, 1991.

68. Silviera, D.J. and Sommers, L.E., Extractability of copper, zinc, cadmium and lead in soils incubated with sewage sludge, *J. Environ. Qual.,* 6, 47, 1977.

69. Bidwel, A.M. and Dowdy, R.H., Cadmium and zinc availability to corn following termination of sewage sludge application, *J. Environ. Qual.,* 16, 438, 1987.

70. Khalid, R.A., Gambrell, R.P., and Patrick, W.H., Jr., Chemical availability of cadmium in Mississippi River sediment, *J. Environ. Qual.,* 10, 523, 1981.

71. Misra, A.K., Sarkunan, V., Das, M., and Nayar, P.K., Transformation of added heavy metals in soils under flooded condition, *J. India Soc. Soil Sci.,* 38, 416, 1990.

72. Dowdy, R.H., Clapp, C.E., Linden, D.R., Larson, W.E., Halbach, T.R., and Polta, R.C., Twenty years of trace metal partitioning on the Rosemount sewage sludge watershed, in

Sewage Sludge Land Utilization and the Environment, Clapp, E.C., Larson, W.E. and Dowdy, R.H., Eds., Am. Soc. Agr., Madison, WI, 1994.

73. Barbarick, K.A., Ippolito, J.A., and Westfall, D.G., Extractable trace elements in the soil profile after years of biosolids application, *J. Environ. Qual,* 27, 801, 1998.

74. LeClaire, J.P., Chang, A.C., Levesque, C.S., and Sposito, G., Trace metal chemistry in arid-zone field soils amended with sewage sludge. IV: Correlations between zinc uptake and extracted soil zinc fractions, *Soil Sci. Soc. Am. J.,* 48, 509, 1984.

75. Banjoko, V.A. and McGrath, S.P., Studies of the distribution and bioavailability of soil zinc fractions, *J. Sci. Food Agri.,* 57, 325, 1991.

76. Murthy, A.S.P., Zinc fractions in wetland rice soils and their availability to rice, *Soil Sci.,* 133, 150, 1982.

77. Singh, M.V. and Abrol, I.P., Transformation and movement of zinc in an alkali soil and their influence on the yield and uptake of zinc by rice and wheat crops, *Plant Soil,* 94, 445, 1986.

78. Gonzalez, R.X., Sartain J.B., and Miller, W.L., Cadmium availability and extractability from sewage sludge as affected by waste phosphatic clay, *J. Environ. Qual.,* 21, 272, 1992.

79. Han, F.X., Binding, distribution and transformations of polluting trace elements in soils of arid and semi-arid regions receiving waste inputs, Ph.D. dissertation, Hebrew University of Jerusalem, 1998.

80. Han, F.X., Banin, A., Triplett, G.B., Redistribution of toxic metals in arid-zone soils under wetting–drying cycle moisture regime, *Soil Sci.* (accepted).

81. Mann, S.S. and Ritchie, G.S.P., Changes in the form of cadmium with time in Western Australian soils, *Aust. J. Soil Res.,* 32, 241, 1994.

82. Ma, L.Q. and Rao, G.N., Chemical fractionation of cadmium, copper, nickel and zinc in contaminated soils, *J. Environ. Qual.,* 26, 259, 1997.

83. Levy, D.B., Barbarick, K.A., Siemer, E.G., and Sommers, L.E., Distribution and partitioning of trace metals in contaminated soils near Leadville, Colorado, *J. Environ. Qual.,* 21, 185, 1992.

84. Han, F.X., Hargreaves, J., Kingery, W.L., Huggett, D.B., and Schlenk, D.K., Accumulation and distribution of copper in catfish pond soils receiving applications of copper sulfate, *J. Environ. Qual.* (accepted).

85. Mandal, L.N. and Mandal, B., Zinc fractions in soils in relation to zinc nutrition of lowland rice, *Soil Sci.,* 142, 141, 1986.

86. Hazra, G.C., Mandal, B., and Mandal, L.N., Distribution of zinc fractions and their transformation in submerged rice soils, *Plant Soil,* 104, 175, 1987.

87. Mandal, B., Chatterjee, J., Hazra, G.C., and Mandal, L.N., Effect of preflooding on transformation of applied zinc and its uptake by rice in lateric soil, *Soil Sci.,* 153, 250, 1992.

88. Han, F.X. and Banin, A., Pathways and kinetics of redistribution of cobalt among solid-phase components in arid-zone soils, *Eur. J. Soil Sci.* (in press).

89. Han, F.X. and Banin, A., Long-term transformations of Cd, Co, Cu, Ni, Zn, V, Mn and Fe in arid-zone soils under saturated conditions, *Commun. Soil Sci. Plant Anal.,* 31, 943, 2000.

90. McLaren, R.G., Lawson, D.M., and Swift, R.S., The forms of cobalt in some Scottish soils as determined by extraction and isotopic exchange, *J. Soil Sci.,* 37, 223, 1986.

91. Jarvis, S.C., The association of cobalt with easily reducible manganese in some acidic permanent grassland soils, *J. Soil Sci.,* 35, 431, 1984.

92. McKenzie, R.M., The mineralogy and chemistry of soil cobalt, in *Trace Elements in Soil–Plant–Animal Systems,* Nicholas, D.J.D. and Egan, A.R., Academic Press, Inc., New York, 1975.

9 Contaminant Transport in the Root Zone

Iris Vogeler, Steven R. Green, Brent E. Clothier, M.B. Kirkham, and Brett H. Robinson

ABSTRACT

The passage through soil and uptake by poplar of copper, a pollutant in the soils of New Zealand, was monitored to determine if poplar can be used in phytoremediation. The chelate ethylenediaminetetraacetic acid (EDTA) was added to the soil to solubilize copper and enhance its uptake by the tree. Another objective was to compare the impact of vegetation on solute leaching beyond the root zone. The movements of the tracer bromide through soil with pasture grass and soil with the poplar tree were compared. The experiment was conducted under greenhouse conditions in two large lysimeters, one with pasture grass and one with a poplar tree. The transports of water and contaminant (Cu, Br) were also modeled using a mechanistic scheme based on Richards' equation for water transport and the convection–dispersion equation to simulate the chemical movement in the root zone. These were linked to a distributed macroscopic sink term for plant upake.

Poplar roots in the surface of the lysimeter took up large amounts of copper, three times ($29.4\ \mu g\ g^{-1}$) the background concentration in the roots ($9.7\ \mu g\ g^{-1}$). EDTA solubilized the copper and changed its distribution in the soil with depth. Without EDTA, the added copper would have remained in the surface 50 mm of the soil. With EDTA, it was distributed through the top 200 mm. Below 800 mm, copper concentrations in the soil with EDTA were at background levels ($15\ \mu g\ g^{-1}$). Nevertheless, a small amount of the solubilized copper (ca. 2% of the total amount of copper added) escaped into the leachate from the lysimeter.

The peak concentration of bromide in the leachate occurred three days earlier under the pasture grass than under the poplar tree due to the higher water uptake of the poplar tree (average of $5.9\ L\ d^{-1}$) compared with the grass (average of $0.84\ L\ d^{-1}$). This shows the impact of vegetation on solute leaching beyond the root zone toward groundwater. Contamination of groundwater would occur earlier under pasture grass than under trees.

The model successfully predicted the movements of copper and bromide through the soil, indicating that mechanistic modeling of the processes of contaminant transport in the root zone will help develop sustainable land-management practices, and will advance effective procedures for remediating contaminated sites.

1-56670-507-X/01/$0.00+$.50

INTRODUCTION

Water can carry contaminants and nutrients such as trace elements through the unsaturated soil of the root zone to the groundwater. Conversely, water taken up by the roots can move the contaminating solutes into the plant, and thereby reduce deep percolation losses of both water and contaminants. The extent of the partitioning of contaminating chemicals into these two transport processes depends on soil-based mechanisms such as adsorption, as well as on uptake by plant roots.

One trace-element contaminant in the soils of New Zealand is copper. It is widely used as a fungicide in orchards. Along with chromium and arsenic, it is also used to preserve wood. The copper–chromium–arsenic (CCA) treatment, which involves impregnation under pressure of a 1–4% solution of copper sulphate, sodium dichromate, and arsenic pentoxide, is the main preservation method used in New Zealand, and 1.4×10^6 m^3 of timber is treated annually in that country.[1] Timber is graded for use according to the amount of treatment solution fixed in it. The copper acts as a fungicide, arsenic as an insecticide, and chromium as a fixative. Serious contamination of soil, surface waters, and groundwater has occurred because of waste disposal practices at timber treatment sites. Total concentrations of copper in the soil at these sites range from 108–8020 mg kg^{-1}.[1] The average total concentration of copper in nonpolluted soils is 20 mg kg^{-1}.[2] The Ministry for the Environment in New Zealand has set the acceptance criteria for copper at 30–100 μg g^{-1} in agricultural soils and 130 μg g^{-1} in residential soils.[1]

Contaminated soils have been carted away at much expense. An alternative and perhaps more environmentally beneficial method is phytoremediation,[3,4] which is the use of green plants to remove pollutants from the soil. The main advantages of phytoremediation are that the procedure is carried out *in situ,* without physical removal of the soil, and it is relatively inexpensive. Classical methods for *in situ* remediation of water-soluble pollutants can cost \$100,000–\$1,000,000 per hectare, and other procedures are even more expensive. Phytoremediation is estimated to cost between \$200 and \$10,000 per hectare.[4,5]

Work in the U.S. has shown that poplar trees are effective in removing contaminants from soil,[6–8] but no work has been done in New Zealand on the use of poplar to remediate soils. Therefore, the first objective of this research was to develop a proof-of-concept prototype for phytoremediation of contaminated soil in New Zealand using poplar.

Recently, the chelating agent ethylenediaminetetraacetic acid (EDTA) has been added to soil to increase the solubility of trace elements for plant uptake during phytoremediation or phytomining.[4,7,9] If plant roots do not take up the solubilized element, it will travel to groundwater and pollute it. Therefore, a second objective of this study was to determine the effect of EDTA on the movement of copper in soil with poplar.

A third objective was to compare measured results with a new mechanistic model that joins water and solute movements through the root zone with uptakes of water and solutes by roots. Measurements combined with mechanistic modeling of the processes of contaminant transport in the root zone will help develop sustainable land-management practices and procedures that can remediate contaminated sites.

As an adjunct, a further objective was considered. This involved comparison of poplar with pasture grass to understand soil-based and plant-based mechanisms that control passage of chemicals and water through these two root zones. This is of concern in New Zealand, because shallow-rooted pasture plants are being replaced by deeper-rooted trees, which leads to the diminution of drainage recharge.

An abbreviated report on this research has been published.[10]

THE MODEL

The model PSILAYER[11] was used. The model and its equations have been described in detail by Vogeler and his colleagues,[12] and we repeat salient information here. It is a deterministic model based on Richards' equation and the convection–dispersion equation linked with sink terms for root water and chemical uptakes. The model assumes that water uptake is dependent on the root distribution. Uptake is also modified by the matric pressure head. Solute uptake by the plants is considered to be passive convection with the water.

Water movement through unsaturated soil is described by Richards' equation:

$$\left(\frac{d\theta}{dh}\right)\left(\frac{\partial h}{\partial t}\right) = \frac{\partial}{\partial z}\left\{[K(h)]\left(\frac{\partial h}{\partial t}\right) - K(h)\right\} - Uw(z,h,t) \tag{9.1}$$

where θ is the volumetric water content (m^3 m^{-3}); h is the soil matric potential (m); t is time (s); z is the depth (m), which is positive downwards; K(h) is the hydraulic conductivity of the soil (m s^{-1}); and U_w is a sink term (s^{-1}) representing the uptake of water by plant roots. PSILAYER allows for three choices for the soil's hydraulic conductivity functions, K(h) or K(θ), which is described below under Parameters in the Model.

For water uptake by the plant roots, we assumed that the macroscopic sink term (U_w) is proportional to the local root density and is modified by the water content of the surrounding soil. In most crops, the local root density L(z) (m^{-1}) declines exponentially with depth,[13] and root effectiveness is considered to be proportional to root density.[14] Thus, the potential water uptake as a function of depth $U_w{}^*(z)$ can be expressed by the exponential equation:

$$U_w{}^*(z) = \frac{U_p L(z)}{L(0)} = U_p \exp\left[-\gamma_R\left(\frac{z}{z_R}\right)\right] \tag{9.2}$$

where U_p is the maximum potential root uptake near the soil surface (s^{-1}), L(0) is the root density at the soil surface (m^{-1}), z_R is the maximum root depth (m), and γ_R is a dimensionless parameter describing the depthwise profile of root volume. Thus, when soil water is nonlimiting, the potential uptake will decrease exponentially with soil depth. However, when soil water becomes limiting at any depth in the root zone, the water uptake there decreases. Hence, local water uptake is modified to reflect the local availability of soil water, and actual root uptake decreases below that of the potential. PSILAYER includes two models to account for the effect of reduced soil water availability, but we use only one of them in this work. In both models the actual root uptake is modeled as

$$U_w(z,h) = f(h)U_w^*(z) \tag{9.3}$$

where f(h) is an empirical function that assumes a value between zero and unity. To account for reduced soil water availability, we chose the model given by Feddes and his colleagues,[15] which assumes that the actual root uptake is limited by the soil water pressure head and is modeled as

$$f = 1 \qquad\qquad h \geq h_d \tag{9.4}$$
$$f = \left[\frac{(h - h_w)}{(h_d - h_w)}\right]^P \qquad h_w \leq h \leq h_d \tag{9.5}$$
$$f = 0 \qquad\qquad h \leq h_w \tag{9.6}$$

where h_d is the lowest matric pressure head at which $U_w = U_w^*$, h_w is the matric potential at which uptake ceases, and P is an assumed constant. A P-value of unity implies a linear reduction in root uptake as the soil dries.

For solute movement, PSILAYER uses a physical nonequilibrium concept that assumes convective–dispersive flow in a mobile domain together with a rate-limited diffusive exchange between the mobile and the complementary immobile soil domains. In our work, we assumed all the soil water was mobile and used the classical convection–dispersion equation describing one-dimensional solute transport during transient flow, as follows:

$$\frac{\partial(\rho S)}{\partial t} + \frac{\partial(\theta C)}{\partial t} = \left(\frac{\partial}{\partial z}\right)\left[(\theta D_s)\frac{\partial C}{\partial z}\right] - \frac{\partial(q_w C)}{\partial z} - S_m(z) \tag{9.7}$$

where C is the local solute concentration in the soil solution (mol m^{-3}), S is the concentration adsorbed onto the soil matrix (mol kg^{-1} of soil solid), D_s is the solute diffusion–dispersion coefficient (m^2 s^{-1}), ρ is the bulk density of the soil (kg m^{-3}), q_w is the soil water flux density (m s^{-1}), and $S_m(z)$ is the solute uptake as a function of depth (mol m^{-3} s^{-1}).

We considered the solute diffusion–dispersion coefficient to be velocity dependent and modeled it using a commonly adopted approach based on the sum of diffusion and dispersion:[16]

$$D_s = \tau D_0 + \lambda |q_w| / \theta \tag{9.8}$$

where τ is a tortuosity factor (0.45), D_0 is the free-water diffusivity (m^2 s^{-1}), and λ is the dispersivity (mm).

For an inert solute, the amount of solute adsorbed, S, is zero. However, for a reactive solute, we need to parameterize the link between S and C. PSILAYER allows for two models, described below, to relate the amount of solute adsorbed onto the soil matrix, S, to the concentration in the liquid phase, C.

1. The Freundlich isotherm:

$$S = K_D C^N \tag{9.9}$$

where K_D is the distribution coefficient and N is a dimensionless constant and usually has a value of $N < 1$. For N equals unity, the Freundlich equation often is referred to as the linear absorption isotherm.

2. The Langmuir isotherm:

$$S = \frac{S_{max} \, \omega C}{1 + \omega C} \tag{9.10}$$

where S and C have the same meaning as in the Freundlich isotherm (Eq. 9.9), S_{max} is the adsorption maximum, and ω is a constant. The Langmuir equation, unlike the Freundlich, provides the advantage of a limiting value for adsorption. The values S_{max} and ω are adjustable parameters that can be obtained by least-squares fitting procedures using equilibrium batch-isotherm data (the procedure is described below under Materials and Methods). Here $\omega (m^3 \, g^{-1})$ is a measure of the bond strength of molecules on the matrix surface, and $S_{max} (g \, kg^{-1})$ is the maximum sorption capacity or total amount of available sites per unit soil mass.

Solute uptake often is assumed to proceed passively. Thus, the local solute uptake, S_m, is proportional to the product of the local water uptake and the local solute concentration. However, experimental evidence also shows that roots can reduce solute uptake whenever the local concentration exceeds a certain critical threshold, C_1. So an active exclusion mechanism needs to be invoked. In the model, the depth-wise pattern of solute uptake, $S_m(z)$ (mmol $L^{-1} s^{-1}$), is given by

$$S_m(z) = U_w(z)C(z) \qquad C \le C_1 \tag{9.11}$$
$$S_m(z) = U_w(z)C_1 \qquad C > C_1 \tag{9.12}$$

where $U_w(z)$ (s^{-1}) is the root water uptake with depth and C_1 (mmol L^{-1}) is the highest concentration at which solute uptake is passive. Thus, at low concentrations, solute moves passively into the plant along with the transpiration stream. At high concentrations some solute is excluded from the uptake process.

The boundary conditions for unsaturated water flow into a soil lysimeter under irrigation are

$$\theta = \theta_i(z) \qquad t = 0 \qquad 0 \le z \le 1 \tag{9.13}$$
$$q_w = q_o(t) \qquad t > 0 \qquad z = 0 \tag{9.14}$$

where θ_i is the initial water content ($m^3 \, m^{-3}$), l is the lysimeter length (m), and q_o is the flux density imposed at the surface ($m \, s^{-1}$).

For modeling, the above equations were solved numerically using a fully implicit Newton–Raphson iteration for the water flow equation, and a time-centered, Crank–Nicholson scheme for the solute flow. The complete equations are given by Green.[11]

MODEL PARAMETERS

Four parameter files were developed to run the model: pasture grass with bromide, poplar with bromide, poplar with copper and no EDTA, and poplar with copper and EDTA. As an example of one of the parameter files, we append the file for poplar with copper and EDTA (see appendix). The soil and plants are described in detail under Materials and Methods. In the numerical model (PSILAYER), all the modeling is done simultaneously, e.g., water flow, solute movement, water uptake, and solute uptake. It is done in time steps so that the values for each parameter (e.g., water content) can be used for calculating the values at the new time step. When a choice of equations was available, we state in this section which one we chose. The parameters that we put into the model to make it run are listed below.

INITIAL INFORMATION

All four parameter files contained the following information: the domain length was 1260 mm (length of the lysimeter); the number of grid points, normally called compartments (the soil profile is divided into compartments), was 21 (3 layers with 7 grids); the number of layers in the soil was one; the n-compartments for the ith layer were 7 7 7; and the orientation was vertical for the soil column.

HYDRAULIC PROPERTIES OF THE SOIL

In PSILAYER, the water-retention characteristics of the soil are described by the following power-law relation:[17]

$$\theta(h) = \theta_r + \frac{\theta_s - \theta_r}{[1 + (\delta|h|)^n]^m} \tag{9.15}$$

where θ_s is the water content at saturation ($m^3\ m^{-3}$), θ_r is a residual water content ($m^3\ m^{-3}$), δ is the inverse of the air entry value (m^{-1}), and n and m are parameters to describe the shape of the $\theta(h)$ curve. Values for each parameter can be determined by fitting Eq. 9.15 to water retention data using least-squares methods.[11] Van Genuchten (1980) assumes the Mualem[18] model, where $m = 1 - (1/n)$.

PSILAYER allows for three choices for the soil's hydraulic conductivity functions:

1. The exponential relation of Gardner:[19]

$$K(h) = K_s \exp(\beta_1 h) \tag{9.16}$$

where K_s is the hydraulic conductivity at saturation (m s^{-1}), when $\theta = \theta_s$ and β_1 is a slope parameter (m^{-1}).

2. The power-law function of Brooks and Corey:[20]

$$K(\theta) = K_s\left(\frac{\theta - \theta_r}{\theta_s - \theta_r}\right)^{\beta_2}$$
(9.17)

where β_2 is a dimensionless empirical constant and the other parameters are as described above. The formulation of Eq. 9.17 is given by Corey (p. 41).[21]

3. The closed-form equation of van Genuchten:[17]

$$K(\Theta) = \Theta^{\frac{1}{2}} - [1 - (1 - \Theta^{\frac{1}{m}})^m]^2$$
(9.18)

where $\Theta = (\theta - \theta_r)/(\theta_s - \theta_r)$, which is the relative water content, and the m-parameter is that of Eq. 9.15.

We chose the power-law function of Brooks and Corey[20] (Eq. 9.17). All four parameter files contained the following information: the value K_s, the saturated hydraulic conductivity for the Brooks and Corey[20] equation, was equal to 2.25×10^{-2} mm s^{-1} (this is the value for the soil under repacked conditions); the value β_2 was equal to 0.00836 (dimensionless); the value δ, which is the inverse of the air entry value, was equal to 0.00162 m^{-1}; the value n was equal to 1.4027 (dimensionless); and the value m was equal to 0.2871 (dimensionless).

SOLUTE TRANSPORT PROPERTIES OF SOIL

In the model, we assumed that bromide was not adsorbed, that the copper was adsorbed by the soil without EDTA following the Langmuir isotherm, that copper was adsorbed by the soil with EDTA following the Freundlich isotherm, and that no immobile water was present. Therefore, all equations in PSILAYER dealing with immobile water were ignored.

The value D_0 was equal to 1×10^{-8} mm^2 s^{-1} (free-water diffusivity or chemical diffusion coefficient in pure water). This value was used for modeling bromide, copper without EDTA, and copper plus EDTA. Its low value indicates that it can more or less be neglected.

The ground surface area of each lysimeter was equal to 0.567 m^2. This value was the same for all four parameter files.

Bromide Modeling for Pasture Grass and Poplar Tree

The gram molecular weight of bromide (bromine) was needed: it is equal to 79.904 g mol^{-1}.

Next, we describe the parameters that needed to be designated at this point in the model. The bulk density of the soil, ρ_b, was equal to 1200 kg m^{-3}. The value m was equal to 0.0 (dimensionless), which means no adsorption of bromide. The value n was equal to 1.0 (dimensionless). The fraction of sorption sites in the mobile phase, f, was equal to 1.00 (dimensionless). The fraction of the mobile water content, θ_m/θ, was equal to 1.00. As noted, in the current modeling all water was considered to be mobile. The dispersivity, λ was equal to 50 mm. The mass transfer coefficient, α, was equal to 0.0×10^{-5} s^{-1} (i.e., zero), as there was no immobile water.

Copper Modeling for Poplar Tree Without EDTA

The gram molecular weight of copper was needed; it is equal to 63.645 g mol^{-1}. A Langmuir isotherm was assumed (Eq. 9.10). This equation was modified to

$$S = \frac{aC}{b + C} \tag{9.19}$$

The values for a and b in this equation come from fitting the equation to data obtained from batch experiments, which are described under Materials and Methods. The value a was equal to 0.08 (dimensionless). The value b was equal to 0.38 (dimensionless). The other parameters were the same as those for the bromide modeling.

Copper Modeling for Poplar Tree with EDTA

A nonlinear Freundlich isotherm was assumed (Eq. 9.9).

The bulk density, ρ_b, was equal to 1200 kg m^{-3}, K_D was equal to 0.124 (m^3 g^{-1}), and N was equal to 3.08 (dimensionless). The other parameters were the same as those for the bromide modeling.

SPECIFICATION OF ROOT UPTAKE PROPERTIES

We assumed that root-length density L declined exponentially with depth. We let the maximum root depth z_R change linearly with time, and the following equation held:

$$z_R = z_{Ri} + (z_{Rf} - z_{Ri})\left(\frac{t}{t_g}\right) \tag{9.20}$$

That is, the maximum depth of the root system, z_R, was assumed to vary linearly with time, with the extent of the root system spanning from a depth z_{Ri} to z_{Rf} in t_g days. Because L declines exponentially with depth, a value for the slope parameter, γ_R, is required. In Eq. 9.19, γ_R was equal to 3.5 (dimensionless). This value, 3.5, a shape factor that describes the exponential decline in root distribution with depth, was determined from the root length density with depth at the end of the experiment. The value t_g was equal to 1 day. The plants were well established before the experiment started, and no change in root distribution with depth occurred during the experiment. Consequently, the value z_{Ri} was equal to 600 mm, and the value z_{Rf} was also equal to 600 mm. Rooting depth was about the same for the pasture grass and the poplar, and it was negligible beyond 600 mm. The solute leaching was performed after the plants were well established, so we could assume that the root distribution did not change during the experiment.

Root Water Uptake Model for Pasture Grass and Poplar Tree with Bromide

We chose the model of Feddes and his colleagues[15] to describe plant water uptake (Eqs. 9.4–9.6). Three reduction parameters had to be put in here: h_d, which was equal to -1×10^4 m; h_w, which was equal to -1×10^{-5} m; and P, which was equal to 1.0 (dimensionless). The value h needed in the model is a variable that changes with each time step and depends upon the water content of the various layers.

SOLUTE UPTAKE MODEL FOR PASTURE GRASS AND POPLAR TREE WITH BROMIDE

Equations 9.11 and 9.12 describe this part of the model. At low concentrations, solute moves passively into the plant along with the transpiration stream. At high concentrations, solute uptake is reduced and some solute is excluded from the uptake process. We assumed that all uptake was passive, so we needed to define only one reduction parameter, C_1, which was arbitrarily set to the high value of 1000. As noted, the value C_1 is the limiting concentration of uptake. If C_1 is high, there is no limitation and passive uptake takes place. Thus, C_1 is the maximum uptake concentration at which uptake changes from active to passive. However, we do not know at what concentration uptake changes from passive to active (exclusion). We propose that multiple tracers, with measurements of water and solute concentrations in the soil and in the plant sap, be used to determine an appropriate input parameter for C_1.[12]

Root water and solute uptake parameters for all four parameter files were identical.

INITIAL WATER AND SOLUTE CONCENTRATIONS

Initial conditions for 21 compartments were then set. The model was divided into 21 n-compartments (see above), so initial conditions for 21 lines of data were needed in the model at this point. The values for the first line were as follows. The volumetric water content, θ, was equal to 0.344 mm^3 mm^{-3}. The water content in the immobile region, θ_i, was equal to 1.0×10^{-2}, which is more or less zero. The concentration in the mobile water region, C_m, was equal to 1.0×10^{-9} mmol L^{-1}. (Units of mmol L^{-1} were used when the model was run. Units of g L^{-1} or mmol L^{-1} can be used, but one must be consistent throughout the model in using the same units.) The concentration in the immobile water region, C_i, was equal to 0 mmol L^{-1}.

The values for C_m and C_i stayed constant with each layer (with depth), except C_m for the copper with EDTA file. In this file, copper concentration was 0.343 mmol L^{-1} in the first layer. The value for C_m in all other layers was equal to 1.0×10^{-9} mmol L^{-1}, and C_i was equal to 0. The water content changed with depth. In three of the four parameter files (all except poplar with EDTA), θ was 0.344 mm^3 mm^{-3} at the surface and 0.363 mm^3 mm^{-3} at the bottom of the lysimeter, and was driest at the seventh and eighth layers (0.307 mm^3 mm^{-3}). For the poplar with EDTA parameter file, θ was 0.397 mm^3 mm^{-3} at the surface, 0.394 mm^3 mm^{-3} at the bottom layer, and 0.393 or 0.394 mm^3 mm^{-3} for all layers except the three surface layers. For the poplar with EDTA parameter file, the model was started with the same conditions as in the poplar with copper parameter file (no EDTA). The model was then run until the date of the application of EDTA. One of the output files is the water content (and solute concentration) as a function of depth. This output file was then used as an input for the different layers in the poplar with EDTA parameter file.

UPPER BOUNDARY CONDITION, FLUX, TRANSPIRATION, EVAPORATION, LOWER BOUNDARY CONDITION, DRAINAGE, AND TIME

Thirty-six experimental days (lines of data) were used for the parameter files for poplar with bromide and copper (no EDTA), 40 for pasture grass, and 20 for poplar with EDTA.

The first line of data for the poplar with bromide parameter file read as follows. The upper boundary condition was the flux type. This is what is called a Type 3 boundary condition in the literature. It remained constant for all four parameter files. The irrigation rate, V_o, was equal to 38.8 mm d^{-1}. This value remained constant for all four parameter files, except for the day when the bromide and copper were washed into the soil and it increased to 42.4 mm d^{-1}. The value, C_o, was equal to 0 mmol L^{-1} on all days except the day when the bromide and copper were washed into the soil. On that day, this value, C_o, was 428 mmol L^{-1} for the pasture grass file, 243 mmol L^{-1} for the poplar with copper (no EDTA) file, and 214 mmol L^{-1} for the poplar with bromide file. EDTA was not considered to affect C_o, so C_o was 0 for all days in the poplar with EDTA parameter file. The value for transpiration rate, Tr, was equal to 18.54 mm d^{-1} (for the poplar tree) on the first day, but changed daily. It was determined using the Penman–Monteith model. The values for transpiration rate of the poplar tree determined using the Penman–Monteith model were in close agreement with the values obtained using the heat-pulse equipment (data not shown). The transpiration rate was much lower for the pasture parameter file (2.37 mm d^{-1} for the first day of the experiment). The value for evaporation, E, which was equal to 2.37 mm d^{-1} on the first day of the poplar experiment, changed daily for the poplar parameter files. It was determined using the Penman–Monteith equation. E was zero in the pasture grass parameter file. The lower boundary condition was considered to be free draining. Under this boundary condition, the water is freely draining through the soil profile, and it implies that the water flux and hydraulic conductivity at the outlet end are equal. This value was constant for all four parameter files. Drainage, D, was equal to 0 mm d^{-1} and was set to 0 on every day of the experiment for all four parameter files. Drainage is relevant in the model only if one chooses flux as the lower boundary condition, i.e., the flux of water is fixed at the outlet end. Because we chose the free drainage situation, however, the model calculated the amount of drainage from the hydraulic properties, and so we entered values of 0 in the D column of the model.

MATERIALS AND METHODS

The experiment was carried out in a shade house with screen doors located at HortResearch, Palmerston North, New Zealand (40°21′S; 17°43′E). Two large cylindrical lysimeters (0.85 m in diameter; 1.15 m deep) with truncated, cone-shaped bottoms were used. The cut-off base of the cone was 0.46 m in diameter, and the cone extended 22.8 cm below the base of the cylinder. The lysimeters were packed with Manawatu fine sandy loam (Dystric Fluventic Eutrochrept) to a bulk density of 1.2 Mg m^{-3}. In November 1997 one lysimeter was planted with a hybrid poplar (*Populus deltoides* Marsh × *Populus nigra L.*) seedling, and the other lysimeter was planted with pasture grass that was a blend of fescue and browntop (Hoddes and Tolley, Ltd., Palmerston North, New Zealand). Browntop is also known as colonial bentgrass (*Agrostis tenuis* Sibth.). The fescue component was probably a mixture of several species that are called "fineleaf fescues" (*Festuca* sp.). These have a very fine leaf blades, in contrast to the most common fescue in the semi-arid region of Kansas, tall fescue (*Festuca arundinacea* Schreb.). Time domain reflectometry (Model 1502C Cable Tester, Textronix, Beaverton, OR, U.S.A.) with multiplexer capability was used to measure the soil's volumetric water content and electrical conductivity. Three-rod TDR probes (6 mm in diameter and 400 mm long with 25-mm spacing between rods) were installed horizontally in pairs on opposite sides of each lysimeter at depths of 150, 350, 600, and 850 mm. Another set of probes was installed horizontally in pairs on opposite sides of each lysimeter, but 90° away from the previously mentioned probes, at the following depths: 250, 450, 750, and 1050 mm. So, a row of TDR probes was installed at each of the four "sides" of the lysimeter. In addition, two probes were installed vertically in the surface of each lysimeter at the 0–400 mm depth to give a total of 18 probes per lysimeter. The water-content measurements were used to infer the depthwise pattern of water uptake by the roots, and the measurements of electrical conductivity were used to monitor the movement of chemicals through the soil. The method is described by Vogeler and her colleagues.[22,23]

The first day of the experiment (DOE) was arbitrarily designated 7 April 1998 (DOE 1), when the pasture grass was well established and the poplar tree was about 2.5 m tall. On 9 April 1998 (DOE 3), pulses of KBr were applied to the soil surface of both lysimeters. For the grass, we sprayed a solution of 42 g KBr dissolved in 2 L water onto the surface of the lysimeter, to give an application rate of 50 g Br m^{-2}. Bromide was used as a conservative tracer (i.e., a tracer that is nonreactive with the soil, as well as nonvolatile, and shows the water-flow pathways). Bromide was chosen rather than chloride because, unlike chloride, it occurs in negligible quantities in New Zealand soils. Bromide can be considered to mimic the movement of nitrate in the soil. For the poplar, to keep the same ionic strength, we applied a mixed pulse of 21 g KBr and 24.4 g $Cu(NO_3)_2 \cdot 3H_2O$ dissolved in 2 L water (3200 $\mu g\ g^{-1}$ Cu or 6.4 g Cu total added). The pulses were washed into the soil by spraying water on the surface (4.5 L water added to the pasture lysimeter and 250 mL water added to the poplar lysimeter).

On 24 April 1998 (DOE 18), 50 g of the chelate EDTA $\{[CH_2 \cdot N(CH_2 \cdot COOH) \cdot CH_2COONa]_2 \cdot 2H_2O$, molecular weight 372.24$\}$ (disodium salt) dissolved in 1 L of distilled water was sprayed on the surface of the soil with the poplar tree. This amount of EDTA gave 2 g kg^{-1} of soil in the upper 50 mm. The EDTA was added to solubilize the copper in the soil, which would otherwise be strongly adsorbed by the soil.

Daily throughout the experiment, 22 L of irrigation water was applied automatically to each lysimeter (11 L at 9:00 a.m. and 11 L at 9:00 p.m.) by 11 drippers about 10 cm above the soil surface. This resulted in a high irrigation rate of about 40 mm d^{-1}, which we wanted to achieve so that the passage of the chemicals could be observed. The irrigation rate was verified at the end of the experiment by catching in buckets and weighing the water delivered by the irrigation equipment.

Each day between 7 April 1998 (DOE 1) and 13 May 1998 (DOE 36), drainage water was collected at 7:00 a.m. A small tension (pressure head) of about 10 cm was kept on the water in the soil at the bottom of each lysimeter. Every three days during the experiment, four poplar leaves were sampled from different locations on the tree. Once a week, the pasture grass was cut and the cuttings were retained for analysis.

Movements of water, bromide, and copper were monitored by the time domain reflectometry probes, and also by analyzing the effluents quitting the base. Only the effluent results will be given here. The heat-pulse technique[24] was used to determine the daily transpiration of the poplar tree from the measurements of sap flow made in the trunk of the tree every 30 min. The transpiration rates of the poplar tree as well as the pasture grass were calculated using the Penman–Monteith equation.[25,26] This is based on the measured leaf area and ambient atmospheric factors. Net radiation, soil-heat flux, wind speed, vapor pressure deficit, and air temperature were monitored automatically throughout the experiment at a weather station located between the two lysimeters, which were about 1.5 m apart in the greenhouse. Transpiration rates determined using the Penman–Monteith equation are given in this chapter.

The poplar tree was harvested on 13 May 1998, 34 days after solute application, by chopping it down at the soil surface. All leaves were plucked off the tree, and total leaf area was measured using a leaf-area meter (LI-3100 Area Meter, Li-Cor, Inc., Lincoln, NE, U.S.A.). The leaves were divided into three groups: senescent, medium-sized, and small. The grass lysimeter was stopped on 17 May 1998, 38 days after solute application. The poplar lysimeter was dismantled on 13 May, and the grass lysimeter was dismantled on 25 May. Soil was removed from both at 100-mm depth increments (12 samples, 0 to 1200 mm), and subsamples of these layers were used to determine the depthwise pattern of the resident concentrations of bromide and copper in the soil (described below). Gravimetric soil water content at each depth was also determined at harvest. In the lysimeter with the poplar tree, large roots were removed for analysis of copper. No attempt was made to extract fine roots, as labor was limiting. Roots were not extracted from the lysimeter with the pasture grass.

Soil, leaves, roots, and drainage water were analyzed for bromide and copper. To analyze the soil for bromide, approximately 5 g of the wet weight of the soil was put in a test tube and 25 mL of distilled water was mixed into it. The soil and solution were filtered, and the extracts were analyzed on an HPLC (high-performance liquid chromatography) instrument (DX500 Ion Chromatograph; Dionex Corp., Sunnyvale, CA, U.S.A.) with a conductivity meter for detection. Drainage water was analyzed directly for bromide using the HPLC. To analyze leaves for bromide, 0.5 g of leaf sample, which had been oven-dried at 80°C, was placed in a flask and 60 mL distilled water was added. The flask was put on a hotplate at 80–90°C for 2 h, and the contents were stirred every 15 min. The extract was filtered through Whatman no. 54 hardened 110 mm filter paper, and the bromide in the extract was analyzed on the HPLC.

The method described by Robinson and his colleagues[27] was used to analyze for copper. Leaves and roots were placed in a drying cabinet at 80°C until a constant weight was reached. Samples were weighed and ground. Approximately 0.2 g from each sample was weighed into a 50-mL Erlenmeyer flask. Concentrated nitric acid (10 mL) was added to each flask, which was placed on a heating block and heated until a final volume of ca. 3 mL was reached. Each sample was then diluted to 10 mL using deionized water and stored in a polythene container. Soil samples were dried at 80°C and sieved to <1 mm size using a nylon sieve. About 0.2 g of each sieved soil was weighed into a boiling tube. Ten mL of concentrated nitric acid was then added, and the mixture was boiled until a final volume of 3 mL was reached. A further 10 mL of concentrated hydrochloric acid was added, and the mixture was again evaporated to 3 mL. After filtration, each solution was diluted to 10 mL with deionized water. Copper concentrations in the plant and soil solutions and the drainage water, which was analyzed directly, were determined using an atomic absorption spectrophotometer (Model 904; GBC Scientific Equipment Inc., Arlington Heights, IL).

To determine the effect of EDTA on the adsorption of copper to the soil, batch experiments were done using the following technique. Five grams of air-dried soil (fresh soil that had not been used in the lysimeters) was put in each of ten glass beakers. Four stock solutions were prepared to give 500, 1000, 5000, and 10,000 ppm Cu, using the copper nitrate salt [$Cu(NO_3)_2 \cdot 3H_2O$]. A fifth solution was distilled water (0 ppm Cu). Then 25 mL of each stock solution was added to the soil to give two beakers at each of the five concentrations of copper. The beakers were placed on a shaker. In addition to the shaking, the solution in each beaker was stirred manually with a glass stirrer for 5 s every 10 min for 2 h. Then 2 h later (4 h after the copper solutions had been added to the soil), the beakers were taken off the shaker, and the solution in each beaker was decanted into a test tube. Then 1 g EDTA (disodium salt) per kg soil was added to half the beakers. The soil and EDTA solutions were shaken and stirred as described above, and then decanted into test tubes. The copper in solution was determined using the atomic absorption spectrophotometer. The amount adsorbed to the soil was the difference between copper added and copper found in the solution. We could then plot the isotherm, which is adsorbed copper (mol kg^{-1}) versus copper in solution (mol L^{-1}). We plotted two isotherms, one for soil without EDTA and one for soil with EDTA.

RESULTS AND DISCUSSION

Figure 9.1 shows the concentrations of bromide and copper in the leachate leaving the base of the lysimeter with the poplar tree. The EDTA added on DOE 18 mobilized the copper, and seven days later (on DOE 25) copper concentrations in the drainage water started to increase. The concentrations got high after DOE 31 and reached a peak on DOE 34 (1.83 ppm or 1.83 $\mu g\ g^{-1}$). Total amount of copper lost via drainage water was 0.12 g, which was about 2% of the copper added. The average value of copper in drainage water without EDTA was 0.24 $\mu g\ g^{-1}$, and the concentration after DOE 31 was 0.89 $\mu g\ g^{-1}$. We do not know why the drainage water on DOEs 3 and 4 had high values for copper but not for bromide. The variability in copper concentration in the leachate between DOEs 3 and 10 indicated that soil pores with copper were not emptying uniformly. The application of EDTA, which resulted in solubilization of the copper, meant that copper could be taken up more easily by the plant (see below). But the solubilization also posed the risk of copper leaching into the groundwater, as can be seen from the copper concentrations measured in the leachate after DOE 25.

Bromide concentrations are shown in Figure 9.1 to compare with copper. Unlike copper, concentrations of bromide in the drainage water followed the typical pattern of a breakthrough curve for a nonreactive tracer. They were zero for 9 days after the pulse of KBr was added on 9 April, peaked 15 days after that date, and then decreased as the bromide was leached out of the lysimeter. Bromide concentrations are discussed below, when leachate under the pasture grass and poplar tree are compared.

Poplar leaves took up small amounts of copper (0.004 g for all leaves), perhaps because the experiment was carried out during the autumn months in New Zealand, when plants are not vigorously growing and transpiring as they are in the spring.

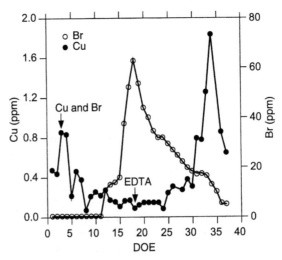

FIGURE 9.1   Copper and bromide concentrations in leachate from lysimeter with poplar tree during experiment. Copper and bromide were added on day of experiment (DOE) 3. Chelate EDTA was added on DOE 18.

Copper concentrations at harvest were 11.1 μg g^{-1} in the senesced leaves, 9.2 μg g^{-1} in the small leaves, and 7.1 μg g^{-1} in the medium-sized leaves. Transpiration rate was on a downward trend throughout the experiment (data not shown). The highest transpiration rate for poplar, 18.5 mm d^{-1}, occurred on the first day of the experiment. We do not know what a maximum transpiration rate of poplar might be. But the maximum that we observed in our experiment was less than half that of seedlings of *Eucalyptus* (39.9 mm d^{-1}), another hardwood species (see Table 13 in Doley[23]).

For the poplar roots, copper concentrations of 23.0, 35.7, 12.0, 12.6, 6.8, 8.4, 7.5, 11.4, and 9.1 μg g^{-1} occurred at the 0–100, 100–200, 200–300, 300–400, 400–500, 500–600, 600–700, 700–800, and 800–900 mm soil depths, respectively. The maximum concentration of copper found in plant tissue under normal conditions is 15 μg g^{-1}.[29] Thus, the copper concentration was elevated in the top 200 mm of the soil (average concentration of 29.4 μg g^{-1} in the roots), but not below 200 mm, where the average copper concentration in the roots was 9.7 μg g^{-1}. At the 100–200-mm depth, the concentration of copper in the roots was more than two times the normal maximum level. Roots took up a total of 0.0147 g copper. More copper might have been found in the roots had we analyzed the fine roots, which are the most active ones in uptake.

At harvest, the measured concentrations of copper in the soil decreased curvilinearly with depth down to a background level of 14 μg g^{-1}, which occurred below the 800-mm depth. Concentrations of 50.6, 34.5, 28.1, 23.1, 21.7, 19.6, 18.2, 16.4, 14.5, 13.5, 13.7, and 14.3 μg g^{-1} were measured at the 0–100, 100–200, 200–300, 300–400, 400–500, 500–600, 600–700, 700–800, 800–900, 900–1000, 1000–1100, and 1100–1200 mm soil depths, respectively. Even with the EDTA added, massive amounts of copper remained in the top 200 mm of the soil, which accounted for essentially all the copper added (6.4 g).

Because of adsorption onto the soil, copper should not move at the same speed as bromide. Given its strong exchange isotherm (discussed below), the added copper should have remained in the top 50 mm according to the model and using the adsorption isotherm for copper without EDTA. The distribution of copper with depth without EDTA, therefore, would have been discontinuous, with the surface 50-mm layer having a high concentration of copper and lower depths having background levels. However, the addition of the EDTA changed the distribution of copper in the profile. At the end of the experiment, copper decreased continuously and curvilinearly with depth. The measured and predicted concentrations of copper in the soil profile at the end of the experiment, with the background level subtracted, are shown in Figure 9.2.

In the model, we assumed that EDTA decreased the adsorption of copper by the soil. The solubilization was considered to result in a different isotherm, which expressed the weak adsorption of copper in the presence of EDTA. As noted above, a Langmuir equation was used to model copper without EDTA, and a Freundlich equation was used to model copper with EDTA. The isotherm of copper with EDTA is shown in the inset of Figure 9.2 as the solid line. The broken line shows the isotherm of copper without EDTA. The agreement between the predicted and measured values for copper distribution with depth at the end of the experiment was good. Although simulation via a changed isotherm might not be an apt mechanistic

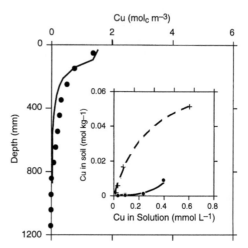

FIGURE 9.2 Measured and predicted concentrations of copper in soil at different depths in lysimeter with poplar tree at end of experiment. Inset shows relationship between adsorbed and dissolved concentrations of copper (adsorption isotherm) in soil with or without chelate EDTA. Crosses show measured data from batch experiment without EDTA, and broken line was fitted using Langmuir equation. Solid circles show measured data from batch experiment with EDTA, and solid line was fit using Freundlich equation.

description (cf., Reference 30), its simplicity and ease of parameterization mean it has some merit. As a first approximation to this process of chelation-induced mobility, it provides a reasonable rendition of our observations.

The breakthrough curves of bromide measured in the effluent from the lysimeters with the pasture grass and the poplar tree are shown in Figures 9.3a and 9.3b, respectively. The timing and magnitude of the peak concentrations of bromide under the two lysimeters differed. The peak concentration with the poplar tree occurred 3 d after that with the pasture grass. Water was applied at the same rate to both, so this difference was due to the higher water uptake of the poplar tree compared to the pasture grass, with averages of 9.2 and 1.3 mm d^{-1} (determined using the Penman–Monteith equation), respectively. The mm d^{-1} calculations are derived by dividing amount of water lost in liters by the surface area of the lysimeter. On a lysimeter basis, the transpiration was 5.9 L d^{-1} and 0.84 L d^{-1} for the poplar tree and pasture grass, respectively. The difference shows the impact of the vegetation on leaching of solutes beyond the root zone. Solutes below the root zone cannot be taken up and contaminate the groundwater. Thus, in the case of the pasture grass, most of the applied chemical would have leached beyond 1 m after 800 mm of rainfall. For the poplar tree, this would occur after 1000 mm of rainfall. Considering the different rooting depths occurring under field conditions, leaching beyond the root zone would occur much earlier under pasture compared to trees. Thus, contamination of groundwater (e.g., from nitrate) might occur more readily under pasture than under trees.

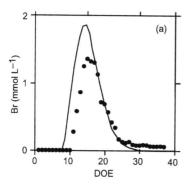
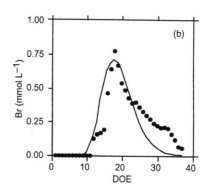

FIGURE 9.3 Measured (circles) and predicted (solid line) concentrations of bromide in leachates from lysimeters with pasture grass (a) and poplar tree (b) during experiment. DOE = day of experiment.

This emphasizes the role of land management and root zone processes in groundwater protection. The size of the water-drive is important in contaminant transport, and the magnitude of this is determined by the size of the plant-water extraction. Because the leaf area of the tree was much greater than the leaf area of the pasture grass, there was an additional 5.06 L d^{-1} available to leach through the root zone of the pasture grass (i.e., 5.9–0.84 L d^{-1}).

Bromide concentrations in the leachate were zero until DOE 12, when the pulse of bromide, added on DOE 3, started to appear in the drainage water. The zero concentration of bromide confirms that the Manawatu fine sandy loam contains no measurable amount of bromide. Worldwide, the normal range of bromide in soils is <0.5 to 515 mg kg^{-1}.[2] This soil, therefore, falls at the low end of the range with bromide concentration of <0.1 mg kg^{-1}. As expected, peak concentrations of bromide in the effluent from the lysimeter with the poplar tree were about half those in the effluent from the lysimeter with the grass, because we added 42 g KBr to the grass lysimeter and 21 g KBr to the poplar lysimeter. Figures 9.3a and 9.3b also show the predictions of the bromide concentrations using the model. The predictions are good. The assumption that all the water in this repacked soil was active in carrying the solute through the soil appeared to be valid.

The changes in bromide concentrations with time in the grass and poplar leaves are shown in Figures 9.4a and 9.4b, respectively. The concentrations first rise and then decrease with the leaching of the bromide beyond the root zone. Over the duration of the experiment, 8% of the applied bromide was taken up by the grass and only 2% by the poplar tree, which suggests exclusion of bromide by the poplar trees. Measurable amounts of bromide are found in most plant tissues and normally vary from trace concentrations to about 0.026% of the dry tissues. Our results confirm that plants differ in their ability to absorb the bromide ion.[30] Bromide has not been shown to be essential for growth, nor has it been regarded as particularly toxic. Residues following fumigation with chemicals containing bromide probably constitute the only cause of bromine toxicity.[31]

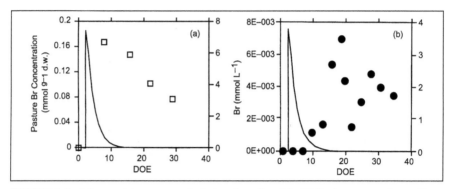

FIGURE 9.4 Measured (squares or circles) and predicted (solid line) concentrations of bromide in pasture grass (a) and poplar tree (b) during experiment. DOE = day of experiment.

Figures 9.4a and 9.4b also show the uptake concentrations predicted by our model. Substantial mismatches occur between our measurements and predictions. The measured concentrations, which were determined from leaves, lag behind the predicted ones, which the model calculates for root concentrations. Thus, the disagreement probably is due to a slow translocation of the bromide from the roots into the leaves. This remains an area of challenging and exciting research.

CONCLUSIONS

Our mechanistic model of water and solute movements was used successfully to describe the movement of copper, which was solubilized by EDTA. However, the simplified assumption that EDTA changes the isotherm might not always be valid. In the future, we plan to measure and model the movements of EDTA, copper, and other trace elements. The work would include treating the trace elements separately and linking their movements to that of the metal-complexed chelate. In addition, we also propose to study the movement of the free chelate itself because we need to understand the impact of EDTA on the soil environment.

The use of EDTA resulted in a small fraction of the applied copper being leached out of the base of the lysimeters, which mimics the arrival of copper in groundwater. Although EDTA solubilized the metal to make it plant-available, the impact of this chelation on mobilizing the contaminant cannot be overlooked. This poses some interesting questions about the use of solubilization reagents for phytoremediation of contaminated sites. We feel that phytoremediation strategies using chelating agents can be developed, but they must acknowledge the mobility of the metal-complexed chelate in the soil as well as the enhanced uptake that chelation induces.

Our mechanistic model of water and solute movements was also used successfully to describe the movement of a conservative tracer, bromide, through the soil. The experimental data and the model showed that leaching of chemicals like bromide (e.g., nitrate) beyond the root zone, with subsequent contamination of groundwater, would occur earlier under pasture than under trees.

ACKNOWLEDGMENTS

We thank Carlo van den Dijssel for help in dismantling the lysimeters, Jack D. Fry for identifying the pasture grass species, Suzanne Clark for the loan of the shaker, and David R. Scotter for helping to measure the total leaf area of the poplar tree and participating in discussions throughout the experiment. This is contribution no. 99-497-B of the Kansas Agricultural Experiment Station.

REFERENCES

1. Roberts, A.H.C., Cameron, K.C., Bolan, N.S., Ellis, H.K., and Hunt, S., Contaminants and the soil environment in New Zealand, in *Contaminants and the Soil Environment in the Australasia–Pacific Region,* Naidu, R., Kookana, R.S., Oliver, D.P., Rogers, S., and McLaughlin, M.J., Eds., Kluwer Academic Publishers, Dordrecht, The Netherlands, 579, 1996.
2. Kirkham, M.B., Trace elements, in *The Encyclopedia of Soil Science and Technology,* Finkl, C.W., Jr., Ed., Chapman and Hall, London, 2000 (in press).
3. Cunningham, S.D., Anderson, T.A., Schwab, A.P., and Hsu, F.C., Phytoremediation of soils contaminated with organic pollutants, *Advances in Agron.,* 56, 55, 1996.
4. Brooks, R.R., General introduction, in *Plants that Hyperaccumulate Heavy Metals,* Brooks, R.R., Ed., CAB International, Wallingford, Oxon, U.K., 1, 1998.
5. Iskandar, I.K. and Adriano, D.C., Remediation of soils contaminated with metals, in *Advances in Environmental Sciences,* Science Reviews, Northwood, England, 1997.
6. Schnoor, J.L., Licht, L.A., McCutcheon, S.C., Wolfe, N.L., and Carreira, L.H., Phytoremediation of organic and nutrient contaminants, *Environ. Sci. Tech.,* 29, 318A, 1995.
7. Watanabe, M.E., Phytoremediation on the brink of commercialization, *Environ. Sci. Tech.,* 31, 182, 1997.
8. Burken, J.G. and Schnoor, J.L., Predictive relationships for uptake of organic contaminants by hybrid poplar trees, *Env. Sci. Tech.,* 32, 3379, 1998.
9. Robinson, B.H., The phytoextraction of heavy metals from metalliferous soils, Ph.D. thesis presented to Massey University, Palmerston North, New Zealand, 1997.
10. Vogeler, I., Green, S.R., Clothier, B.E., Kirkham, M.B., and Robinson, B., Contaminant transport in the root zone, in *Best Soil Management Practices for Production, Proc. Workshop,* Currie, L.D., Hedley, M.J., Horne, D.J., and Loganathan, P., Eds., Occasional Report No. 12, Fertilizer and Lime Research Centre, Massey University, Palmerston North, New Zealand, 1999, 219.
11. Green, S.R., PSILAYER. *A User's Manual of the PSILAYER Model for Water and Solute Movement in Layered Soils and Uptake by Distributed Roots,* HortResearch Internal Report no. IR97-65, The Horticultural and Food Research Institute of New Zealand, Ltd., A Crown Research Institute, Palmerston North, New Zealand, 1997.
12. Vogeler, I., Green, S.R., Scotter, D.R., and Clothier, B.E., Measuring and modelling the transport and root uptake of chemicals in the unsaturated zone, *Plant and Soil,* 2000 (in press).
13. Gerwitz, A. and Page, E.R., An empirical mathematical model to describe plant-root systems. *J. Appl. Ecol.,* 11, 773, 1974.
14. Feddes, R.A., Bresler, E., and Neuman, S.P., Field test of a modified numerical model for water uptake by root systems, *Water Resources Res.,* 10, 1199, 1974.

15. Feddes, R.A., Kowalik, P., Kolinska–Malinka, K., and Zaradny, H., Simulation of field water uptake by plants using a soil water dependent root extraction function, *J. Hydrol.*, 31, 13, 1976.

16. Nielsen, D.R., van Genuchten, M.Th., and Biggar, J.W., Water flow and solute transport processes in the unsaturated zone, *Water Resources Res.*, 22 (no. 9, Supplement), S89, 1986.

17. van Genuchten, M.Th., A closed-form equation for predicting the hydraulic conductivity of unsaturated soils, *Soil Sci. Soc. America J.*, 44, 892, 1980.

18. Mualem, Y., A new model for predicting the hydraulic conductivity of unsaturated porous media, *Water Resources Res.*, 12, 513, 1976.

19. Gardner, W.R., Some steady state solutions of the unsaturated moisture flow equation with application to evaporation from a water table, *Soil Sci.*, 85, 228, 1958.

20. Brooks, R.H. and Corey, A.T., *Hydraulic Properties of Porous Media,* Hydrology Papers no. 3, Colorado State University, Fort Collins, Colorado, 1964.

21. Corey, A.T., Pore-size distribution, in *Indirect Methods for Estimating the Hydraulic Properties of Unsaturated Soils,* van Genuchten, M. Th., Leij, F.J., and Lund, L.J., Eds., University of California, Riverside, CA, 37, 1992.

22. Vogeler, I., Clothier, B.E., Green, S.R., Scotter, D.R., and Tillman, R.W., Characterizing water and solute movement by time domain reflectometry and disk permeametry, *Soil Sci. Soc. America J.*, 60, 5, 1996.

23. Vogeler, I., Clothier, B.E., and Green, S.R., TDR estimation of the resident concentration of electrolyte in the soil solution, *Australian J. Soil Res.*, 35, 515, 1997.

24. Green, S.R. and B.E. Clothier, Water use of kiwifruit vines and apple trees by the heat pulse technique, *J. Exp. Bot.*, 39, 115, 1988.

25. Green, S.R., McNaughton, K.G., Greer, D.H., and McLeod, D.J., Measurement of the increased PAR and net all-wave radiation absorption by an apple tree caused by applying a reflective ground covering, *Agri. Forest Meteorol.*, 76, 163, 1995.

26. Moreno, F., Fernández, J.E., Clothier, B.E., and Green, S.R., Transpiration and root water uptake by olive trees, *Plant and Soil,* 184, 85, 1996.

27. Robinson, B., Mills, T., Petit, D., Fung, L., Green, S., and Clothier, B., Phytoremediation of cadmium-contaminated soils using poplar and willow, in *Function and Management of Plants in Land Treatment Systems,* Proceedings of the Technical Session 19, New Zealand Land Treatment Collective, Christchurch, New Zealand, 19–20 April 1999, Wang, H., and Tomer, M., Eds., Forest Research, Rotorua, New Zealand, 3, 1999.

28. Doley, D., Tropical and subtropical forests and woodlands, in *Water Deficits and Plant Growth. Volume 6. Woody Plant Communities,* Kozlowski, T.T., Ed., Academic Press, New York, 209, 1981.

29. Kirkham, M.B., Trace elements in sludge on land: Effect on plants, soils, and ground water, in *Land as a Waste Management Alternative,* Loehr, R.C., Ed., Ann Arbor Science, Ann Arbor, MI, 209, 1977.

30. Kedziorek, M.A.M., Dupuy, A., Bourg, A.C.M., and Compere, B., Leaching of Cd and Pb from a polluted soil during the percolation of EDTA: Laboratory column experiments modeled with a non-equilibrium solubilization step, *Environ. Sci. Tech.*, 32, 1609, 1998.

31. Martin, J.P., Bromine, in *Diagnostic Criteria for Plants and Soils,* Chapman, H.D., Ed., Quality Printing Co., Inc., Abilene, TX, 62, 1973.

APPENDIX

Data file for PSI-1D.FOR program.

1260.0	!	domain-len:	domain length (mm)
21	!	n-intervals :	number of r-grid points < 100
3	!	n-layers :	number of layers in the soil
7 7 7	!	n-compartments for the i-th layer	
2	!	orientation :	$< 1 >$ = horizontal; ,2 $>$ = vertical

. . . Set the soils hydraulic properties

. . . Input max and min soil-water contents [m3/m3] for each layer
---

0.525 0.000	!	és, én for the 1st layer
0.525 0.000	!	
0.525 0.000	!	és én for the n-th layer

. . . Select the hydraulic conductivity model K(h) . . .

. . .	1=Brooks & Corey	:	$K(Ks,B)=Ks. [(é-én)/(és-én)]^B$ $B=(5n + 4)/n$
. . .	2=van Genuchten	:	$K(Ks,*)=Ks.Se .5(1-(1-Se^ 1/m))^m)^2$
. . .	3=Gardner (1958)	:	$K(Ks,B)=Ks.exp(Bh)$
. . .	4=Gardner (1960)	:	$K(Ks,B)=Ks.(é/ém)^B$

2.0	!	Select K(h) model
2.25e-2 0.00836	!	K-sat, á parameters for the 1st layer
2.25e-2 0.00836	!	
2.25e-2 0.00836	!	K-sat, á parameters for the nth layer

. . . Select the water retention model é(h) . . .

. . .	1=Brooks & Corey	:	$é(à, n, *)=én + (és-én)/(àh)^n$
. . .	2=van Genuchten	:	$é(à, n, m)=én + (és-én)/[1+(àh)^n]^m$

2.0	!	
0.00162 1.4027 0.2871	!	à, n, m parameters for the 1st layer
0.00162 1.4027 0.2871	!	
0.00162 1.4027 0.2871	!	à, n, m parameters for the n-th layer

---
. . . Set up the solute transport model

1	!	Solve for S?	$<1>$ = solve S, $<0>$ = dont solve S
10.0	!	Smax	used for plotting only
1.0E-8	!	Input chemical diffusion coefficient in pure water [mm2/s]	
63.546	!	Input the gram molecular weight (g/mol)	
0.567	!	ground surface area [m2]	

. . . Parameters rho-b, Kd-mod, m, n, and f for each layer
. . . Model 1 = Freundlich   S = $Kd.C^n$ (if n=1.0 then isotherm is linear)
. . . Model 2 = Langmuir   S = $a.b.C/(1+bC)$

1200 1.000 0.124 3.08 1.00	!	rho-b, Kd-mod, Kd or a, n or b, f for 1st layer
1200 1.000 0.124 3.08 1.00		
1200 1.000 0.124 3.08 1.00	!	rho-b, Kd-mod, Kd or a, n or b, f for nth layer

---
. . . Input the mobile fraction [ém/é], the dispersivity [lambda, mm]
. . . and the mass tx. coeff [s-1] for each layer. mass tx coeff. Note: the
. . . mass tx. coeff is normally 1e-9 to 1e-6 [s-1]

```
1.00 50.0 0.0E-5 0.0e-5      !    ém/é, lambda [mm], à-coefficient, æ-decay [s-1]
1.00 50.0 0.0E-5 0.0e-5      !    ém/é, lambda [mm], à-coefficient, æ decay [s-1]
1.00 50.0 0.0E-5 0.0e-5      !    ém/é, lambda [mm], à-coefficient, æ-decay [s-1]
```

```
 . . . specify root uptake properties
```

```
1                            !    Do you solve for root uptake: <1>=yes, <0>=no
```

```
 . . . Assume root-length density L declines exponentially with depth
 . . . Let max root depth Zr change with time Zr = Zri + (Zrf-Zri)*t/tg
```

```
3.5                          !    Input ç parameter for L = Lm.exp(-çz/zR)
600 600 1                    !    Input parameters ZRi, ZRf, tg
```

```
 . . . Select model for root water uptake
 . . . We assume that U(é) = f1(L).f2(soil) such that ET = int[U(é,z)dz]
 . . . 1 > U(éw,éd,P)  [=1.0 if é>éd]   [=0.0 if é<éw]  ELSE  [=[(é-éw)/(és-én)]^P
 . . . 2 > U(hw,hd,P)  [=1.0 if h>hd]   [=0.0 if h<hw]  ELSE  [=[(h-hw)/(hs-hn)]^P
 . . . 3 > U(*,*,*)   = ET.RLD.K(h)/[int RLD.K(h).dz]
 . . . 4 > U(*,*,*)   = ET.RLD.D(é)/[int RLD.D(é).dz]
```

```
2.0                          !    Water uptake model: 1 = é, 2 = í, 3 = K(í), 4 = D(í)
-1E4 -1E5 1.0    !           Reduction parameters
```

```
 . . . Select a model for solute uptake
 . . . 1 > S(*,C1,*,*) = U(é) times min[C,C1]
 . . .       passive uptake: C1=large; active uptake:  C1=small
 . . . 2 > S(Kc,C1,*,*) = [Kc.C(z)/(C1+C(z)).U(é)] Mikaelis-Menten type
```

```
1.0                          !    Solute uptake model: 1=passive, 2=active
1.0                          !    Ke efficiency [g/mm]  S-tot = Ke times ET-tot
1.0 1000 0.0 0.0  !          Reduction parameters, Kc, C1 (rest pars redundant)
```

```
 . . . Set the initial water and solute concentrations
```

```
0.397    0.343    1.0E-9  0.0    !   ém or hm; Cm,1; Ci,1
0.397    1.0E-9   1.0E-9  0.0
0.395    1.0E-9   1.0E-9  0.0    !!!   Enter -ve values for hm
0.394    1.0E-9   1.0E-9  0.0
0.394    1.0E-9   1.0E-9  0.0
0.393    1.0E-9   1.0E-9  0.0
0.393    1.0E-9   1.0E-9  0.0
0.393    1.0E-9   1.0E-9  0.0
0.393    1.0E-9   1.0E-9  0.0
0.393    1.0E-9   1.0E-9  0.0
0.393    1.0E-9   1.0E-9  0.0
0.393    1.0E-9   1.0E-9  0.0
0.393    1.0E-9   1.0E-9  0.0
0.394    1.0E-9   1.0E-9  0.0
0.394    1.0E-9   1.0E-9  0.0
0.394    1.0E-9   1.0E-9  0.0
0.394    1.0E-9   1.0E-9  0.0
0.394    1.0E-9   1.0E-9  0.0
0.394    1.0E-9   1.0E-9  0.0
0.394    1.0E-9   1.0E-9  0.0
0.394    1.0E-9   1.0E-9  0.0    !   ém,n Cm,n Ci,n    :    é initial from i=1,n
```

```
------------------------------------------------------------------------
. . . ADD IN THE TIME OFFSET (D)
0.0
------------------------------------------------------------------------
. . . Set the UPPER  b.c.  <1>=flux;  <2>=conc. Note: Vo, Tr, E & D [mm/d]
. . . Set the LOWER b.c.  <1>=flux;  <2>=free draining
```

ubc	Vo,éo	Co	Tr	E	lbc	D	Day	Hr	Min
1	38.8	0	14.02	1.84	2	0	1	12	0
1	38.8	0	7.61	0.79	2	0	2	12	0
1	38.8	0	6.82	1.00	2	0	3	12	0
1	38.8	0	9.89	1.45	2	0	4	12	0
1	38.8	0	11.97	1.69	2	0	5	12	0
1	38.8	0	9.31	1.22	2	0	6	12	0
1	38.8	0	8.40	1.16	2	0	7	12	0
1	38.8	0	3.89	0.58	2	0	8	12	0
1	38.8	0	8.63	1.13	2	0	9	12	0
1	38.8	0	6.64	0.89	2	0	10	12	0
1	38.8	0	6.35	0.93	2	0	11	12	0
1	38.8	0	8.99	1.36	2	0	12	12	0
1	38.8	0	8.60	1.27	2	0	13	12	0
1	38.8	0	8.50	1.22	2	0	14	12	0
1	38.8	0	8.49	1.34	2	0	15	12	0
1	38.8	0	2.48	0.26	2	0	16	12	0
1	38.8	0	7.18	0.98	2	0	17	12	0
1	38.8	0	6.96	0.92	2	0	18	12	0
1	38.8	0	2.03	0.33	2	0	19	12	0
1	38.8	0	7.05	1.04	2	0	20	12	0
1	0.00	0	7.14	1.02	2	0	21	12	0
1	0.00	0	3.62	0.43	2	0	22	12	0
0	0	0	0	0	0	0	0	0	0

10 Partitioning and Reaction Kinetics of Cd-109 and Zn-65 in an Alum Shale Soil as Influenced by Organic Matter at Different Temperatures

Åsgeir Almås and Bal Ram Singh

ABSTRACT

An alum shale soil, developed on sulfide-bearing rocks, was treated with organic matter at the rate of 0 and 4% and was placed in temperature-controlled climate chambers at 9 and 21°C to study the partitioning and reaction kinetics of [109]Cd and [65]Zn. Soil subsamples, collected at time intervals ranging from 0.5 to 8760 h (1 yr) after spiking the soil with [109]Cd and [65]Zn, were subjected to a seven-step sequential extraction procedure. Soil adsorption of [109]Cd and [65]Zn was rapid, but the subsequent transfer of the metals toward irreversibly sorbed fractions was a much slower process. The [109]Cd and [65]Zn concentrations decreased in fractions F1 to F3 and increased in fractions F5 and F6 with time. The concentrations of [109]Cd and [65]Zn determined in the seven extracts were used in a three-component soil-water kinetic model to calculate the kinetics of [109]Cd and [65]Zn partitioning between the water-soluble, reversibly sorbed, and irreversibly sorbed fractions. The addition of organic matter reduced the rate of [109]Cd and [65]Zn transfer into the irreversibly sorbed fraction, and the potential mobility of [109]Cd and [65]Zn were thus increased by the addition of organic matter. Increased temperature, on the other hand, enhanced the rate of transfer and the potential mobility was reduced. The interaction between organic matter and temperature affected both the rates of transfer and the pseudo equilibrium constants.

INTRODUCTION

The bioavailability of trace metals in the complex and dynamic soil–plant–water system changes with time and shifting environmental conditions. The alum shale soil used in this study was weathered from sulfide-bearing rocks formed in the latter part

1-56670-507-X/01/$0.00+$.50

of the Cambrian era. Such soils are naturally rich in a number of trace metals due to the geochemical composition of soil.[1] Among the trace metals found, Cd and Zn are the dominant ones from the point of view of their environmental effects. Despite the limited regional distribution of these soils in Norway, they are located in important agricultural districts around Oslo and in the Mjøsa region in Hedmark County with a high production potential.

When the mobility and biological uptake of trace metals from soils are studied, chemical or biological methods are used to quantify the fraction of soil contaminants available for biological uptake. According to Pickering,[2] multistep extraction procedures provide more detailed information about the status of trace metals in soils than do single-extraction methods. In addition, the early stages in the extraction sequence provide information comparable to that obtained from single-extraction methods, so the initial fractions can be used to assess the potential short-term biological uptake of soil metals. A complementary use of partitioning schemes, where the same soil is repeatedly sampled and studied using the same method over a period of time, can provide information on the relative variation in elemental phases over time, irrespective of the method applied. Such information can be used for predictive modeling in systems where the soil–plant pathway is a key contributor to potentially harmful effects to plants and animals. All information on processes regulating plant uptake and leaching of trace metals to groundwater are relevant to risk assessment.

Organic substances, in addition to mineral surfaces, are responsible for the dissolution and retention of metals in the soil–water system. Increased concentrations of soil organic matter increase the retention of trace metals in soil. However, at higher soil pH levels, dissolved organics can increase the solubility of metal ions by the formation of soluble organo-metallic complexes, which compete with the solid phase for the metal ions.[3–5] Naidu and Harter[6] reported that the release of Cd into soil solution as metal–organic complexes was important when soil pH > 5.5. The extent of release, however, varied considerably with the Cd-complex-forming ligand. Treating organic matter as a single variable may complicate the role of soil organic matter, because the composition and nature may vary considerably by origin. For instance, fresh organic matter is chemically different as compared with residual humus in soils. At higher soil pH, soluble organic matter (e.g., fulvic acid) can suppress trace metal adsorption on silicates and oxides and promote dissolution of trace metals from adsorption sites on clay minerals.[7] Because the metal reaction with a ligand is endothermic,[8,9] formation of metal complexes with organic and inorganic ligands is generally favored at higher temperatures. The relationship between temperature and reaction rate, known as the Arrhenius equation, postulates that the reaction rate increases with temperature. Hence, the rate and extent of metal sorption in soil increases with increasing temperature.[10] The trace metal reaction with soil particles has previously been reported to increase with temperature,[10,11] and, according to Barrow[10] and Bergseth,[12] temperature also influences the metal affinity for surfaces.

It has previously been reported that diffusion is involved in the reaction of metals with soil components.[10,13–15] It is suggested that the rapid surface adsorption reactions are followed by a subsequent but slow transfer of the metals into soil fractions with a tendency toward irreversibility. An adequate model designed to estimate the partitioning rates of trace metal interaction with soil can provide valuable estimations on the mobility and retention of trace metals in the soil system, but these calculations do not necessarily provide a unique identification of the reaction mechanism.

This chapter is based on a study issued in two publications.[3,16] Basically, the study aimed to (1) investigate the geochemical partitioning of ^{109}Cd and ^{65}Zn over time, (2) use results from sequential extraction analysis to study the exchange kinetics of ^{109}Cd and ^{65}Zn partitioning in the soil by setting up a simple three-component kinetic model, and (3) investigate the effect of temperature and the addition of organic matter on the rate of ^{109}Cd and ^{65}Zn sorption and partitioning into irreversibly sorbed soil fractions.

MATERIALS AND METHODS

SOIL AND SOIL TREATMENT

The soil we used for these investigations (a clay loam) was collected from the A horizon of an alum shale soil located at an agricultural school farm in southeastern Norway. Alum shales are sulfur-bearing rocks naturally rich in a number of trace metals and radionuclides. Detailed descriptions of the mineralogical and chemical properties of this soil are presented by Jeng and Bergseth[17] and Singh et al.[18] The most important properties of the soil are presented in Table 10.1. The soil was either untreated or mixed with organic matter at the rate of 4%. The organic matter source was fresh pig manure (*Sus scrofa,* "Norwegian Landrace"); the chemical composition is also shown in Table 10.1. The soil and organic material mixture was spiked with a tracer solution(from DUPONT, Belgium) of ^{109}Cd^{2+} and ^{65}Zn^{2+} (specific activity: 1.24×10^8 and 1.45×10^8 Bq mg^{-1}, respectively) diluted with MilliQ-water to 25 mL, mixed thoroughly with 2.2 L soil (soil volume in pots), and moistened immediately to 60% of field capacity. The spiked solution was acidified to pH 2 to prevent adsorption of the tracers to the equipment surfaces, but the acid spike did not change the soil pH noticeably. Tracers of Cd and Zn (^{109}Cd and ^{65}Zn) were used to separate the metals added from those naturally existing. The method is very precise, and the gamma emission from the radioactive metals enables detection of extremely low metal concentrations.[19] The specific activities of ^{109}Cd and ^{65}Zn in soil were estimated to be 500 and 600 Bq g^{-1} soil, respectively. This soil was utilized for a pot experiment conducted in temperature-controlled climate chambers (phytotrons) at 9 and 21°C, and the moisture content of the soil in pots was maintained at 60% of the field capacity throughout the experimental period.

SOIL SAMPLING AND SEQUENTIAL EXTRACTIONS

The soil samples were collected from the pots in a moist state to be as much as possible in agreement with field conditions.[20] About 3 g moist subsamples of soil from the pot experiment at 9 and 21°C were collected from the phytotron for sequential extraction after 0.5, 1, 3, 24, ≈ 168 (7 d), ≈ 720 (1 mo) and ≈ 8760 h (1 yr) contact time between ^{109}Cd and ^{65}Zn and the soil used. The data presented are corrected to dry weight soil (oven dried at 105°C). The solid-phase fractionation of ^{109}Cd and ^{65}Zn was

TABLE 10.1

Selected soil properties and the composition of organic matter added

Parameters		Soil[a] Alum shale Typic Cryoboroll	Organic Matter[b] Pig manure Sus scrofa
pH	H_2O, 1:2.5	6.5	
TOC	%	5.0	
CEC	cmol kg^{-1}	44.0	
C	%		44.6
N	%		2.8
Fe-acid-oxalate	%	0.8	
Fe-dithionate-citrate	%	3.3	
Grain size distribution	Sand	30	
(%)	Silt	41	
	Clay	29	
Total metal	Fe		659.0
concentrations	Mn		81.0
(mg kg^{-1} d.w.)	Cd	1.9	0.1
	Cu	97.0	15.0
	Ni	115.0	4.2
	Zn	169.0	158.0

a From Almås et al.[3]

b From Narwal and Singh[31]

done according to the procedure of Salbu et al.,[21] which is a modified version of that of Tessier et al.[22] In this procedure,[21] the reagents were chosen to differentiate between binding mechanisms:

Reversible physical sorption by providing H_2O (F1), followed by 1 M NH_4OAc at soil pH 7 (F2).
Reversible electrostatic sorption using the pH effect by extracting with 1 M NH_4OAc at pH 5 (pH lower than soil pH) (F3).
Irreversible chemisorption by using redox systems, which was provided sequentially by hydroxylamine (F4),
H_2O_2 in 1 M HNO_3 (F5),
7 M HNO_3 (F6).

The detailed extraction procedure is presented in Table 10.2. For slowly reversible electrostatic sorption, time is needed to dissolve metals. Due to the experimental conditions (limited time of extraction), this fraction is also included when redox reagents are applied.[23] It is expected that ^{109}Cd and ^{65}Zn will be released from soil oxides in F4, and that in F5 organic matter will be oxidized so the ^{109}Cd and ^{65}Zn complexed

TABLE 10.2
Summary of the sequential extraction steps for ^{109}Cd and ^{65}Zn fractionation (adapted from Salbu et al.[21])

Fraction	Extracts	Contact Time (h)	Solution Volume (ml)	Treatment	Temp.
F1	H2O, pH 5.5	1	20	Rolling table	20°C
F2	1-M NH4OAc, pH 7	2	20	Rolling table	20°C
F3	1-M NH4OAc, pH 5	2	20	Rolling table	20°C
F4	0.04-M NH2OH • HCl in 25% HAc	6	20	Warm water	80°C
F5	30% H_2O_2 at pH 2 (HNO$_3$),	5.5	15	Warm water	80°C
	added 3.2 M NH$_4$OAc in 20% HNO$_3$	0.5	5	Rolling table	20°C
F6	7-M HNO$_3$	6	20	Warm water	80°C
F7	Residue				

with these organic ligands will be released. Overlapping of these fractions will occur, however. All extractions were carried out by shaking the samples in 50-mL centrifuge tubes with 20 mL of each extract in the sequence shown in Table 10.2.

The solid phases were shaken and washed with 10 mL MilliQ-water. The supernatant was separated from the solid by high-speed centrifugation for 30 min at $11000 \times g$ and evacuated for analysis. The concentrations of ^{109}Cd and ^{65}Zn recovered in the washing step were added to the extracts, from the step prior to washing. A Packard Minaxi 3-in. through-hole NaI Gamma Counter, 5000 Series, was used to determine the activity of ^{109}Cd and ^{65}Zn in the extracts and in the residue left after the 7-M HNO$_3$ extraction step. The stabile Cd and Zn concentrations were also determined in the same extracts by using a Perkin–Elmer graphite furnace Atomic Absorption Spectrophometer and a Jarrell Ash ICP, respectively. All funnels and containers used were of polycarbonate, and all the equipment was soaked in 10% HCl for days and thoroughly rinsed in MilliQ-water prior to use. A Shimadzu TOC-5000 analyzer was used for the analysis of total organic carbon (TOC) in the water extracts.

MODELING THE RATE OF METAL EXCHANGE BETWEEN FRACTIONS

To study the ^{109}Cd and ^{65}Zn transformation rate in this soil with increasing contact time, the results from the sequential extractions were used in a three-component soil–water kinetic model (Eq. 10.1). The calculations, performed by ModelMaker 2.0,[24] were based on the relative distribution among the three components. Since plant uptake of metal ions occurs through soil solution, the water-soluble fraction (F1) made of the first component. Reversibly sorbed metals recovered in the extracts from steps F2 and F3 in the sequence made up the second component of the model (reversibly sorbed). The apparent irreversibly sorbed fractions (F4, F5, F6, and F7), from which sorbed metals are released by using redox agents, made up the third component of the model (irreversibly sorbed). Due to the experimental conditions (limited time of extraction), this model component also includes the slowly reversibly sorbed metal fractions. It is expected that mechanisms of fixation can include chemisorption, (co)precipitation, and solid-state diffusion of metals or a combination thereof. The effect of organic matter and temperature on the exchange kinetics of ^{109}Cd and ^{65}Zn among the three components was thus studied in this model. Ideally, the initial concentrations used in the model should have been 0 for the reversibly sorbed and the irreversibly sorbed components, and 100 for the water component at the moment of spiking. Therefore, maximum adsorption of ^{109}Cd and ^{65}Zn was nearly completed before the first withdrawal, the adsorption rate could not be calculated satisfactorily, and the initial values used in the calculations refer therefore to those obtained after the first withdrawal (0.5 h). The results from the calculations are therefore valid only for the time span of 0.5–8760 h. As outlined below, the apparent rate constants k_1, k_2, k_3, and k_4, calculated by ModelMaker 2.0,[24] can be used to define the time-dependent distribution coefficients K_r and K_i (L kg^{-1}) between the phases. Rates of metal transfer between the phases followed a first-order reaction kinetics and, according to Børretzen,[25] Eqs. 10.2, 10.3, and 10.4 can be used to describe the metal flux between the three components (N_w, N_r, and N_i) in Eq. 10.1 below.

$$N_w \underset{k_2}{\overset{k_1}{\rightleftarrows}} N_r \underset{k_4}{\overset{k_3}{\rightleftarrows}} N_i \qquad (10.1)$$

$$\frac{dN_w}{dt} = -k_1 \bullet N_w + k_2 N_r \qquad (10.2)$$

$$\frac{dN_r}{dt} = k_1 \bullet N_w - k_2 N_r - k_3 \bullet N_r + k_4 \bullet N_i \qquad (10.3)$$

$$\frac{dN_i}{dt} = k_3 \cdot N_r - k_4 \cdot N_i \qquad (10.4)$$

where t is the time (h) and N_w, N_r, and N_i denote the relative metal concentrations in the water, reversibly sorbed, and irreversibly sorbed fractions, respectively. The k_1, k_2, k_3, and k_4 factors are apparent rate constants (h^{-1}) and describe the fluxes between components in the model. The exchange reactions are at equilibrium when $dN_w/dt = 0$, $dN_r/dt = 0$, and $dN_i/dt = 0$, and the pseudo equilibrium constants K_r and K_i (L kg^{-1}) can then be expressed in Eqs. 10.5 and 10.6, respectively, by adjusting the k_1/k_2 and k_3/k_4 ratios for solution volume (V) and weight of soil (m). The V/m ratio in this investigation was ≈ 8.

$$K_r = \frac{N_r}{N_w} \bullet \frac{V}{m} = \frac{k_1}{k_2} \bullet \frac{V}{m} \qquad (10.5)$$

$$K_i = \frac{N_i}{N_r} \bullet \frac{V}{m} = \frac{k_3}{k_4} \bullet \frac{V}{m} \qquad (10.6)$$

Because the soil environmental conditions influence all sorption reactions, the calculated values may be regarded as typical for the soil investigated.

Statistical Analysis and Graphical Presentations

The experiment was designed as a two-factorial design with two levels of temperature (9 and 21°C) and two levels of organic matter addition (0 and 4%). The observations were hence treated as a 2^2 factorial design with two levels for each of the factors (high and low) and three replicates of each factor combination. Statistical analysis was performed using JMP 3.1[26] for Windows, and the level of significance is 95% if not otherwise specified in the text. Graphical presentations and curve fitting were done by Origin 5.0,[27] and the rate constants were calculated by SB ModelMaker 2.0.[24]

RESULTS AND DISCUSSION

Transformation of ^{109}Cd and ^{65}Zn within Soil Fractions with Time

The relative distribution shows that the ^{109}Cd and ^{65}Zn concentrations decreased in the water and the reversibly sorbed fractions and increased in the irreversibly sorbed fractions with time (Figure 10.1). Approximately 30% and 20% of ^{109}Cd and ^{65}Zn, respectively, were reversibly sorbed after 0.5 h of contact time. However, at the end of the experimental period (1 yr), these values were reduced to 20% and 10%, respectively. The largest amount of solid-phase ^{109}Cd and ^{65}Zn was recovered in the F4 fraction (40–55% of the total), but the slow accumulation of ^{109}Cd and ^{65}Zn occurred mainly in the F5 fraction. Zinc-65 also accumulated in the F6 fraction with time. Metals associated with exchangeable sites (reversibly sorbed) at organic and inorganic colloids are expected to be extracted by the weaker extracting reagents, as used in F1–F3.[3] Christensen[28] previously reported that the sorption equilibrium for Cd was attained for 95% of the total Cd within 10 min of contact time with soil in 10^{-3}-M $CaCl_2$ solution, and from a study where ^{109}Cd was spiked to a loam soil, Riise and her colleagues[29] reported that the ^{109}Cd distribution in soil did not significantly change beyond 1 h. In their report, however, it was shown that the amount of ^{109}Cd decreased slightly in the 1-M NH_4OAc extract and increased slightly in the H_2O_2 extract with time. Other investigations of Cd and Zn sorption[13] reported that solid diffusion resulted in increased mineral fixation of Cd and Zn (among other metals) with time.[14]

Rate of Reaction

The rates of ^{109}Cd and ^{65}Zn transfer among the three components in the model (Eq. 10.1) were adequately described by Eqs. 10.2, 10.3, and 10.4 (p < 0.001). The ^{109}Cd and ^{65}Zn reaction kinetics in this soil have been described in more detail in Almås et al.[16] The reversible sorption was nearly completed before the first withdrawal, and the respective kinetic parameters (k_1, k_2, and K_r) are thus discussed only briefly in this chapter. The k_1 and k_2 rates reflect, however, the effect of temperature and the addition of organic matter on ^{109}Cd and ^{65}Zn sorption. The loss of reversibly sorbed ^{109}Cd and ^{65}Zn being partitioned by the irreversibly sorbed fractions (Figure 10.2) was described satisfactorily. All the kinetic constants are presented in

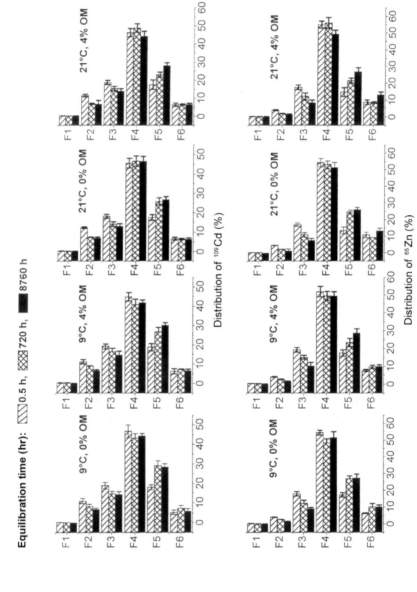

FIGURE 10.1 Distribution of ^{109}Cd and ^{65}Zn among fractions F1–F6 is shown for soil untreated and treated with 4% (w/w) organic matter, with increasing contact time, at 9 and 21°C. Bars represent mean of n = 3. (after Almås et al.[3]).

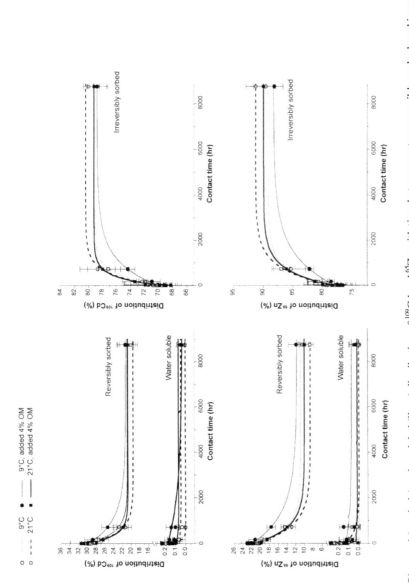

FIGURE 10.2   Observed (symbols) and modeled (lines) distribution of [109]Cd and [65]Zn with time between water, reversibly sorbed, and irreversibly sorbed fractions as affected by temperature (9 and 21°C) and addition of organic matter (4% OM). Each symbol represents mean of n = 3 (after Almås et al.[16]

TABLE 10.3
Effect of temperature and addition of organic matter on apparent rate constants (k_1–k_4) and pseudoequilibrium constants (K_r and K_i) for ^{109}Cd and ^{65}Zn interactions in soil–water system model (from Almås et al.;[16] reproduced by permission of Soil Science Society of America)

		^{109}Cd				^{65}Zn			
Temperature(°C)		9		21		9		21	
Organic Matter Addition (%)		0	4	0	4	0	4	0	4
Rates (h^{-1})	k_1	1.8×10^{-2}	1.6×10^{-2}	5.9×10^{-2}	1.2×10^{-1}	2.0×10^{-2}	1.7×10^{-1}	3.7×10^{-2}	6.8×10^{-2}
	k_2	5.6×10^{-5}	9.7×10^{-5}	2.6×10^{-4}	5.3×10^{-4}	1.1×10^{-4}	1.0×10^{-3}	1.8×10^{-4}	2.5×10^{-4}
	k_3	1.7×10^{-3}	7.7×10^{-4}	1.7×10^{-3}	2.2×10^{-3}	1.5×10^{-3}	6.5×10^{-4}	1.3×10^{-3}	1.4×10^{-3}
	k_4	4.5×10^{-4}	2.1×10^{-4}	4.1×10^{-4}	5.8×10^{-4}	1.8×10^{-4}	8.4×10^{-5}	1.2×10^{-4}	1.5×10^{-4}
Pseudo-	K_r	2571	1320	1815	1811	1454	1360	1644	2176
equilibrium	K_i	30	29	33	30	67	62	87	75
constants ($L\ kg^{-1}$)	r^2	0.95	0.98	0.97	0.96	0.90	0.94	0.99	0.98

r^2 indicates model fitness to variation of observations ($p < 0.001$).

Table 10.3, and the observed and calculated distribution of ^{109}Cd and ^{65}Zn with time among the three components in the model are shown in Figure 10.2. The rate constants for ^{109}Cd and ^{65}Zn were tested statistically in a 2^2 factorial design, and both organic matter and temperature significantly affected the rates. Since the interaction effect of temperature and organic matter was significant, the effect of organic matter application was dependent on temperature. Although the effect of organic matter appeared to be more important for ^{65}Zn than for ^{109}Cd, the effect of temperature was found to be the most important effect for both.

EFFECT OF ORGANIC MATTER

Both ^{109}Cd and ^{65}Zn were increasingly sorbed with time, but since the rate of transfer into the irreversibly sorbed fractions was reduced by the addition of organic matter, the addition of organic matter increased the potential mobility of ^{109}Cd and ^{65}Zn (Figure 10.2). The concentrations of ^{109}Cd and ^{65}Zn were higher in the F1–F3 extracts withdrawn from the soil treated with organic matter when compared with the untreated soil. Organic matter significantly affected the sorption process at both temperature levels. Because the organic matter-induced reduction of the forward rate (k_3) was counteracted by the reverse reaction (k_4; recall Eqs. 10.5 and 10.6), the shift in K_i was hardly detectable, and the K_i decrease of only 2–7% was not significant. The decrease in K_i was, on the other hand, significant for ^{65}Zn, both at 9 and 21°C. Addition of organic matter may thus have increased the potential mobility of the metals. Analysis of the water extracts showed that the concentrations of both ^{109}Cd and ^{65}Zn and the total organic carbon (TOC) were significantly higher in the extracts withdrawn from the soil treated with organic matter. This was found to be significant during the whole experimental period (paired t-test), and dissolved organic acids may thus have increased the concentration of ^{109}Cd and ^{65}Zn by formation of dissolved metal-organic complexes.

EFFECT OF TEMPERATURE

Although the k_3 values either remained the same or decreased with increasing temperature, the K_i value increased because the rate coefficient that describes the reverse reaction (k_4) decreased even more (Table 10.3). Paired t-tests were conducted on observations from the different fractions, testing the response from different temperature treatment levels. The test showed that the concentrations of both ^{109}Cd and ^{65}Zn were significantly lower in the F1 and F3 fractions and higher in the F4, F5, and F6 fractions at 21°C compared with those at 9°C. The rate of ^{109}Cd and ^{65}Zn transfer into the irreversibly sorbed fractions (k_3) was thus increased with temperature. Irrespective of the organic matter levels, the increased temperature also significantly shifted the K_i values to higher levels for both metals (Table 10.3). This resulted in increased concentrations of ^{109}Cd and ^{65}Zn in the irreversibly sorbed fractions at 21°C and decreased concentrations of reversibly sorbed metals and those in the water fractions (Figure 10.2). Other investigators have reported similar observations for metal reaction with soils[13] and oxides.[14,30] It also appears from the K_i values that the effect of temperature was more pronounced for ^{65}Zn than for ^{109}Cd. The temperature-induced increase in the K_i values was in the range of 9 and 3% for ^{109}Cd, and 23 and 19% for ^{65}Zn in the untreated and in the organic-matter-treated soils, respectively. Bergseth[12] showed that increasing temperature not only increased the sorption of Cd^{2+} and Zn^{2+} by vermiculite, but it also increased the selectivity for Zn^{2+} over Cd^{2+}. As discussed previously, the addition of organic matter increased the potential mobility of the metals, but the effect was found to be more pronounced at 9 than at 21°C; the effect of organic matter was therefore more effective at the lower temperature. Since the concentration of TOC was also found to be significantly higher at 9 than at 21°C in the same period, it is also possible that the aggregation of suspended metal–organic complexes has increased with temperature, which would subsequently increase the solid-phase fraction of ^{109}Cd and ^{65}Zn. Complexation is known to be more important for metal retention in soil than exchange sites on organic matter,[31] so it is also likely that as temperature increased, solid organic and inorganic matter increasingly sorbed labile metal ions previously associated with dissolved organics.

CONCLUSIONS

In the present study, a soil was spiked with ^{109}Cd and ^{65}Zn to determine the kinetics of reactions affecting the geochemical partitioning of trace metals with time as influenced by a temperature regime and the addition of organic matter. The initial sorption of ^{109}Cd and ^{65}Zn was nearly completed before the first withdrawal (after 0.5 h), but

[109]Cd and [65]Zn were transferred from the reversibly sorbed into the irreversibly sorbed fractions by a much slower process. On one hand, increasing temperature facilitated the forward transfer rate (k_3) and, because this lowered the K_i, the potential mobility of [109]Cd and [65]Zn was thus reduced at the higher temperature. The addition of organic matter, on the other hand, increased the mobility of [109]Cd and [65]Zn, and the rate of the irreversible sorption (k_3) was therefore reduced. These results indicate that the potential mobility of [109]Cd and [65]Zn is *reduced* with increasing temperature, but it is *increased* with the addition of organic matter. The effect of these two treatment factors on the exchange rates and pseudoequilibrium constants for the [109]Cd and [65]Zn distribution in the model used were, hence, antagonistic.

ACKNOWLEDGMENT

The Research Council of Norway supported this research through a fellowship to the senior author as well as through research funding. This assistance is gratefully acknowledged.

REFERENCES

1. Jeng, A.S., Weathering of some Norwegian alum shales, *Acta Agric. Scand., Sect. B, Soil and Plant Sci.,* 41, 13, 1991.
2. Pickering, W.F., Metal ion speciation—Soils and sediments, *Ore Geol. Rev.,* 1, 83, 1998.
3. Almås, Å.R., Singh, B.R., and Salbu, B., Mobility of cadmium-109 and zinc-65 in soil influenced by equilibration time, temperature and organic matter, *J. Environ. Qual.,* 28, 1742, 1999.
4. Harter, R.D. and Naidu, R., Role of metal-organic complexation in metal sorption by soils, *Adv. Agron.,* 55, 219, 1995.
5. Sauvé, S., McBride, M., and Hendershot, W., Soil solution speciation of lead (II): Effects of organic matter and pH, *Soil Sci. Soc. Am. J.,* 62, 618, 1998.
6. Naidu, R. and Harter, R.D., Effect of different organic ligands on cadmium sorption by and extractability from soils, *Soil Sci. Soc. Am. J.,* 62, 644, 1998.
7. Zachara, J.M., Resch, C.T., and Smith S.C., Influence of humic substances on Co^{2+} sorption by a subsurface mineral separate and its mineralogic components, *Geochim. Cosmochim. Acta,* 58, 553, 1994.
8. Hodgson, J.F., Geering, H.R., and Fellows, M., The influence of fluoride, temperature, calcium, and alcohol on the reaction of cobalt with montmorillonite, *Soil Sci. Soc. Am. J.,* 28, 39, 1964.
9. Taylor, R.W., Hassan, K., Mehadi, A.A., and Shuford, J.W., Zinc sorption by some Alabama soils, *Commun. Soil Sci. Plant Anal.,* 26, 993, 1995.
10. Barrow, N.J., A brief discussion on the effect of temperature on the reaction of inorganic ions with soil, *J. Soil Sci.,* 43, 37, 1992.
11. Pehlivan, E., Ersoz, M., Pehlivan, M., Yildiz, S., and Duncan, H.J., The effect of pH and temperature on the sorption of zinc (II), cadmium (II), and aluminum (III) onto new metal-ligand complexes of sporopollenin, *J. Coll. Inter. Sci.,* 170, 320, 1995.
12. Bergseth, H., Effect of increasing temperature on the selectivity of a suspended vermiculite for Cu^{2+}, Zn^{2+} and Cd^{2+}, *Acta Agric. Scand., Sect. B, Soil and Plant Sci.,* 32, 373, 1982 (in German).

13. Barrow, N.J., Testing a mechanistic model, II. The effect of time and temperature on the reaction of zinc with a soil, *J. Soil Sci.*, 37, 277, 1986.
14. Bruemmer, G.W., Gerth, J., and Tiller, K.G., Reaction kinetics of the adsorption and desorption of nickel, zinc and cadmium by goethite, I. Adsorption and diffusion of metals, *J. Soil Sci.*, 39, 37, 1988.
15. Gerth, J. and Tiller, K.G., Reaction kinetics of the adsorption and desorption of nickel, zinc and cadmium by goethite, I. Adsorption and diffusion of metals, *J. Soil Sci.*, 39, 37, 1988.
16. Almås, Å.R., Salbu, B., and Singh, B.R., Changes in partitioning of ^{109}Cd and ^{65}Zn in soil as affected by organic matter addition and temperature, *Soil Sci. Soc. Am. J.*, 2000 (in press).
17. Jeng, A.S. and Bergseth, H., Chemical and mineralogical properties of Norwegian alum shale soils, with special emphasis on heavy metal content and availability, *Acta Agric. Scand., Sect. B, Soil and Plant Sci.*, 42, 88, 1992.
18. Singh, B.R., Narwal, R.P., and Almås, Å.R., Crop uptake and extractability of cadmium in soils naturally high in metals at different pH levels, *Commun. Soil Sci. Plant Anal.*, 26, 2123, 1995.
19. Salbu, B., Radioactive tracer techniques in speciation studies, *Env. Tech. Lett.*, 8, 381, 1987.
20. Kennedy, V.H., Sanchez, A.L., Houghton, D., and Rowland, A.P., Use of single and sequential chemical extractants to assess radionuclide and heavy metal availability from soils for root uptake, *Analyst*, 122, R89, 1997.
21. Salbu, B., Krekling, T., and Houghton, D., Characterisation of radioactive particles in the environment, *Analyst*, 123, 843, 1998.
22. Tessier, A., Campbell, P.G.C., and Bisson, M., Sequential extraction procedure for the speciation of particulate trace metals, *Anal. Chem.*, 51, 844, 1979.
23. Salbu, B., *Encyclopedia of Analytical Chemistry*, 2000 (in press).
24. SB Technology Ltd., SB ModelMaker, V. 2.0b, Basingstoke, U.K., 1994.
25. Børretzen, P.E., Mobility of trace elements in sediments: An overview of the physico-chemical forms of mainly the trace elements Cd and Zn, in sediments from the Karah Sea area, Thesis, Agricultural University of Norway, Laboratory of Analytical Chemistry, 1995 (in Norwegian).
26. SAS Institute, Inc., JMP, Statistical Discovery Software, V. 3.1. Cary, NC, 1995.
27. Microcal Software, Inc., Origin, Data Analysis and Technical Graphics, V. 5, Northampton, MA, 1997.
28. Christensen, T.H., Cadmium soil sorption at low concentrations, I. Effect of time, cadmium load, pH, and calcium, *Water Air Soil Poll.*, 21, 105, 1984.
29. Riise, G., Salbu, B., Singh, B.R., and Steinnes, E., Distribution of ^{109}Cd among different soil fractions studied by a sequential extraction technique, *Water Air Soil Poll.*, 73, 285, 1994.
30. Barrow, N.J., Gerth, J., and Brümmer, G.W., Reaction kinetics of the adsorption and desorption of nickel, zinc and cadmium by goethite, II. Modelling the extent and rate of reaction, *J. Soil Sci.*, 40, 437, 1989.
31. Basta, N.T., Pantone, D.J., and Tabatabai, M.A., Path analysis of heavy metal adsorption by soil, *Agron. J.*, 85, 1054, 1993.
32. Narwal, R.P. and Singh, B.R., Effect of organic materials on partitioning, extractability and plant uptake of metals in an alum shale soil, *Water Air Soil Poll.*, 103, 405, 1998.

11 Solid-Phase Speciation of Cd, Ni, and Zn in Contaminated and Noncontaminated Tropical Soils

Abul Kashem and Bal Ram Singh

ABSTRACT

Twelve samples collected from 0–15 cm depths at different locations in Bangladesh were designated as contaminated, noncontaminated, and background soils to investigate the solid-phase speciation of Cd, Ni, and Zn with widely differing soil properties. Total Cd, Ni, and Zn in the soils ranged from 0.01–0.69, 21–52, and 32–939 mg kg^{-1}, respectively. Total contents of Ni except in two hilly soils and Zn in the city sewage (939 mg kg^{-1}) and pharmaceutical soils (162 mg kg^{-1}) exceeded the Dutch limit for clean soil. Sequential extraction was used to speciate Cd, Ni, and Zn from 12 Bangladeshi surface soils to assess metal mobility. The sequences of extractions were six operationally defined groups: water soluble (F1), exchangeable (F2), carbonate (F3), oxide (F4), organic (F5), and residual (F6). The contaminated, noncontaminated (control), and other background soils contained 42, 32, and 26% of Cd; 4, <1, and 1% of Ni; and 11, 3, and 2% of Zn in mobile (F1–F3) fractions, respectively. Higher proportions of metals in mobile fractions in the contaminated soils indicate higher mobility of anthropogenically added metals. Cadmium was found predominately associated with the oxide (44%) fraction irrespective of soil types. In contrast, higher percentages of Ni (83%) and Zn (72%) were found in the residual fraction. The relative extractability, expressed as the ratio of DTPA to aqua regia-extractable contents, was 29% for Cd, 2% for Ni, and <5% for Zn in an average of 12 soils. Significant correlations between mobile and immobile fractions of metals (p = 0.01 and 0.05, respectively) with soil properties were found only with organic carbon. Total contents and mobile fractions of metals were much higher in the city sewage soil than in other soils, so this soil should be of greater concern for the contamination of agricultural products and groundwater.

INTRODUCTION

Rapid population growth, uncontrolled urbanization, unregulated industrialization, and growing agricultural activities in Bangladesh have created a problem of heavy metal contamination. About 9000 metric tons of different pesticides (including those that are metal based) and more than 2 million metric tons of various fertilizers are used in Bangladesh annually.[1] The chemical pollution scenario is not well known, and analytical capabilities are limited. Kashem and Singh[2] found that the total concentrations of Cd, Cu, Ni, Pb, and Zn in soils were several times higher than the natural background levels in four industrial sites around the city of Dhaka. Heavy metal contamination in the terrestrial ecosystems is the result of geochemical processes as well as anthropogenic activities. Geochemical forms of heavy metals in soils affect their solubilities and thus directly influence their bioavailability.[3] The degree of association with different forms depends on soil properties such as soil pH, organic matter, redox conditions, and grain size distribution.

Total concentration of trace metals in soils provides only limited information when considering their toxic effects[4] and the chemical reactivity of the different geochemical forms of the metals found in soils.[5] Single chemical extractions are generally used to determine "available" amounts of soil metals, and usually aim to extract the water-soluble, easily exchangeable, and (some of the) organically bound metals. Other metal fractions may become available over time through chemical weathering or the decomposition of organic matter, but metals occluded by stable secondary minerals may not become available in the short or medium term. Multistep extraction procedures can provide more information about the heavy metal status in soils relative to single extraction methods.[6] More commonly used to quantify the different fractions of metal retention in soils are sequences of different chemical extractions, usually starting with the weakest and least aggressive chemicals and ending with the strongest and most aggressive. The chemical forms determine metal behavior in the environment and its remobilization ability.[7] The sequential procedures used here are based on operationally defined mobile (F1–F3) and immobile (F4–F6) fractions. The reversible fractions where metals are extracted by H_2O or by exchange reactions are defined as mobile. The immobile fractions refer to inner-sphere complexes where strong acids or digestion are required to release the fractions of metals.[8,9] These fractions are not entirely specific, and there will be overlapping between them. The term mobile fraction mostly denotes a negative aspect of environmental "speciation," in that faster transfer from one environmental medium into another usually involves greater reactivity and bioavailability of potentially toxic elements. The term speciation is used here to describe the distribution and transformation of metal species in various media. Methods for assessing the bioavailability of metals in the field are dependent on chemical extraction techniques.[10] Extraction with 0.005 mol L^{-1} DTPA (diethylenetriamine pentaacetic acid)[11] has been used by a number of workers to measure available metals in soils.[12–14] Information on metal fractionations into different phases is almost nonexistent for the soils of Bangladesh. Therefore, this study was planned to investigate the solid-phase speciation of Cd, Ni, and Zn in contaminated and noncontaminated soils with widely differing soil properties.

MATERIALS AND METHODS

Twelve soil samples were collected from the surface layer (0–15 cm) at different locations in Bangladesh. Samples 1a, 2a, 3a, and 4a were collected from areas contaminated by a tannery, city sewage, pharmaceuticals, and a paper mill, respectively, and samples 1b, 2b, 3b, and 4b are the same soils from noncontaminated areas (as controls). Samples 5 and 6 were collected from hilly areas, and 7 and 8 from basin (flood plain) areas (Table 11.1). The last four samples were collected from sites that are far from the contamination sources and that experience intensive agricultural practices throughout the year; as such, they can be treated as background soils. Before use, subsamples of each soil sample were air dried and ground to pass through a 2-mm stainless steel screen. Locations, soil classification, and some characteristics of the soils used in this study are presented in Table 11.1. Soil pH was measured in a 1:2.5 soil/water suspension. The soil suspension was allowed to stand overnight before pH was determined. Soil organic carbon (OC) was found by combustion in an EC-12 LECO-carbon analyzer. Cation exchange capacity (CEC) was estimated by extraction with neutral 1-M NH_4OAc (pH 7.0).[15] Particle size distribution was measured by using the pipette method.[16] Total metals (Cd, Ni, and Zn) in the soils were

TABLE 11.1
Characteristics of Soils Used

Soil	Nature and Location of Samples (District)	Series	FAO Soil Unit	Texture	pH	OC (mg kg^{-1})	CEC (cmol kg^{-1})
1a	Hazaribagh tannery cont. soil, Dhaka	Khaler char	Chromi-Eutric Fluvisols	Silt loam	6.55	15.1	23.6
1b	Control, Dhaka				7.80	3.9	10.1
2a	City-sewage cont. soil, Kawran bazar, Dhaka	Karil	Chromi-Mollic Gleysols	Silty clay loam	5.54	60.6	50.8
2b	Control, Dhaka			Clay	5.78	33.7	31.2
3a	Beximco Pharmaceutical, Tongi, Dhaka	Khailgaon	Chromi-Eutric Gleysols	Silty clay loam	6.77	22.8	20.8
3b	Control, Tangail			Clay	5.13	13.9	11.6
4a	Paper mill cont. soil, Pubna	Gopal pur	Chromi-Calcaric Gleysols	Silty clay loam	7.37	57.6	54.7
4b	Control, Pubna				7.72	14.2	86.9
5	Sylhet	Beani Bazar	Haplic Alisols	Sandy loam	5.71	2.3	6.3
6	Sylhet	Baralekha	Haplic Alisols	Sandy clay loam	5.22	3.8	9.7
7	Barisal	Barisal	Chromi-Eutric Gleysols	Silty clay loam	4.91	7.2	31.0
8	Jalakhati	Barisal	Chromi-Eutric Gleysols	Silt loam	4.62	7.0	25.8

TABLE 11.2
Procedure of Sequential Extraction Scheme

Step	Fraction	Extract	Reaction Time	Device	Centrifuge/ Filtrate
1	F1: Water soluble	20 mL deionized water	1 h in 20°C	Rolling table	10,000 rpm in 30 min
2	F2: Exchangeable	20 mL 1-M NH_4OAc (pH 7)	2 h in 20°C	Rolling table	10,000 g in 30 min
3	F3: Carbonate bound	20 mL 1-M NH_4OAc (pH 5)	2 h in 20°C	Rolling table	10,000 g in 30 min
4	F4: Fe & Mn oxide bound	20 mL 0.04-M $NH_2OH \cdot HCl$ in 25% (v/v) Aac (pH 3)	6 h in 80°C	Shaking water bath	10,000 g in 30 min
5	F5: Organically bound	15 mL 30% H_2O_2 (adj. pH 2) 5 mL 3.2-M NH_4OAc in 20% (v/v) HNO_3	5.5 h in 80°C 0.5 h in 20°C	Shaking water bath and rolling table	— 10,000 g in 30 min
6	F6: Residual	7 M HNO_3	6 h in 80°C	Shaking water bath	Filtrate

determined after digestion with aqua regia.[17] The same metals in these soils were also determined by extracting with DTPA.[11] The sequential extraction method of Salbu and his coworkers,[9] modified from that of Tessier and his coworkers,[18] was used to determine the solid-phase speciation of Cd, Ni, and Zn. The experimental details of this procedure are presented in Table 11.2. Two-gram soil samples were used throughout the experiment. The reagents used for analysis were of analytical grade. All equipment and containers were soaked in 10% HCl and rinsed thoroughly in deionized water before use. Metals were determined by flame atomic absorption spectrophotometry (AAS). Graphite furnace AAS was employed to determine metal concentrations when they were too low to be detected accurately by flame AAS. One duplicate sample was run for every five samples. The results were evaluated graphically as percentage of total content using Microsoft Excel,[19] and stepwise multiple regression analyses was by Minitab Inc.[20]

RESULTS AND DISCUSSION

TOTAL METAL CONTENTS

The total (aqua regia-extractable) Cd, Ni, and Zn in the soils ranged from 0.01–0.69, 21–52, and 32–939 mg kg^{-1}, respectively (Table 11.3). Total concentrations of Ni, except in two hilly soils and Zn in the city sewage (939 mg kg^{-1}) and pharmaceutical soils (162 mg kg^{-1}), were found above the Dutch limit for clean soil. The Dutch A^{20} reference values are 0.8, 35, and 140 for mg kg^{-1} of Cd, Ni, and Zn in soils,

TABLE 11.3
Concentration of Cd, Ni, and Zn in Individual Fractions and Totals of Soils

Soil No.	F1	F2	F3	F4	F5	F6	Sum[a]	Total	Recovery (%)
				Cadmium ($\mu g\ kg^{-1}$)					
1a	4	13	32	54	4	12	119	126	94
1b	n.d.	5	16	43	4	16	84	98	86
2a	25	74	148	335	47	30	659	693	95
2b	n.d.	8	21	53	10	31	123	127	97
3a	3	30	44	61	7	9	154	163	95
3b	n.d.	11	36	47	7	12	113	127	89
4a	6	24	62	71	45	29	237	249	95
4b	3	7	50	45	7	36	148	162	91
5	n.d.	n.d.	6	13	n.d.	3	22	19	116
6	n.d.	n.d.	n.d.	12	n.d.	6	18	16	113
7	3	4	7	17	5	9	45	46	98
8	6	5	8	14	5	10	48	47	102
				Nickel ($mg\ kg^{-1}$)					
1a	0.1	0.2	0.9	2.6	2.3	34.3	40.37	42	96
1b	—	—	0.2	2.3	2.0	32.2	36.7	40	92
2a	0.8	1.2	3.3	9.5	6.8	24.6	46.2	52	89
2b	n.d.	0.3	0.1	7.0	6.3	21.3	34.07	40	86
3a	0.1	0.1	0.6	2.3	4.1	32.5	39.7	40	99
3b	n.d.	n.d	0.5	2.7	2.3	36.9	42.4	42	101
4a	n.d.	n.d.	0.2	1.4	4.7	32.7	39.0	39	100
4b	n.d.	n.d.	0.2	1.4	1.4	33.6	36.6	37	99
5	n.d.	n.d.	0.1	1.1	0.3	16.3	17.8	21	85
6	n.d.	n.d.	n.d.	0.4	0.3	18.4	19.1	23	83
7	0.1	0.2	0.4	0.7	1.1	38.5	41.0	38	108
8	0.2	0.3	0.6	0.7	1.1	38.3	41.2	39	106
				Zinc ($mg\ kg^{-1}$)					
1a	0.4	0.7	5.5	20.5	15.4	71.8	114.3	120	95
1b	0.1	0.2	0.4	6.6	4.4	71.0	82.7	89	93
2a	3.0	72.8	144.8	373.2	144	117	854.8	939	91
2b	0.1	0.6	4.4	10.2	5.1	58.4	78.8	80	99
3a	0.2	0.9	14.5	42.2	14.0	87.4	159.2	162	98
3b	0.1	0.4	2.9	9.4	5.6	65.5	83.9	93	90
4a	0.1	0.2	3.3	13.6	3.7	95.6	116.5	125	93
4b	0.1	0.2	0.8	4.2	0.6	67.5	73.4	79	93
5	0.1	0.3	0.8	5.1	1.7	24.2	32.2	32	101
6	ND	0.2	0.4	3.6	1.3	22.8	28.3	32	90
7	0.4	0.4	0.8	13.5	5.8	78.0	101.9	96	106
8	0.9	0.4	1.1	10.5	3.2	80.7	96.8	91	106

[a]Sum of F1–F6: F1—Hydrosoluble, F2—Exchangeable, F3—Carbonate bound, F4—Oxide bound, F5—Organically bound, and F6—Residual.

Recovery %: (F1 + F2 + F3 + F4 + F5 + F6)/Aqua regia-extractable × 100.

n.d.: Not detectable

respectively. In soils from all contaminated areas, total Cd, Ni, and Zn concentrations were higher than in the control soils, except for Ni in the Beximco pharmaceutical (3a) area soil. The overall order of contamination was Zn > Cd > Ni. Cadmium and Zn concentrations were about 5 and 12 times higher in the city sewage-contaminated soils than in the control. The degrees of contamination were found in the order: city sewage > paper mill > Beximco pharmaceutical > tannery area soils for Cd and city sewage > Beximco pharmaceutical > paper mill > tannery area soils for Zn. The level of contamination was higher in the city sewage soil due to discharging of liquid waste and flocculated sludge with high metal contents coming from different sources.

PARTITIONING AND DISTRIBUTION OF METALS IN SOILS

Sequential extraction procedures indirectly assess the potential mobility and bioavailability of metals in soils. Bioavailability of metals decreases in the order: water soluble > exchangeable > carbonate > oxides > organic > residual.[7] Recovery of Cd, Ni, and Zn was calculated as the ratio of the sum of the fractions to the aqua regia (total), expressed in percent. The recovery exceeded total Cd, Ni, and Zn in only three samples, indicating that enhanced recovery through the multistep extraction procedure was not a serious problem. Higher amounts of metals extracted by sequential extraction procedures were also found by other investigators.[4,22] However, there were good correlations (r^2 = 1, 0.85, and 1 for Cd, Ni, and Zn) between the sum of all fractions and the amount extracted by aqua regia. Miller and McFee[23] also reported high correlation coefficients between the total metals (acid digested) and the sum of the fractions.

Cadmium

The distribution of various fractions of Cd in the investigated soils indicate that on average, 34% of total Cd was associated with the mobile (F1–F3) fractions and 66% with the immobile (F4–F6) fractions (Figure 11.2). There were variations for all Cd fractions in different soils. Metals extracted with H_2O and NH_4OAc were treated as mobile fractions. These fractions ranged from 38 to 50% in contaminated soils (1a–4a), 24 to 42% in noncontaminated (1b–4b) control soils, and NT to 40% in background (5–8) soils, respectively. The higher proportion of mobile Cd in contaminated soils was due to anthropogenic contamination, generally low clay content, and low pH values as compared to control soils. In the background floodplain soils, Cd was present in all three mobile fractions (F1–F3) because of low pH and light-texture alluvial soils. Soil pH is considered the single most important factor controlling mobility and availability of trace metals in soils, as reported by Witter.[24] High pH favors sorption and precipitation of heavy metals as oxides, hydroxides, and carbonate. Soils that are high in clays and CEC provide sorption sites for metals and strongly retain them in the lattice. It was reported that for soils with the same metal total Cd content, Cd was more soluble and plant-available in sandy soil than in clay soil.[25,26]

The mean percentages of Cd distribution in the F1 + F2, F3, F4, F5, and F6 fractions were 7, 21, 45, 6, and 21% in noncontaminated soils, and 16, 26, 42, 8, and 8% in contaminated soils. These results show that contamination increased the values of

Cd in F1 + F2 and F3 fractions. The higher proportion of Cd in the F1 to F3 fractions in contaminated soils indicates that Cd from contaminated sources could account for higher bioavalability and leaching. Among the mobile fractions, the carbonate fraction of Cd was found to be higher. Cadmium percentage in organic fractions was also found to be higher in contaminated soils because of high organic carbon content, but the contaminated soils contained lower values of residual and oxide Cd compared with noncontaminated soils. In the most highly contaminated city sewage soil, the residual fraction only accounted for 5% of the total. Lower levels in the residual fraction of a sequential extraction in contaminated soils were also reported by other investigators.[12,27] The carbonate form of Cd is susceptible to pH changes, mainly in the rhizosphere during plant growth, so it may be regarded as potentially phytoavailable. Cadmium added to soil as carbonate is relatively mobile in acidic conditions and within a few years may convert to exchangeable form.[28] In groupings, contaminated, noncontaminated (control), and background soils contained about 42, 32, and 26% of Cd that was found in mobile fractions (Figure 11.1). Higher percentages of Cd in this fraction in the contaminated soils than in the other groups indicate that the anthropogenically added Cd remains in the mobile fraction and has not been incorporated within the crystal lattice of minerals.

The association of Cd in different fractions was in the order: F4 (29–67%) > F3 (NT–34%) > F6 (4.6–33%) > F2 (NT–19%) and > F5 (NT–19%) and > F1(NT–13%) (Figure 11.2). A similar distribution order of Cd was observed by Chlopecka et al.[29] and Ramos et al.[30] in their investigations. Cadmium was present in all six fractions in the contaminated and floodplain soils. The proportion of F1 fraction was higher in the floodplain soil, which was lighter in texture, had low pH, and was relatively young, and where silt deposition by flood water occurs every year. Mobile and organic fractions were not traceable in a sandy clay loam hilly soil, probably due to leaching, surface erosion, low content of organic carbon, and low CEC.

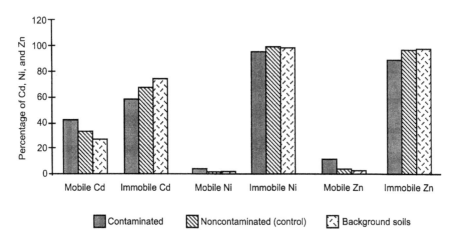

FIGURE 11.1 Distribution of mobile and immobile fractions of Cd, Ni, and Zn in contaminated, noncontaminated (control), and background soils.

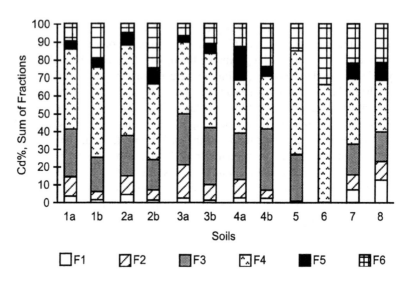

FIGURE 11.2   Distribution of Cd in various fractions based on total contents in soils.

Nickel

The mobile fraction of Ni was only 2% of the total (Table 11.3). Mobile fractions varied among the soil groups, and the values were about 4, <1, and 1% in the contaminated, control, and background soils, respectively (Figure 11.1). Comparing the mean percentage of Ni in the mobile fractions among the soil groups, the differences, as with Cd, were also due to the level of contamination, low pH, and low clay content. In the city sewage soil, Ni was higher (12%) in the mobile fraction and lower (88%) in the immobile fraction as compared with all other soils. Higher percentages of Ni in the mobile fraction in contaminated soils were also observed for Cd; the explanation could be same for Ni distribution in different soils. The mean percentage of Ni distribution in noncontaminated soils was 1% in F1–F3, 6% in F4, 5% in F5, and 88% in F6 fractions, and in contaminated soils it was 4, 9, 11, and 76% in the F1–F3, F4, F5, and F6, respectively (Figure 11.3). This implies that the contamination increased the oxide, organic, and mobile fractions but decreased the residual fraction, as this study found for Cd.

The distribution among the fractions was in the order: residual (53–96%) > Fe–Mn oxide (2–21%) ≈ organic (2–18%) > carbonate (NT–7%) > exchangeable (NT–3%) > water soluble (NT–2%) (Figure 11.3). Higher values were found in all fractions except F6 in city sewage-contaminated soil. The F4 (oxide) fraction in the floodplain and F5 (organic) fraction in the hilly soils were low when compared with other soils. The distribution order of Ni among fractions coincided well with the results reported by Pengxin and his colleagues[36] and Hickey and Kittrick.[32] Similar to the results of this study, the concentration of Ni in the residual fraction has also been found by other investigators.[33–35] Metals bound to residual fraction may be those bound with detrital silicate minerals, resistant sulphides, and refractory organic materials.[36,37] Metals in this fraction are expected to be chemically stable and biologically

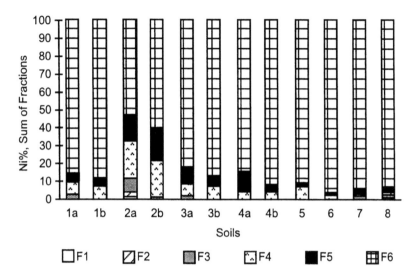

FIGURE 11.3   Distribution of Ni in various fractions based on total contents in soils.

inactive. Another explanation for this may be that Ni^{2+} has the highest crystal field stabilization energy of the common divalent metals and thus has a high potential to be enriched in clay minerals.[38] The next Ni-containing fractions were Fe–Mn oxide and the organically bound; these two fractions both showed similar distribution of Ni (7%).

Zinc

The proportion of Zn associated with the mobile fractions was 6% of the total (Table 11.3). The sum of Zn in the mobile fractions ranged from 3 to 26% in the contaminated soils, <1–7% in the controls, and 2–4% in the background soils. The city sewage soil showed higher amounts of Zn in the mobile fractions than the other soils did because of higher Zn contamination. The mean percentages of mobile Zn were 11, 3, and 2 in the contaminated, control, and background soils, respectively (Figure 11.1). The same trend was observed for the other metals (Cd and Ni) in this study. The fractions of Zn in the noncontaminated soils were 5, 9, 5, and 81% in the F1–F3, F4, F5, and F6 fractions, respectively, but in the contaminated soils they were 11% in F1–F3, 25% in F4, 11% in F5, and 53% in the F6 fractions. The amount of Zn in the residual fraction was lower (53%) in the contaminated soils than in the controls (83%), as was the case with Cd; the explanation given for Cd also applies for the behavior of Zn.

Zinc showed a distribution similar to that of Ni (Table 11.3). Among the fractions, the order of Zn association was F6 (14–92%) > F4 (6–44%) > F5 (<1–17%) > F3 (0.5–17%) > F2 (0.2–9%) and > F1 (NT–0.9%) (Figure 11.4). In the city sewage-contaminated soil, the proportion of Zn in the F2–F5 fractions was higher than in the other soils, and in F6 it was lower. Higher proportions of Zn in the residual (72%) and oxide (15%) fractions are in agreement with results reported by

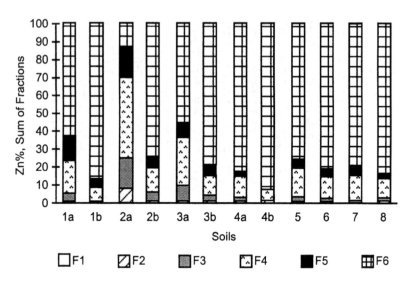

FIGURE 11.4 Distribution of Zn in various fractions based on total contents in soils.

other investigators for varying types of soils.[4,39,40] The organic matter had little effect (7%) on Zn partitioning, although the contaminated soils contained higher amounts of organic matter. Lower percentages of soil Zn in the organic fraction among the immobile fractions studied here indicates that the Zn–organic matter association may be relatively unstable. In city sewage-contaminated soil (2a), a relatively larger fraction (44%) of Zn was associated with F4, partially due to the high stability constant of Zn oxides. This has been observed by other investigators.[30,41] The amount of Zn present in the mobile fraction ranged from 0.8% (1b) to 26% (2a). Among the mobile fractions, the F3 (carbonate) fraction contained the highest amount of Zn. Tessier and his colleagues[18] found a high percentage of Zn (13–16%) in the carbonate fraction; the same was true for the city sewage soil (17%).

DTPA-extractable metals

Metals removed by the chelating agent DTPA are thought to represent the plant-available fractions.[11,42] DTPA forms soluble complexes with metals, reducing the activity of the metals in solution. Therefore, ions desorb from the surface and enter into the solution. The relative extractability, expressed as the ratio between DTPA and aqua regia-extractable contents, were 29% for Cd, 2% for Ni, and <5% for Zn in an average of 12 soils (Table 11.4). We found similar results in the mobile fraction (F1–F3) of sequential extraction procedures in this study: solubility was higher for Cd, intermediate for Zn, and least for Ni. Cadmium is reportedly more soluble than other metals.[42–44] Nickel is the least extractable, which might be due to its high affinity to soil minerals and oxides.[33,45] Relative extractability for Cd, Ni, and Zn were 36, 4, and 11% in contaminated soils; 27, 1, <1% in noncontaminated (control) soils; and 26, 1.5, and 2% in other background soils, respectively. The extractability was higher for

TABLE 11.4
DTPA-Extractable Cd, Ni, and Zn in Soils
with Percentage of Their Total Contents
(% in parentheses)

Soil	Cadmium (mg kg^{-1})	Nickel (mg kg^{-1})	Zinc (mg kg^{-1})
1a	0.050 (40)	1.22 (2.9)	8.6 (7.2)
1b	0.033 (34)	0.20 (0.5)	0.6 (0.7)
2a	0.215 (31)	4.40 (8.5)	210 (22)
2b	0.041 (32)	0.62 (1.6)	1.0 (1.3)
3a	0.063 (39)	1.22 (3.1)	19 (11.4)
3b	0.025 (20)	1.02 (2.4)	0.9 (1.0)
4a	0.082 (33)	0.62 (1.6)	3.2 (2.7)
4b	0.035 (22)	0.16 (0.4)	0.6 (0.8)
5	0.006 (32)	0.10 (0.5)	0.6 (1.9)
6	0.004 (25)	0.12 (0.5)	0.4 (1.3)
7	0.009 (20)	0.90 (2.4)	1.8 (1.9)
8	0.012 (26)	1.02 (2.6)	2.6 (2.8)

Cd, Ni, and Zn in the contaminated soils than in noncontaminated soils, as was observed in the mobile fractions of sequential extractions. The high ratio between DTPA and total concentrations of Cd in the contaminated soils indicate that this metal is more available for plant uptake and leaching as the mobile (F1–F3) fraction of Cd. Mellum and his coworkers[42] found that 52, 13, and 6% of Cd, Ni, and Zn, respectively, of the total concentrations in 50 Norwegian soils developed on alum shale geological material were extractable by DTPA. Average DTPA-extractable concentrations of 15% for Cd, 12% for Ni, and 18% for Zn in an acid (pH 5.6) silt loam soil amended with municipal sewage sludge was reported by Schauer and his coworkers.[46] Singh and his colleagues[12] found 38% Cd, 26% Zn, and <10% Ni by DTPA-extractant of their total concentrations.

Availability, as denoted by DTPA-extractant, is predictable when comparing with the results of mobile fractions of sequential extraction methods. In our study, we found strong positive correlations (r^2 = > 0.95 for Cd, Ni, and Zn) between DTPA-extractable concentrations and the concentration of mobile fractions (sum of F1–F3). This implies that both DTPA-extractable concentration and the mobile fractions of sequential extractions could be considered to represent the labile pool of heavy metals. In stepwise multiple regression with soil properties, DTPA-extractable Cd, Ni, and Zn were positively correlated with organic carbon (Table 11.5). Organic matter has both a positive and a negative role in heavy-metal mobility, depending on its complexing ability. Brummer and Herms[47] concluded that organic matter in soils immobilizes heavy metals in strongly acidic conditions and mobilizes metals in weakly acid to alkaline (pH 6–8) conditions by forming insoluble and soluble complexes, respectively. The addition of organic matter increased the Cd extracted by DTPA in

TABLE 11.5
Stepwise Multiple Regression between Mobile
and Immobile Fractions with Soil Properties

Metals	Equation	R^2
Mobile Cd	= 0.1 + 2.71 OC	0.68[a]
Immobile Cd	= 3.2 + 4.4 OC	0.69[a]
Mobile Ni	= 0.21 + 0.04 OC	0.42[b]
Immobile Ni	= 8.7 + 0.33 silt + 0.28 clay	0.83[a]
Mobile Zn	= −18.6 + 2.00 OC	0.42[b]
Immobile Zn	= 18.8 + 5.52 OC	0.47[a]
DTPA Cd	= 4.2 + 2.26 OC	0.66[a]
DTPA Ni	= 0.22 + 0.04 OC	0.42[b]
DTPA Zn	= −17.6 + 1.89 OC	0.41[b]

OC—Organic carbon

[a] $p = 0.01$

[b] $p = 0.05$

soils of varying textures.[25] This was explained by increased soil CEC and decreased soil pH in soils treated with organic matter.

Relationships among Mobile, Immobile, and DTPA-Extractable Metals with Soil Properties

Stepwise multiple-regression procedures were used to evaluate the possible relationships between mobile, immobile, and DTPA fractions with soil: organic carbon (OC), silt, and clay (Table 11.5). Cadmium, Ni, and Zn were below the detection limit in different fractions of sequential extraction in some soils, so we used only mobile and immobile fractions for this purpose. The results showed that among the properties, only OC showed a relationship with the metal fractions. Significant positive correlation between mobile and immobile fractions with organic carbon suggests that these metals have strong affinity with OC, except Ni in the immobile fraction. Almås and his colleagues[8] reported that in a pig manure-treated alum shale soil, ^{109}Cd and ^{65}Zn increased significantly in mobile fractions, with a corresponding reduction in immobile fractions. Low-molecular-weight organic acids may inhibit the retention of metals in the solid phase and hence increase solubility. McBride[48] reported that soluble organics can increase the solubility of metal cations bound to organic molecules. The gradual decomposition of organic matter could release the available binding sites of these metals and thus could affect retention of the metals.[49] Immobile Ni was influenced by silt and clay, indicating that most of the Ni resided predominately in the clay and silt fractions. In this study, pH did not show a significant role with any of the metals, although it is the single main factor controlling the mobility and solubility of heavy metals.

CONCLUSIONS

Total Cd and Zn in the city sewage soil were 6 and 11 times higher than in the control soil, and the contents varied among the soils with soil properties, level of contamination, and agricultural practices. The proportion of DTPA-extractable fractions was higher in the contaminated soils, and the extractability of Cd was 15 and 6 times that of Ni and Zn, respectively. The extraction sequence (F1–F6) followed the order of decreasing solubility. The order of Cd association was F4 > F3 > F6 > F2 > F5 > F1. Nickel and Zn showed a similar distribution, and their proportions were in the order F6 > F4 > F5 > F3 > F2 > F1. The percentage of mobile fractions was highest for Cd and lowest for Ni. The proportion of mobile fractions (F1–F3) for all three metals was higher in the contaminated soils as compared with other soils. This implies that metals added through contamination remained in the mobile fractions. Strong significant correlations of mobile and immobile fractions of metals were found with organic carbon only. Among the soils, the city sewage-contaminated soil was characterized by higher total metal contents and mobilities and should therefore of be concern because it is used intensively for agriculture.

ACKNOWLEDGMENTS

This study is part of the Ph.D. thesis of Abul Kashem, associated with the project *Accumulation, speciation, and geochemical processes and heavy metal budgets in agricultural systems,* supported by the Research Council of Norway. The authors gratefully acknowledge the Research Council of Norway for financial support.

REFERENCES

1. Matin, M.A., Environmental pollution and its control in Bangladesh, *Anal. Chem.,* 14, 468, 1995.
2. Kashem, M.A. and Singh, B.R., Heavy metal contamination of soil and vegetation in the vicinity of industries in Bangladesh, *Water Air Soil Pollut.,* 115, 347, 1999.
3. Xian, X., Chemical partitioning of cadmium, zinc, lead, and copper in soils near smelters, *J. Environ. Sci. Health, A,* 527, 1987.
4. Jeng, A.S. and Singh, B.R., 1993. Partitioning and distribution of cadmium and zinc in selected cultivated soils in Norway, *Soil Sci.,* 156, 240, 1993.
5. Spevackova, V. and Kucera, J., Trace element speciation in contaminated soils studied by atomic absorption spectrometry and neutron activation analysis, *Int. J. Environ. Anal. Chem.,* 35, 241, 1989.
6. Pickering, W.F., Metal ion speciation—Soils and sediments, *Ore Geol. Rev.,* 1, 83, 1998.
7. Lena, Q.M. and Gade, N.R., Chemical fractionation of cadmium, copper, nickel, and zinc in contaminated soils, *J. Environ. Qual.,* 26, 259, 1997.
8. Almås, Å., Singh, B.R., and Salbu, B., Mobility of [109]Cd and [65]Zn in soil influence by equilibration time, temperature and organic matter, *J. Environ. Qual.,* 28, 1742, 1999.
9. Salbu, B., Krekling, T., and Oughton, D.H., Characterisation of radioactive particles in the environment, *Analyst,* 123, 843, 1998.

10. Bryan, C.W. and Langston, W.J., Bioavailability, accumulation and effects of heavy metals in sediments with special reference to United Kingdom estuaries: A review, *Environ. Pollut.*, 76, 89, 1992.
11. Lindsay, W.L. and Norvell, W.A., Development of DTPA soil test for zinc, iron, manganese, and copper, *Soil Sci. Soc. Am. J.*, 42, 421, 1978.
12. Singh, S.P., Tack, F.M., and Verloo, M.G., Heavy metal fractionation and extractability in dredged sediment derived surface soils, *Water Air Soil Pollut.*, 102, 313, 1998.
13. Tack, F.M.G. and Verloo, M.G., Chemical speciation and fractionation in soil and sediment heavy metal analysis: A review, *Int. J. Environ Anal. Chem.*, 59, 225, 1995.
14. Alloway, B.J. and Jackson, A.P., The behaviour of heavy metals in sewage sludge amended soils, *Sci. Total Environ.*, 100, 151, 1991.
15. Page, A.L., Miller, R.S.H., and Keeney, D.R. (Eds), *Method of Soils Analysis, Part 2. Chemical and Microbiological Properties*, 2nd edition, American Society of Agronomy, Inc., Madison, WI, 1982.
16. Elonen, P., Particle size analysis of soil, *Acta Agric. Fenn.*, 122, 1, 1971.
17. Jeng, A.S., Weathering of some Norwegian alum shales, I. Laboratory simulations to study acid generation and the release of sulphate and metal cations (Ca, Mg & K), *Acta Agric. Scand., Sect. B, Soil and Plant Sci.*, 41, 13, 1991.
18. Tessier, A., Campbell, P.G.C., and Bisson, M., Sequential extraction procedure for the speciation of particulate trace metals, *Anal. Chem.*, 51, 844, 1979.
19. Excel Inc., Microsoft Excel for Windows 95, 1985–1996 Microsoft Corporation, 1996.
20. Minitab Inc., *Minitab Handbook*, 2nd ed., 1992.
21. VROM, Dutch A-reference value Leidraad bodembescherming, Staatsuitgeverij, s-Gravenhage, 6th ed., 1990.
22. Harrison, R.M., Laxen, D.P.H., and Wilson, S.J., Chemical associations of lead, cadmium, copper, and zinc in street dusts and roadside soils, *Environ. Sci. Technol.*, 15, 1378, 1981.
23. Miller, W.P. and McFee, W.W., Distribution of cadmium, zinc, copper, and lead in soils of Northwestern Indiana, *J. Environ. Qual.* 12, 29, 1983.
24. Witter, E., Agricultural use of sewage sludge, Controlling metal contamination of soils, National Environment Protection Board, Stockholm, Sweden, Rep. 3620, 1989.
25. He, Q.B. and Singh, B.R., Cadmium distribution and extractability in soils and its uptake by plants as affected by organic matter and soil types. *J. Soil Sci.*, 44, 641, 1993.
26. Eriksson, J.E., The influence of pH, soil type, and time on adsorption and uptake by plants of Cd added to the soil, *Water Air Soil Pollut.*, 48, 317, 1989.
27. Zhang, J., Huang, W.W., and Wang, Q., Concentration and partitioning of particulate trace metals in the Changiang (Yangze river), *Water Air Soil Pollut.*, 52, 57, 1990.
28. Chlopecka, A., Forms of trace metals from inorganic sources in soils and amounts found in spring barley, *Water Air Soil Pollut.*, 69, 127, 1993.
29. Chlopecka, A., Bacon, J.R., Wilson, M.J., and Kay, J., Forms of cadmium, lead and zinc in contaminated soils from southwest Poland, *J. Environ. Qual.* 25, 69, 1996.
30. Ramos, L., Hernandez, L.M., and Gonzalez, M.J., Sequential fraction of copper, lead, cadmium and zinc in soils from or near Donana National Park, *J. Environ. Qual.*, 23, 50, 1994.
31. Pengxin, W., Erfu Q., Zhenbin L., and Shuman, L.M., Fractions and availability of nickel in loessial soil amended with sewage sludge, *J. Environ. Qual.*, 26, 795, 1997.
32. Hickey, M.G. and Kittrick, J.A., Chemical partitioning of cadmium, copper, nickel and zinc in soils and sediments containing high levels of heavy metals, *J. Soil Sci.*, 156, 240, 1984.
33. Li, Z. and Shuman, L.M., Redistribution of forms of zinc, cadmium and nickel in soils treated with EDTA, *Sci. Total Environ.*, 19, 95, 1996.

34. Sims, J.T. and Kline, J.S., Chemical fractionation and plant uptake of heavy metals in soils amended with co-composted sewage sludge, *J. Environ. Qual.*, 20, 387, 1991.
35. Emmerich, W.E., Lund, L.J., Page, A.L., and Chang, A.C., Solid phase forms of heavy metals in sewage sludge-treated soils, *J. Environ. Qual.*, 11, 178, 1982.
36. Narwal, R.P. and Singh, B.R., Effect of organic materials on partitioning, extractability and plant uptake of metals in an alum shale soil, *Water Air Soil Pollut.*, 103, 405, 1998.
37. Norrish, K., Geochemistry and mineralogy of trace elements, in *Trace Elements in Soil–Plant–Animal Systems,* Nicholas, D.J.D., and Egan, A.R. (Eds.), Academic Press, San Diego, CA, 55, 1975.
38. Bruemmer, G.W., Gerth, J., and Tiller, K.G., Reaction kinetics of the adsorption and desorption of nickel, zinc, and cadmium by geothite: I. Adsorption and diffusion of metals, *J. Soil Sci.*, 39, 37, 1988.
39. Calvet, R., Bourgeois, S., and Masky, J.J., Some experiments on extraction of heavy metals present in soil, *Int. J. Environ. Anal. Chem.*, 39, 1990, 31–45.
40. Shuman, L.M., Effect of organic matter on the distribution of manganese, copper, iron, and zinc in soil fractions, *Soil Sci.*, 146, 1988, 192–198.
41. Kuo, S., Heilman, P.E., and Baker, A.S., Distribution and forms of copper, zinc, cadmium, iron and manganese in soils near a copper smelter, *Soil Sci.*, 135, 1983, 101–109.
42. Mellum, H.K., Arnesen, A.K.M., and Singh, B.R., Extractability and plant uptake of heavy metals in alum shale soils, *Commun. Soil Sci. Plant Anal.*, 29, 1183, 1998.
43. McGrath, S.P. and Cegarra, J., The effects of soil organic matter levels on soil solution concentrations and extractabilities of manganese, zinc and copper, *Geoderma*, 42, 177, 1992.
44. Alloway, B.J. and Morgan, H., The behaviour and availability of Cd, Ni and Pb in polluted soils, in *Contaminated Soil,* van den Brink, J.W. (Ed.), Martinus Nifhoff Publishers, Dordrecht, The Netherlands, 101, 1986.
45. Wang, P., Erfu Q., Li, Z., and Shuman, L.M., Fractions of availability of nickel in loessial soil amended with sewage or sewage sludge, *J. Environ Qual.*, 26, 795, 1997.
46. Schauer, P.S., Wright, W.R., and Pelchat, J., Sludge-borne heavy metal availability and uptake by vegetable crops under field conditions, *J. Environ. Qual.*, 9, 69, 1980.
47. Brummer, G. and Herms, U., Influence of soil reaction and organic matter on the solubility of heavy metals in soils, in *Effects of Accumulation of Air Pollutants in Forest Ecosystems,* Ulrich, B. and J. Pankrath, J. (Eds.), D. Reidel Publishing Co., Dordrecht, The Netherlands, 233, 1983.
48. McBride, M.B., *Environmental Chemistry of Soils,* Oxford University Press, New York, 1994.
49. Zhang, M., Alva, A.K., Li, Y.C., and Calvert, D.V., Chemical association of Cu, Zn, Mn, and Pb in selected sandy citrus soils, *Soil Sci.*, 162, 181, 1997.

12 Quality of Estimated Freundlich Parameters of Cd Sorption from Pedotransfer Functions to Predict Cadmium Concentrations of Soil Solution

Günther Springob, Dörthe Tetzlaff, Angela Schön, and Jürgen Böttcher

ABSTRACT

To predict the cadmium concentrations of soil solutions C_{Cd} under varying properties of soil matrix and electrolyte, we established Cd sorption isotherms for 225 samples from sandy, northern German arable and forest soils, and from these derived the Freundlich parameters k and M. Standard electrolyte was 5 mM $Ca(NO_3)_2$. As the initially (native) sorbed fraction of Cd, we used the amount extracted with 0.025 M Na_2–EDTA at 20°C, 2 h shaking. The average value of parameter M was 0.815 ($-$). There was some correlation of M with pH: samples above pH 6 had an average M of just 0.730. The main information about the sorption properties of the soils was contained in k, which could be predicted by multiple regressions from pH, organic carbon (OC,%) and clay content (%) for one subset of Ap horizons ($r^2 = 0.96$). When all 225 samples were combined, no more statistical influence was found for the variable "clay" on the multiple regression models. Clay, therefore, is not included in the final model in which Freundlich k ($mg^{1-M} L^M kg^{-1}$) is given by $-0.993 \, pH^{0.537} \, OC^{0.783}$. The resulting values are valid for 5 mM $Ca(NO_3)_2$, 20°C and were used, together with the mean M of 0.815 and the Freundlich equation, to predict the Cd concentrations of the soil solutions (C_{Cd}) of the 225 investigated sites, both for the current load of Cd and assuming higher contamination. In a large number of samples, the estimated C_{Cd} exceeded current drinking water threshold values and other solution-based critical limits when a total load of 1 mg Cd kg^{-1} soil was assumed. In a final step, we corrected the predicted C_{Cd} for the strength of the electrolyte, here defined by the

Ca^{2+} concentration, and the proportion of the complexing Cl^- among the accompanying anions. The approach appears to be promising, but there are still some clear deficiencies concerning the prediction of the Freundlich exponent M and the influence on k of the independent variables DOC, time, and temperature on the one hand, and the contents of clay or oxides on the other.

INTRODUCTION

In the context of assessing its bioavailability and ecotoxicology, the mobility of Cd in soils has been expressed in a number of ways. Sequential extraction techniques with agents of different strengths as well as single extractions with diluted salt solutions (e.g., NH_4NO_3, $CaCl_2$) are widely used. Another approach, and a more basic one, is the application of relationships that relate the Cd concentration of the soil solution (C_{Cd}) to the sorbed fraction that sustains the equilibrium. If relatively simple relationships are used, e.g., the Freundlich equation, then this approach is still an empirical and operational one. There is the advantage that the resulting indicator of mobility is a Cd concentration of soil solution C_{Cd} that can be used for different purposes in a more general way. The Freundlich sorption isotherm is given by

$$S = k \, C_{Cd}{}^{M} \tag{12.1}$$

where S is the sorbed fraction and k (slope) and M (shape) are the Freundlich parameters. C_{Cd} for an individual soil can be derived if S, k, and M are known. If a number of different soils are considered, M turns out to be not so closely correlated with basic soil properties, and average values around 0.8 were frequently used (e.g., Streck and Richter[1]). The more important parameter k is clearly determined by soil properties (e.g., Buchter et al.,[2] Gray et al.,[3] more references below) and thus can be derived by multiple regression techniques. A first step would be to extend the basic Freundlich function in the following way:

$$S = k^* \, x_1{}^{a} x_2{}^{b} \ldots x_n{}^{z} C_{Cd}{}^{M}. \tag{12.2}$$

The factors x_1 to x_n represent the soil variables that predict k. The constant k^* contains the unpredictable rest and might not be required in all cases. The exponents a, b . . . z are implicated to allow a nonlinear contribution of each independent variable. In the log domain, Eq. 12.2 can be fitted to data by the classical procedures of multiple linear regression analysis:

$$S = k^* + a \log x_1 + b \log x_2 \ldots + z \log x_n + M \log C_{Cd}. \tag{12.3}$$

Such operational models, based on such easily available, general soil properties as organic carbon (OC), pH, CEC, and others, and mostly somehow modified, were parameterized and used by several researchers.[4–10] Related ways of generalizing the Freundlich equation, which include more sophisticated physicochemical considerations, have been proposed.[11–13]

Soil pH is the main predicting variable for k. OC was also applied in the models cited above. A final term used by Tiktak and his colleagues[10] included pH and the CEC. The theoretical requirement to include dissolved organic carbon (DOC) and other constituents of the soil solution, such as ionic strength, concentrations of Ca^{2+}, or of the Cd-complexing anion Cl^-, were frequently stressed in the studies cited above.

Here, we focus on the simpler multiple regression models. We evaluate the quality of the Freundlich parameters estimated from soil properties by pedotransfer functions and the range of the resulting Cd concentrations of the soil solutions. This is done for the actual state of the soils and for assumptions about future higher Cd loads, e.g., those caused by sewage sludge application or other contaminations. We only present data on sandy soils without much aggregation because any of these equations, irrespective of the degree of sophistication, still predicts average Cd concentrations of bulk soil solutions. In sandy soils this restriction will not cause many problems, but the ecotoxicological implications of mobilized Cd can be severe in these substrates. For example, does further soil acidification occur after converting ploughed and limed agricultural land into forest, a process that has recently been discussed in the European Union? In addition to properties of the soil matrix (OC, clay content, amorphous Fe oxides) and pH, we include the modifying influence of Ca^{2+} and Cl^- in soil solutions on C_{Cd}.

There are a number of possible applications of this kind of pedotransfer functions (see also the literature cited above). These include modeling of Cd transport and transfer into plants in spatially variable soils and translation of geodata from data banks, as well as soil maps into maps showing areas of potential ecological risks for the evaluation of threshold values for metals in soils that are based on total Cd or Cd fractions in terms of Cd mobility. Finally, and basically important, the mobility of Cd is defined in a more general way by the Cd concentration of the soil solution than by any conventionalized extraction or equilibration.

SOILS AND METHODS

SOILS

The parent materials are Pleistocene sand deposits of northern Germany. The amounts of weatherable silicates are generally low, but the scale of contents is still sufficient to influence—in combination with climatic gradients, levels of groundwater, and land use—the soil development towards podsolic, cambic, and gleyic characteristics. Clay contents are mostly low. Higher contents up to 10% are sometimes found near rivers due to alluvial deposits from occasional flooding that are incorporated into the Ap horizon. Some of the arable soils have unusually high contents of OC because of earlier use as grassland when the groundwater level was higher. For these reasons, there is some range in the data in humus properties as well. The soils are practically free of $CaCO_3$, but some chalky grains may be present due to cultivation. The pH ranges from extremely acid (forest) to near-neutral (a few Ap horizons).

There are six sets of data from three regions (Fuhrberg near Hannover, Braunschweig, and Berlin) and one soil from the Fuhrberg area (pH 5.8, 3.3 % clay, 3.4 % OC) that were used for the experiments described below under "recovery of soil solution." The sets Fuhrberg 1 and 2 originate from a drinking water catchment of 300 km² in the north of Hannover. There is no specific contamination except some sewage sludge. The Fuhrberg 1 set comprises 24 samples used earlier to establish conventional Cd sorption isotherms with a number of data points on each function.[8,12] We refer to these functions as standard isotherms. The Fuhrberg 1 samples are not fully representative of the region, as we intended to include some samples with higher clay contents (max. about 10%). The Fuhrberg 2 dataset comprises 95 Ap horizons from the same region and, in fact, represents the variability found in the soils of the catchment (94 samples plus one laboratory standard sample). The Fuhrberg 2 data as well as all other sets described below are derived from what we define as minimal isotherms. The Fuhrberg Garden set originates from the same region but includes 11 topsoil and subsoil samples from a site contaminated by ashes. The Fuhrberg Forest set was collected mainly under pine forests; some samples were taken under beech. The original forest sample collection includes a much broader range of samples. Here, we restricted our investigation to samples from mineral horizons with EDTA-extractable Cd > 100 µg kg^{-1} (57 samples), because the samples from humus layers and humus-poor subsoil horizons were too remote from the arable soils, both in Cd load and soil properties.

The samples from 31 Ap horizons from the Braunschweig area were provided by T. Streck and J. Ingwersen (compare listed literature[1,6,9]). The samples include a higher proportion of soils with cambic properties and less podsolic features, because the precipitation is somewhat lower in that region and the parent materials contain more weatherable minerals. The municipal wastewater of the city of Braunschweig is distributed by sprinkling, meaning it is used for agricultural production after biological clearing (for details, see Streck and Richter[1]). The metal contents of these soils are considerably higher than common background levels. Some spots that received higher quantities of water and sludge approach or even exceed the current German limit for the application of sewage sludge, which is 1 mg Cd kg^{-1} dry soil as measured by aqua regia extraction (referred to as total Cd in this study).

The Berlin samples have a similar but more extreme history of municipal wastewater disposal. The amounts of water as well as the cumulative metal loads were much higher than in the Braunschweig area. Disposal ceased in 1985, and the soils are not presently used for agricultural purposes. A first attempt of afforestation failed due to high levels of contaminants, and probably also because of problems of the water regime. We included seven samples in our measurements from topsoils and subsoils that were provided by C. Hoffmann and his coworkers.

SAMPLE TREATMENT AND DETERMINATION OF SOIL CHARACTERISTICS

The soils were homogenized by sieving (2 mm) and were air-dried at about 25°C. The basic soil properties were determined according to Schlichting et al.[14] The oxides were not removed prior to particle size analysis. OC is the organic carbon from dry combustion. The pH was measured in 0.02 M $CaCl_2$. Feo is noncrystalline Fe extracted by oxalic acid in the dark. The CEC is the effective value at soil pH.

CD EXTRACTIONS AND EQUILIBRATIONS

Total Cd is the aqua regia fraction. The sorbed fraction here is defined by an extraction with 0.025 M Na_2–EDTA (1 h, 20°C).

For the 24 standard isotherms interpreted in earlier papers,[8,12,15] we used 10 g of dry soil that was equilibrated for 24 h at 20°C with 25 mL 0.005-M $Ca(NO_3)_2$ containing increasing additions of Cd. The supernatant obtained from centrifuging was filtered to remove particles >0.45 μm. For the minimal isotherms, the basic procedure was the same, but only two solutions with Cd additions of 200 μg L^{-1} and 2500 μg L^{-1} were used. To obtain the sorbed fraction of Cd, S, for plotting the isotherms, the initially sorbed Cd, S_0, as given by the EDTA extraction described above, was added to the newly sorbed Cd obtained from the experiments. It is important to note that the EDTA extraction used here yielded about 30% less Cd than the earlier procedure used in Springob and Böttcher.[12] Here, the lower actual EDTA fraction was also applied to recalculate the 24 standard isotherms from the original sorption data. Thus, the derived Freundlich k and M data for the Fuhrberg 1 set is not exactly the same as those tabled in Springob and Böttcher[12] because k and M markedly depend on the applied fraction of S_0, as they discuss in detail. Nonetheless, to compare all 225 isotherms in this present study, it was necessary to apply a uniform fraction of sorbed Cd.

It should also be noted that our background solution is only 5 mM $Ca(NO_3)_2$, which we found to be closer to the original soil solutions of our soils than the 10 mM $Ca(NO_3)_2$ used in many other investigations. The lower concentration of the electrolyte is a major reason that the derived k values might appear to be generally higher than in other studies (e.g., Filius et al.[7]).

To obtain Cd data for different soil solutions, Cd-free solutions containing $Ca(NO_3)_2$ and $CaCl_2$ in different concentrations were equilibrated with the soil (1:2.5) for 12 h. These soils were centrifuged two times: once to obtain the supernatant, and a second time at higher speed to remove all suspended particles > 0.45 μm.

Measurements

Most measurements were carried out on a Perkin Elmer 1100B/HGA 700 atomic absorption spectroscopy (AAS) instrument using individual temperature programs and standards for each type of solution matrix (water for fresh soil solution, 0.01-M_0 $Ca(NO_3)_2$, 0.025-M Na_2–EDTA, and aqua regia). The lowest calibration curve was 0.1 μg L^{-1} with readings involving one decimal. The average deviation between replicated measurements was less than 5%, even at the lower end. Most of the Cd concentrations in the EDTA solutions were measured by flame AAS.

Curve Fitting and Multiple Regression Analysis

A linear regression of the log data yielded k and M for the 24 standard isotherms. The parameters of the minimal isotherms were also derived from log data from which M is directly available as the slope of the function log S over log C_{Cd}. The parameter k was obtained by inserting M, S, and C_{Cd} into the Freundlich equation (12.1) and rearranging, with S being given by the EDTA fraction, S_0, plus the newly sorbed amount δS for one of the two isotherm points. C_{Cd} is the related equilibrium concentration measured for the same point.

To obtain the pedotransfer function, Eqs. 12.4 and/or 12.5, which originate from 12.3 and 12.4, were fitted to the k and soil data using a range of combinations of soil properties as independent variables. Just the best results are presented.

$$k = k* x_1^a x_2^b \ldots x_n^z \qquad (12.4)$$

$$\log k = \log k* + a \log x_1 + b \log x_2 \ldots + z \log x_n \qquad (12.5)$$

Equation 12.4 was fitted by an iterative procedure to original (nonlog) data that stresses the influence of higher k values (results of Figure 12.1d only). Equation 12.5 was fitted as a multiple linear regression of log k and log soil data, thus giving more influence to lower k values.

Theoretically, it would also be possible to fit the whole extended Freundlich equation (12.3) and not just k to measured data of S, C_{Cd}, and soil properties. This was done by Streck and Richter[6] using Cd in laboratory suspensions as C_{Cd}, the EDTA fraction as S, and the vertical and horizontal variability in one field (S, C_{Cd}, OC, or pH) for regression statistics. The procedure also yields an estimate for M, but just one average for the group of soil samples used to derive the parameters. Here, we intended to have M values for each sample, and therefore preferred the two-point isotherms.

RESULTS AND DISCUSSION

BASIC DATA

Table 12.1 groups the data according to pH, region, and analytical quality. Quality here means that Freundlich parameters M larger than 1, as obtained from fitting the Freundlich equation to the isotherms, were considered to have a high probability of being erratic and thus were excluded from some of the calculations. Almost all M > 1 occurred in the contaminated sites of Fuhrberg Garden and Berlin. As mentioned above, M depends on the fraction used to correct for S_0. We used the EDTA fraction, which might have been too weak to give reliable results in the higher contaminated soils, thus giving an M > 1. For later calculations with mean M values, we used the value of M = 0.815 (Table 12.1, M > 1 excluded) if complete data were concerned. For some purposes, it may also be useful to apply individual M values for different groups of soils, even within the sandy substrates. Here, the samples above pH 6 had a significantly lower mean (0.730) than those above pH 6 (0.827). Correlations

TABLE 12.1
Mean values and standard deviations (in parentheses) for soil properties, Cd loads, and isotherm-derived Freundlich parameters k and M. C_{Cd-800} is estimated Cd concentration of soil solution for assumed sorbed fraction S of 800 μg Cd kg⁻¹ soil and fixed k value of 207 μg^{1-M} L^M kg⁻¹. Size of k was chosen to obtain Cd concentration of 5 μg L⁻¹ for large basic group no. 2. Current German limit for drinking water is 5 μg L⁻¹

No.	Group	n	OC (%)	pH in CaCl₂	Cd_{EDTA} (μg kg⁻¹)	k (μg^{1-M} L^M kg⁻¹)	M [-]	C_{Cd-800} (μg L⁻¹)
1	All data	225	2.63 (1.75)	4.92 (1.27)	205 (383)	174 (220)	0.839 0.132	4.78
2	M > 1 excluded	212	2.67 (1.77)	4.85 (1.24)	197 (372)	162 (210)	0.815 0.077	5.00
3	Fuhrberg 1,2 arable	119	2.12 (0.99)	5.44 (0.51)	134 (43)	204 (208)	0.817 0.081	5.01
4	Braunschweig arable	31	1.31 (0.91)	5.83 (0.45)	276 (175)	218 (184)	0.798 0.067	5.19
5	Fuhrberg forest	56	4.60 (1.71)	2.92 (0.23)	152 (49)	13.8 (7.2)	0.828 0.085	4.89
6	Contaminated sites	19	2.27 (2.05)	5.99 (0.82)	697 (1205)	393 (354)	1.050 0.280	3.48
7	Samples with pH >6	25	1.92 (1.28)	6.30 (0.15)	240 (183)	491 291	0.730 0.065	6.05
8	Samples with pH <6	185	2.78 (1.81)	4.63 (1.18)	192 (393)	108 117	0.827 0.071	4.90
9	Samples with pH <4	55	4.61 (1.72)	2.92 (0.23)	154 (50)	14.0 7.2	0.825 0.082	4.92

n = Number of samples in each group

between M and pH were sometimes found (e.g., Buchter et al.[2]), but usually were weak (compare Figure 12.5).

Table 12.1 also provides an estimated Cd concentration of the soil solution assuming a sorbed fraction S of 800 μg kg^{-1} soil, which would be the status of the soil if the current limit for the application of sewage sludge would be reached. It can be seen that C_{Cd}, for a uniform value of k (207 μg^{1-M} L^M kg^{-1}), ranges from 3.48 to 6.05 μg L^{-1} if the individual mean M values of each group are applied, and the overall mean M gives a concentration of 5 μg L^{-1} for the same k and S. On the other hand, the differences between the groups of arable soils are small, and even C_{Cd} in the forest soils does not differ significantly from the overall mean. Therefore, if contaminated sites or soils with high pH are regarded, an improper mean M value may be responsible for deviations of about 20% between the mean of estimates and the mean of the real values of C_{Cd}. For individual samples, such deviations can be larger.

The overall ranges found in the data, both for soil properties and sorption parameters, were considerable. In particular, parameter k showed standard deviations higher than the mean values (Table 12.1). As the data were only approximately normally distributed, conventional statistical figures should be taken with some caution. Here, statistical procedures are used to parameterize the pedotransfer functions, not primarily to derive significances. The interpretation should be based only on the final linear relationships between measured and estimated parameters (e.g., Figure 12.2 and Figure 12.3). Significances are only mentioned where there is no doubt that the figure can be used.

MULTIPLE REGRESSIONS BETWEEN FREUNDLICH k AND SOIL PROPERTIES AND PREDICTED Cd CONCENTRATIONS OF SOIL SOLUTIONS

Generally, pH and OC had the strongest influence on k. The clay content, or the content of amorphous Fe that was closely correlated with clay in these soils, contributed significantly to the fits when just the parameters of the 24 complete isotherms were included in multiple regressions (Figure 12.1), because these soils were selected to provide some range in clay contents. When all 225 soils were combined, no more influence of the independent variable clay was observed, probably because there were not enough samples with higher clay contents coming from the investigated regions. The variable "clay," therefore, is not contained in the final multiple model (Figure 12.2), but may have influence in other constellations. Figure 12.1 also shows that the effective CEC, which already combines the influences of clay, OC, and pH as far as the sorption of cations is concerned, does not sufficiently explain the variation of k. The highest r^2 (0.956) is obtained in Figure 12.1d, using iterative curve fitting instead of multiple linear regression of the log data. If the parameters are optimized this way to give a high r^2, by forcing higher k values to move toward the 1:1 line, the arrangement within the low k values gets worse, as can be seen in Figure 12.1. Low k values are ecologically more important as they stand for high solute Cd concentrations. Thus, the models of Figure 12.1b or 12.1c must be considered the best, even with lower r^2.

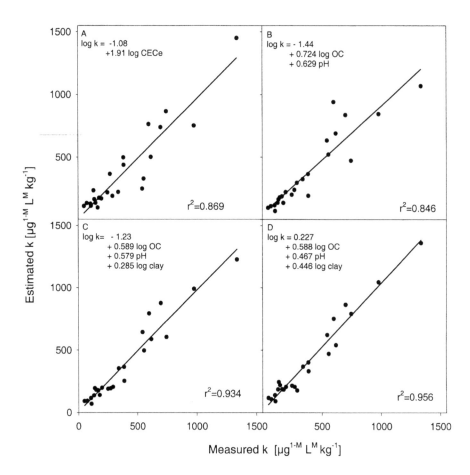

FIGURE 12.1 Freundlich sorption parameter k for Cd as estimated from soil properties vs. isotherm-derived, measured k for Fuhrberg 1 dataset (Ap horizons). (a–c) Multiple linear regression of log data (Eq. 12.4), (d) Iterative fit to original data (Eq. 12.5). Data and functions are recalculated from original isotherm and soil data[12] using different fractions of S_0 (initially sorbed Cd) to define S in preceding fits of Eq. 12.1 to laboratory isotherms.

Figure 12.2 combines the k data from all 225 samples. The scattering around the 1:1 line is fairly high. The coefficient of determination is still 0.882, but the number of soils where the differences between measured and estimated k exceed 100% is considerable. The conclusion is that individual k values estimated from the multiple model used in Figure 12.2 would have a high probability of being wrong by a factor of 2 or 3 (but not much more). This may be a problem when Cd translocation in soil columns is modeled, but not if an evaluation of range of resulting Cd concentrations over a variety of soils of a region, or otherwise grouped samples, is intended.

Figure 12.3 shows the resulting Cd concentrations of the soil solutions, calculated for the original, actual Cd loads of the soils using Eq. 12.1 (S_0 = actual EDTA fraction). For k, estimated and measured values were used. With the estimated

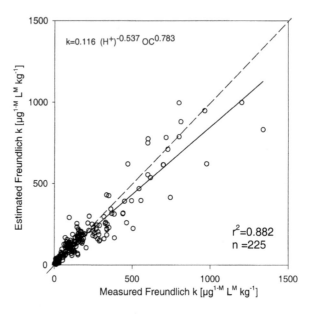

FIGURE 12.2 Freundlich sorption parameter k for Cd as estimated from soil properties vs. isotherm-derived, measured k for 225 sandy north German soils.

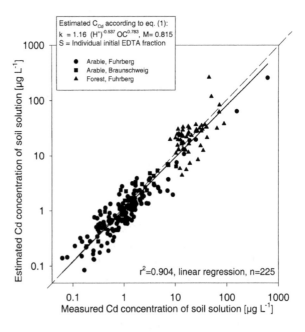

FIGURE 12.3 Estimated Cd concentrations of soil solutions as based on Freundlich k from pedotransfer function (in graph) vs. measured data for all 225 sandy north German soils.

k values, an exponent M of 0.815 was used for all soils. With the measured k data, we also used the measured M values. The resulting concentrations span the range of three orders of magnitude. Considering this enormous range and using a log scale, the remaining error of 100 or 200%, given by the scattering around the 1:1 line in Figure 12.3, can still be taken as moderate. Most of the samples actually sustain Cd concentrations below 2 μg L^{-1}. Only the acid forest soils range between 10 and 100 μg L^{-1}. As mentioned before, the data may still be affected by problems of defining S_0. Therefore, there should be at least some doubt whether the extremely high concentrations around 100 μg L^{-1} really occur in nature.

PREDICTION OF SOLUTE Cd AT INCREASED LEVELS OF SORBED Cd AND REMAINING PROBLEMS FROM INAPPROPRIATE VALUES OF FREUNDLICH EXPONENT M

In Figure 12.4 we calculated solute Cd in the same manner as in Figure 12.3, but for a uniform total Cd load of 1000 μg kg^{-1} in all soils. This is the current German soil limit for the application of sewage sludge. We assumed that 80% of this total amount fell into the EDTA fraction, so all values were derived for a sorbed fraction S of 800 μg kg^{-1}. The Freundlich parameters k were the measured ones. Plotting C_{Cd} vs. k using log scales would, according to Eq. 12.1, result in a function describing a perfect line if the individual values of M were used. Here, we used the average M of 0.815 from Table 12.1 together with the individual k values. Any deviation from the ideal function is due to deviations between mean and individual Ms. It can be seen that there are, in fact, noticeable deviations, although they do not appear to be serious if the whole scale of data is taken into account. Some points, however, mainly in the middle and upper part, are fairly remote from the regression line, meaning that using mean or individual M, respectively, may cause differences higher than 100% in the resulting value of C_{Cd} (compare also the small graph in Figure 12.4b). It is, however, unclear whether the problem is due to problems of the individual M values, which might contain experimental errors, e.g., caused by inappropriate S_0, or whether the use of average M values is generally a problem. In this context, Figure 12.5 shows that some of the variability of M seems to be determined by pH.

Figure 12.4b illustrates that a strong increase of C_{Cd} can be expected if k falls below about 200 μg^{1-M} L^M kg^{-1}. To compare the expected values of C_{Cd} if the total Cd loads are equal to the sewage sludge limit, we added some lines indicating a range of critical liquid Cd concentrations in Figure 12.4a. Value b (2 μg L^{-1}) is taken from Ingwersen et al.,[9] who found unacceptably elevated (modeled) Cd concentrations in wheat grain above this level. Value c (3 μg L^{-1}) is the WHO[16] limit for drinking water, and value d (5 μg L^{-1}) is the corresponding German limit.[17] Values a and e originate from a comprehensive literature study by Bird and his colleagues,[18] who provided levels of Cd that affect biota in aquatic and terrestrial ecosystems. They converted the basic data where necessary to free metal activities of solutions and, on this basis, defined ENEVs—expected-no-effect-values. The final value for Cd in soil solutions was 8 μg L^{-1}; the corresponding level for aquatic systems was only 0.2 μg L^{-1} (value

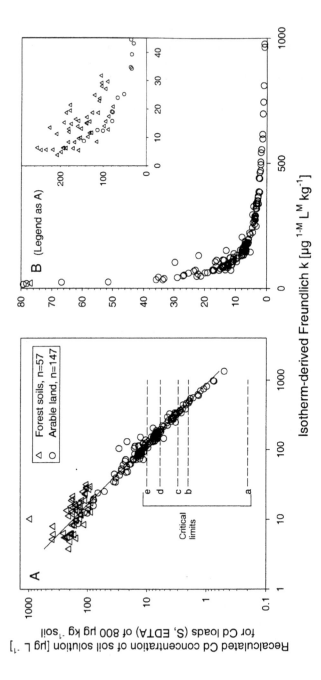

FIGURE 12.4 Cd concentrations of soil solutions calculated from measured individual k values but using one mean M value of 0.815, plotted vs. individual sorption parameter k. All scattering originates from deviations between original, individual M and mean M. Soils are grouped according to samples from forest and arable land. Highly contaminated sites Fuhrberg Garden and Berlin are not included. (a) See text for critical values (a–e) for plant uptake of Cd, drinking water, and effects on biota. (b) Same data, linear scales, and increased resolution in the inset.

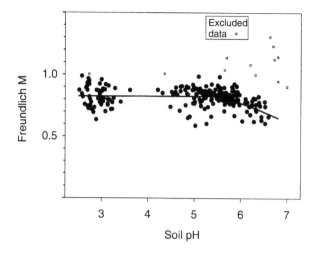

FIGURE 12.5 Freundlich exponent M vs. soil pH for all 225 samples. Samples indicated are those regarded as suffering from analytical problems (definition of S_0) and that were excluded to obtain final overall average of M of 0.815 used to derive estimated Cd concentrations of Figures 12.4 and 12.5.

a). For soils, the authors used different models to derive the free metal activities from measured concentrations. Considering their approaches and our technique, we think that their activity-based ENEV of 8 μg L^{-1} would best be approximated by a Cd concentration of 10 μg L^{-1} (value e in Figure 12.4a). However, Bird and his colleagues[17] also cited literature sources in which effects were found at much lower activities than 8 μg L^{-1} in terrestrial systems. Our data suggest that a large number, probably the majority of the investigated soils, would fall into the range of negative effects, both on soil biota and on soil–plant transfer, if total Cd approached 1 mg kg^{-1}.

INFLUENCE OF IONIC STRENGTH AND PRESENCE OF Cl$^-$ ANION

The above predictions of C_{Cd} are, within some other constraints, restricted to the presence of one electrolyte—5 mM Ca(NO$_3$)$_2$—and are valid for exactly 20°C. The same is true for the measured and estimated k values, which are high in this chapter because the electrolyte is relatively dilute. Soil solutions vary in composition and ionic strength. Especially the presence of Cl$^-$ increases the solubility of Cd and thus increases k. The same can be assumed for dissolved organic carbon (DOC), which is not as abundant in German arable soils as it may be in acid forest soils. Here, we evaluate the influence of the ionic strength and of the Cl$^-$ concentration for a typical sandy soil material from one Ap horizon. The soil was not dried before the experiment, and 12 h were allowed for equilibration. The results are illustrated in Figure 12.6. As expected, C_{Cd} increases markedly if the concentration of the electrolyte increases, and is generally higher if Cl$^-$ is the accompanying anion. The soil solution ratio was 1:2.5, which may have caused some dilution of C_{Cd}. Nonetheless, the

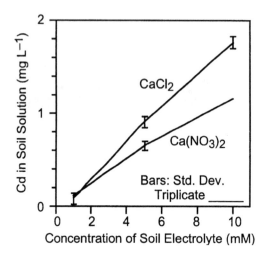

FIGURE 12.6 Cd concentrations of soil solutions of one soil sample (Ap horizon, undried material) as influenced by concentration of background electrolyte and accompanying anion. Soil:solution ratio was 1:2.5.

functions of Figure 12. 6 can be used to derive the concentration of C_{Cd} relative to a basis 1 (C_{Cd}rel) from the concentrations of Ca^{2+} and Cl^- (C_{Ca}, C_{Cl}). The basis, C_{Cd}rel = 1, is given by C_{Cd} for the standard electrolyte, here for 5 mM $Ca(NO_3)_2$:

$$C_{Cd}rel\ (-) = 0.0474 + 0.0948\ C_{Ca}\ (mM) + 0.0448\ C_{Cl}\ (mM). \qquad (12.6)$$

Equation 12.6 is empirical and valid for total ionic strength ranging from about 1 to 10 mM and for Ca^{2+} as dominant cation. The function now allows predictions of C_{Cd} to be corrected for assumptions about the ionic strength and the presence of the complexing anion Cl^-. This was done in Figure 12.7, from which the influence of different combinations of matrix and solution variables on C_{Cd} becomes obvious. For Cd load, we used the total fraction, again assuming that 80% falls into the EDTA fraction, which gives S. Eq. 12.1. The calculations reach into the range of heavy Cd contamination, here up to 8000 μg kg^{-1}. Currently, food production is still taking place at this level on a number of German soils. The application of further sludge is generally forbidden above 1000 μg kg^{-1}. The necessary parameter k to apply Eq. 12.6 was derived from OC and pH using the pedotranfer model given in Figure 12.3, and M is the mean of 0.815 (Table 12.1, M > 1 excluded). From the resulting functions, it can be seen that C_{Cd} can get very high, even if the total load remains below 1000 μg kg^{-1}. Even the relatively high ENEV of Bird and his colleagues[17] of 10 μg L^{-1} Cd, here transformed to concentrations, can be exceeded, even below the current sewage sludge limit. This becomes obvious from Curve 3 of Figure 12.7. Slight decreases in pH plus a decrease in OC, e.g., due to a new equilibrium in a future warmer climate or simply because of the actual level of OC, do not reflect the equilibrium for ploughed soils (soil that was formerly used as grassland), and

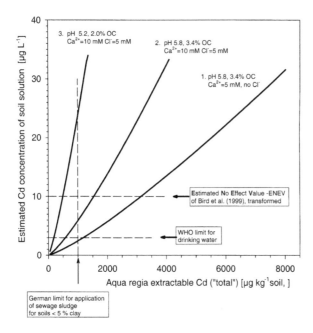

FIGURE 12.7   Examples combining variables pH, organic carbon, concentration of electrolyte, and proportion of Cl$^-$ among anions to estimate Cd concentrations of soil solution of sandy soil as function of total (aqua regia) soil Cd. Indicated are current German limit for disposal of sewage sludge, WHO drinking water limit of 3 µg L^{-1}, and estimated-no-effect-value (ENEV) for soil solution. ENEV is transformed from metal activity of 8 µg L^{-1} (basic value by Bird et al.[17]) to metal concentration 810 µg L^{-1}.

would cause much C_{Cd} for a given Cd load. In addition, many soils in the region presently provide such properties. It is clear from Figure 12.7 that changes of the electrolyte are of major importance if final solute concentrations are used as a criterion to evaluate threshold values defined in terms of total Cd. Curve 2 further shows that even the low WHO limit for drinking water (3 µg L^{-1}) may be exceeded below the sewage sludge limit of 1000 µg kg^{-1} if the composition of the soil solution changes.

CONCLUSIONS

Pedotransfer functions to estimate Freundlich k are promising, at least for the studied sandy soils. If the dataset is well selected, meaning that genetic differences between soils and horizons are not too large even as pH, OC, and clay vary considerably, then about 96% of the variability of k can be explained from these three variables (Figure 12.1). If more heterogeneous data are regarded or less analytical effort is spent on the isotherms, then the influence of variables other than pH and OC remains unclear. There was no significant influence of the clay content in Figure 12.2, where 225 soils were combined.

The Freundlich exponent M should be subjected to more detailed studies, especially as far as its dependence on pH is concerned, because the use of average values does not appear to be justified in every case. Using averages may result in errors in the final prediction of C_{Cd} for individual soils of 100% or more. On the other hand, to get an impression of the variability of C_{Cd} and its ability to predict soil and solution variables over a broad range of soil samples, a mean M of 0.815 appears to be satisfactory, but it must be modified if any other fraction than Cd_{EDTA} is applied as sorbed fraction in the Freundlich equation or if other electrolytes are used. The value of 0.815 is only valid for 5 mM $Ca(NO_3)_2$ at 20°C.

The observed Cd concentrations of the soil solution were relatively high, especially the estimated ones, if the Cd load approached the current German limit for the application of sewage sludge. A number of critical limits for Cd in solution will be exceeded by a large number of soils at this Cd level. Our data and calculations suggest that the limit of 1 mg kg^{-1} total Cd, to say the least, contains no real safety margin for the protection of the full ecological functioning of soils, especially in soils that naturally contain smaller concentrations. Biota probably are affected, and it has to be decided whether this is acceptable in the context of the disposal of sewage sludge.

There is a clear lack of knowledge, however, about the ecotoxicological implications effective at defined levels of Cd in solution because this is no longer a question of orders of magnitude, but of differences in the range of some µg Cd L^{-1}. To deal with this issue, Cd concentrations (or metal activities) of soil solutions will have to be measured with increased precision. This will only be possible with improved techniques to extract "natural" soil solutions. These will also be useful to derive improved pedotransfer functions based on the extended Freundlich equation. At present, such functions are valuable for modeling Cd translocation in spatially variable soils and thus facilitate groundwater protection, but the prediction of C_{Cd} for one spot still contains a possible error of at least 100 to 200%. The same holds true if breakthrough times of Cd toward, e.g., the groundwater surface, are predicted.

Not only pH and sorbents (OC, clay, oxides) influence k and the finally predicted level of C_{Cd}. The electrolyte (strength, presence of Cl^-) has a similarly strong impact. Therefore, the evaluation of threshold values defined by, e.g., total Cd or by the EDTA fraction, cannot simply be based on C_{Cd} derived from extended Freundlich equations via pH and OC, but has to include at least a compensation for ionic strength and Cl^-. Further, variables (DOC, temperature) should be also tested.

ACKNOWLEDGMENTS

J. Griebel, A. Heitkamp, and B. Reuschenberg participated in performing some of the laboratory work. T. Streck, J. Ingwersen, and C. Hoffmann provided a number of processed soil samples from their own study areas.

REFERENCES

1. Streck, T. and Richter, J., Heavy metal displacement in a sandy soil at the field scale, II. Modelling, *J. Environ. Qual.*, 26, 56, 1997.

2. Buchter, B., Davidoff, B., Amacher, M.C., Hinz, C., Iskandar, I.K., and Selim, H.M., Correlation of Freundlich K_d and n retention parameters with soils and elements, *J. Soil Sci.*, 148, 370, 1989.

3. Gray, C.W., McLaren, R.G., Roberts, A.C., and Condron, L.M., Solubility, sorption and desorption of native and added cadmium in relation to properties of soils in New Zealand, *European J. Soil Sci.*, 50, 127, 1999.

4. Chardon, W.J., *Mobiliteit van cadmium in de bodem*, Thesis, Agricultural University, Wageningen, The Netherlands, 1984. (In Dutch)

5. Van der Zee, S.E.A.T.M. and van Riemsdijk, W.H., Transport of reactive solute in spatially variable soil systems, *Water Resource Res.*, 23, 2059, 1987.

6. Streck, T. and Richter, J., Heavy metal displacement in a sandy soil at the field scale. I. Measurement and parameterisation, *J. Environ. Qual.*, 26, 49, 1997.

7. Filius, A., Streck, T., and Richter, J., Cadmium sorption and desorption in limed topsoils as influenced by pH: Isotherms and simulated leaching, *J. Environ. Qual.*, 27, 12, 1998.

8. Springob, G. and Böttcher, J., Parameterisation and regionalisation of Cd sorption in sandy soils, II. Regionalisation: Freundlich k estimates by pedotransfer functions, *J. Plant Nutr. Soil Sci.*, 161, 689, 1998.

9. Ingwersen, J., Streck, T., and Richter, J., Verfahren zur regionalen Bewertung der Cadmiumeinträge in die Böden des Abwasserverregnungsgebietes Braunschweig, in *Bodenökologie und Bodengenese*, 26, Renger, M., et al., Eds., 152, 1998 (in German).

10. Tiktak, A., Leijnse, A., and Vissenberg, H., Uncertainty in a regional-scale assessment of Cd accumulation in the Netherlands, *J. Environ. Qual.*, 28, 461, 1999.

11. Schulte, A. and Beese, F., Adsorptionsdichte-Isothermen von Schwermetallen und ihre ökologische Bedeutung, *J. Plant Nutr. Soil Sci.*, 157, 295, 1994 (in German).

12. Elzinga, E.J., van Grinsven, J.J.M., and Swartjes, F.A., General purpose Freundlich isotherm for cadmium, copper and zinc in soils, *European J. Soil Sci.*, 50, 139, 1999.

12. Springob, G. and Böttcher, J., Parameterisation and regionalisation of Cd sorption in sandy Soils. I. Parameterisation (Freundlich), *J. Plant Nutr. Soil Sci.*, 161, 315, 1998.

13. Welp, G. and Brümmer, G.W., Adsorption and solubility of ten metals in soil samples of different composition, *J. Plant Nutr. Soil Sci.*, 162, 155, 1999.

14. Schlichting, E., Blume, H.-P., and Stahr, K., *Bodenkundliches Praktikum*, Berlin, 1995. (In German)

15. Böttcher, J., Use of scaling to quantify variability of heavy metal sorption isotherms. *Euro. J. Soil Sci.*, 48, 379, 1997.

16. WHO, World Health Organization, *Guidelines for Drinking Water Quality*, Vol. 1, Recommendations, 2nd ed., Geneva, 1993.

17. TVO, *Verordnung über Trinkwasser und Wasser für Lebensmittelbetriebe vom 22.05.1986*, BGBl. I, last change 1993, 1993, p. 760.

18. Bird, G.A., Sheppard, M.I., and Sheppard, S.C., *Effects Characterization: Cd, Cu, Ni, Pb, Zn and As*, prepared in support of PSL Assessment of Releases from Copper and Zinc Smelters and Refineries, Pinawa, Manitoba, Canada, 1999.

13

Effects of Sorbed and Dissolved Organic Carbon on Molybdenum Retention by Iron Oxides

Friederike Lang and Martin Kaupenjohann

ABSTRACT

Molybdenum occurs in soil solution mostly as MoO_4^{2-} and $HMoO_4$. At low pH it is sorbed very strongly to iron oxides. Studies conducted with acid forest soils imply that dissolved organic carbon (DOC) and organic substances sorbed to iron oxides increase the mobility of Mo. The aims of the experiments presented here were to elucidate the role of dissolved and sorbed organic C in the retention of Mo by iron oxides.

In our study we used pure goethite and goethite covered with organic carbon., The goethite samples were analyzed by XRD and BET measurement. Afterward, we incubated C free and C coated iron oxide samples with Mo for 12, 24, and 48h at pH 4. Molybdenum desorption kinetics from the incubated iron oxides were analyzed by anion exchange resin. In addition, XPS analyses of incubated iron oxides were conducted. DOC treatment did not affect the crystallinity of the iron oxides; in contrast, pore volume of the goethite samples clearly decreased after C covering. The desorption kinetics of Mo were biphasic. A fast desorption was followed by only slow Mo mobilization at longer desorption times. Desorbability of Mo from C-coated iron oxides, as well as mobilization rate, were higher than from pure oxide samples. We suggest the fast mobilization represents Mo desorption from the outer surfaces of iron oxides, while slow mobilization is due to diffusion out of the micropores and interdomains of iron oxides. XPS analyses show that, in the presence of C, the percentage of Mo sorbed at outer surfaces is increased. Carbon coverage seems to reduce the accessibility of the pores and may thus decrease the diffusion of Mo into the iron oxides.

To prove whether there are chemical interactions between organic substances and Mo, dialysis experiments with DOC and p-jump experiments with pure and coated iron oxides were performed. The results imply that Mo is complexed by DOC. Molybdenum/DOC co-sorption seems to be possible.

From our results we conclude that Mo fixation to iron oxides is mainly due to the diffusion of Mo into the pores of iron oxides. Organic substances seem to clog the pores and may thus act as diffusion barriers for oxyanions. In addition, Mo-DOC co-sorption and competition for sorption sites may be relevant for the Mo mobility in soils of high DOC concentration.

MOLYBDENUM BIOAVAILABILITY: EXCEPTIONAL POSITION OF ACID FOREST SOILS

Several studies have identified the major factors governing molybdenum sorption to soils and iron oxides. The influence of pH,[1,2] concurring anions,[3] or the ionic strength[2] have been examined. Many experiments also deal with Mo bioavailability in agricultural soils. However, knowledge of Mo availability at acid forest sites, where Mo nutrition might be crucial for NO_3 nutrition of trees,[4] is very scarce.

Studying the Mo supply at forest sites, we determined that there was, in general, low Mo availability in acid soils due to strong Mo sorption at low pH.[5] On the other hand, Mo seems to be leached from topsoil horizons very quickly when compared with agricultural soils, resulting in Mo enrichment in the illuvial horizons.[6] In a Mo fertilization experiment at a spruce site, we found that 50% of applied Mo was leached into the subsoil, in spite of low pH (4.1–4.5). This is in contrast to other studies from nonforest soils, where no Mo leaching could be observed even at higher pH (e.g., Smith and Leeper,[7] pH 5–6).

For the most part, organic anions dominate the soil solution of forest soils. Dissolved organic carbon (DOC) has been shown to increase the mobility of inorganic cations[8] and anions.[9] Batch experiments conducted with DOC, Mo, and a synthetic iron oxide show that this is also true for Mo.[10] We suggest, therefore, that strong sorption of Mo to iron oxides as well as high contents of organic anions limit the validity of conclusions drawn from agricultural studies for acid forest soils.

The aims of the model experiments presented here are (1) to identify rate-limiting steps for Mo mobilization from iron oxides, (2) to study the effect of carbon coatings on Mo desorption kinetics, and (3) to examine whether Mo–DOC cosorption may occur in soils.

EFFECTS OF CARBON COATING ON Mo FIXATION BY IRON OXIDES

Sorption experiments with soils and iron oxides show that Mo sorption is biphasic.[11] Initial rapid Mo sorption is followed by only slow sorption, resulting in decreased Mo desorbability or bioavailability, respectively. This is in agreement with other oxyanions such as phosphate[12] or arsenate.[13] It is widely accepted that slow anion sorption is due to the diffusion of anions into the pores and interdomains of iron oxides (e.g., Strauss et al.[14]). Consequently, the porosity of iron oxides seems to determine ion fixation.[15]

Pure iron oxides are usually used for sorption experiments, neglecting carbon coatings that are mostly present at the surface of soil iron oxides.[16] However, carbon coatings will probably be relevant to ion sorption because they might change the diffusion pathways of Mo, P, or As into iron oxide pores, and thus might affect Mo mobility in soils.

INFLUENCE OF CARBON COATING ON CRYSTALLINITY AND POROSITY OF IRON OXIDES

To prove whether C coating changes the porosity of iron oxides, we compared the x-ray diffraction spectra of coated and uncoated goethite samples and determined the N_2 sorption/desorption hysteresis of the goethites. We used a pure synthetic goethite (G13), with a specific surface of 13 $m^2 g^{-1}$, and goethite G13 coated with organic matter as described elsewhere by Lang and Kaupenjohann.[17] Organic matter coatings were achieved as follows: 2.5 g of the respective iron oxides were equilibrated with 1 L of DOC solution (80 mg C L^{-1}, pH 4.5). The DOC solution was obtained by mixing 1 kg organic forest floor sample with 5 L deionized water for 24 h. The organic soil was extracted just after sampling. The extracted solution was filtered (0.45-μm cellulose acetate membrane filter, Sartorius) before use. After mixing with DOC solution for 6 h, the iron oxides were filtered, washed with deionized water, and freeze-dried. As a result, we obtained goethite samples with a C content of 10 mg g^{-1}.

Coated and uncoated iron oxides were analyzed using N_2 sorption/desorption hysteresis to determine pore size distribution via the BJH method[18] (BET analyzer, Nova 1200, Quantachrome). The iron oxides were further characterized by x-ray diffraction (Röntgendiffraktiometer, Siemens) with Cu K α radiation ($\lambda = 0.15406$ nm) and with a 0.05°-step scanning (1.5 s per step).

The x-ray diffraction spectra of coated and uncoated G13 samples were identical (not shown). Thus, incubation with DOC solution does not seem to influence the crystallinity of G13. In contrast, carbon coating clearly affects pore volume, as determined by N_2 sorption (Table 13.1). Carbon adsorption obviously reduces the accessibility of iron oxide pores for N_2. Since BET analyses were conducted with dried

TABLE 13.1
Porosity of Uncoated (G13−) and Coated (G13+) Synthetic Goethites.

	Average Pore Diameter (nm)	Pore Volume ($\varnothing < 50$ nm) ($mn^3 g^{-1}$)
G13−	18	60
G13+	7	29

goethite samples, real pore-clogging effects may even be more relevant. This is in general agreement with results from BET analyses of soil samples, which show that pore volume increases after removing organic C by H_2O_2 treatment.[19]

MOLYBDENUM DESORPTION KINETICS

Given the concept of diffusion-controlled Mo fixation, reducing pore volume by carbon coating will change Mo desorption characteristics. Thus, studying the Mo desorption kinetics of coated and uncoated goethite samples enabled us to further prove the diffusion hypothesis.

Experimental Setup

The sorbents were incubated with 0.2 mM Mo solution at a solids concentration of 5 g L^{-1}. Molybdenum was provided as Na_2MoO_4, and the pH was adjusted to 4.0. After 12, 24, and 48 h of incubation, Mo desorption kinetics were examined. The samples were shaken gently during incubation. For a detailed description of the preincubation, see Lang and Kaupenjohann.[17]

Desorption of Mo sorbed to the iron oxides during preincubation was studied by anion exchange resin extraction (strong basic anion exchange resin, Cl form, type I, LAB III, Merck). We used 20 mL deionized water per 100 mg of iron oxide and 2 mL of resin, which equals 1.6 mmol$_C$.

The incubated iron oxides were divided into subsamples that were exposed to ion exchange resin for different desorption periods from 0.5 h to 48 h. The samples were shaken throughout the desorption period at 293 K. At the end of the desorption period, the resin was separated from the iron oxides by wet sieving. Desorption of Mo from the resin was achieved within 2 h by adding 30 mL 2-M HNO_3. The desorbability of Mo sorbed to ion exchange resin was tested separately and was about 98%.[20] Mo concentrations in the HNO_3 exchange solution were determined by furnace atomic absorption spectroscopy (AAS) (Varian SpectrAA 800 Z).

Role of Carbon Clogging for Molybdenum Desorption Kinetics

There was no clear effect of incubation time or C coating on Mo sorption (Table 13.2). Most of the Mo was sorbed after just 12 h of incubation. However, the desorbability of Mo decreased with increasing residence time. Much more Mo was desorbable from carbon-coated G13 than from C-free G13. This agrees with experience derived from P desorption experiments with soils. Zhou and his colleagues[21] found that in spite of similar amounts of PO_4 sorbed to Bs and Bh horizons, desorbability of P was higher from Bh than from Bs samples, where coating-free iron oxides are to be expected.

Desorption kinetics of anions from soils and iron oxides have been found to be biphasic. Fast desorption is followed by only slow mobilization.[22-24] To account for this, we combined the first-order equation and the parabolic diffusion equation, which has been used for modeling diffusion limited processes,[25-27] to describe the time dependence of Mo desorption:

TABLE 13.2
Amount of Mo sorbed to uncoated (G13−) and coated (G13+) goethite samples and desorbability of Mo after different incubation times

		\multicolumn{3}{c}{Incubation Time (h)}		
		12	24	48
Mo sorbed (mg g^{-1})	G13−	3.0	3.0	3.0
	G13+	3.0	3.3	3.4
Mo desorbability	G13−	21	4	4
(% of sorbed)[a]	G13+	48	25	20

[a]Determined after 48 h of extraction

$$Mo_{des} = a_0 - a_0 \times e^{-kt} + mt^{0.5} \qquad (13.1)$$

where Mo_{des} is the Mo desorbed at time t (mg g^{-1}), a_0 is the amount of Mo desorbable by the fast reaction, k is the rate constant of the fast reaction (h^{-1}), and m is the rate constant of the slow reaction (mg g^{-1} h$^{-0.5}$). Parameters a_0, k, and m were determined by fitting Eq. 13.1 to the Mo desorption data (SigmaPlot, Jandel). First-order kinetics was accepted for the fast desorption. Mobilized Mo is immediately sorbed to the resin. Hence, readsorption of mobilized Mo can be neglected.

The model is based on the idea that desorption of Mo from outer surfaces is responsible for the fast step, while mass transport phenomena of Mo within mineral pores account for slow mobilization. The rate constant m is related to the diffusion constant of the suggested anion diffusion.[28]

$$D/r^2 = \pi m^2 \times (6x_m)^{-2} \qquad (13.2)$$

where r is the radius of diffusion; x_m is the sum of Mo mobilizable via diffusion (mg g^{-1}); and D/r^2 is the apparent diffusion constant (h^{-1}).

Corresponding to the diffusion concept, Mo desorption kinetics could in general be modeled by Eq. 13.1. This is shown in Figure 13.1, for 24 h of preincubation time. Molybdenum mobilization is more rapid in the presence of C. Diffusion constants determined according to Eq. 13.2 are bigger for C-coated than for pure G13 samples (Figure 13.2a).

With increasing incubation time, Mo anions may penetrate deeper into the pores of iron oxides, getting smaller with increasing distance from the surface.[29] This seems to be reflected by the decrease in diffusion constants with increasing incubation time. The pore-clogging effect of carbon coatings may hinder Mo ions from penetrating into iron oxide pores. Accordingly, diffusion constants increase as a consequence of C coating, as does a_0, representing the portion of Mo sorbed to outer surfaces of G13 (Figure 13.2b).

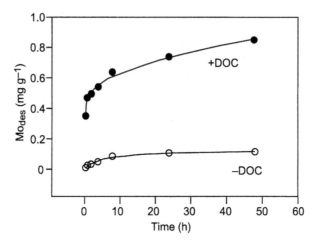

FIGURE 13.1   Molybdenum mobilization kinetics from coated and uncoated G13 samples after 24 h incubation.

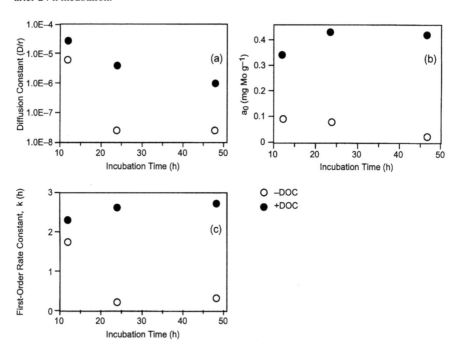

FIGURE 13.2   Parameters of Mo mobilization kinetics from G13 as function of preincubation time.

Following the idea that the desorption model is based on, the rate constant of the first-order term of Eq. 13.1 characterizes the sorption of Mo to outer surfaces of iron oxides. This constant was greater for +DOC treatments than for −DOC treatments at all preincubation times (Figure 13.2c). Different types of bonding to outer surfaces may be the reason. Without C coatings, Mo is bound directly to the iron oxide via ligand exchange; with C coatings, Mo may be complexed by organic molecules sorbed to the iron oxide. However, the parameters of the fast reaction should not be overinterpreted because, in the batch experiments conducted here, desorption and complex solution kinetics may be overlapped by mass transport phenomena.

Pore-Clogging Effect: Spectroscopic Evidence

X-ray photoelectron spectroscopy (XPS) is suitable for determination of molar compositions of solids.[30] This method is sensitive only for the outer layers of minerals (2 nm). The penetration depth of XPS is limited by the ability of electrons, scattered by x-ray, to penetrate solids. Thus, the molar ratio of outer surfaces of goethites can be analyzed via XPS.

To be able to examine the Mo/Fe ratio at the surface of iron oxides by photoelectron spectroscopy, iron oxides with a higher Mo surface concentration were needed. Therefore, we also conducted sorption studies with the respective goethite samples, using a Mo concentration of 1 mM and incubation times of 12 h, 48 h, and 21 d. Incubation of G13 and analyses of supernatant solutions were conducted as described above. Filtered and washed goethite samples were freeze-dried prior to XPS analyses.

The Mo/Fe molar ratio at the surfaces of incubated iron oxides was determined by XPS (ESCAlab 220i vacuum generator) with a constant analyzer energy analyzator, the x-ray source was Mg $K\alpha_{1,2}$ radiation. The measurements were conducted without charge compensation.

In contrast to the experiment with low Mo concentration, in the case of 1-mM Mo sorption experiments, Mo sorption to G13 increased with increasing incubation time and C covering decreased the amount of Mo sorbed to G13 (Table 13.3). However, the Mo/Fe ratio at the uncoated iron oxide surface was constant throughout incubation. Thus, XPS analyses seem to prove true the concept of diffusion-controlled Mo sorption. Slightly increasing the Mo/Fe ratio at the surface of coated samples might be due to a slow diffusion of Mo through the organic carbon coating.

The element ratio after 48 h of preincubation is nearly the same for coated and uncoated G13 samples (Table 13.3). Providing a mean Fe–Fe distance of 0.3 nm at the iron oxide surface, as suggested by Schwertmann and Cornell,[31] results in a Mo sorption density at the goethite surface between 5 and 9 (Table 13.3). Modeling Mo sorption envelopes to G13 gives a total sorption site density from 6 to 8,[32] so the surface covering of Mo seems to correspond to the sorption capacity of the outer surface of G13. Consequently, differences in sorbed Mo with and without carbon might be due mainly to the different accessibility of inner surfaces.

In the case of lower surface concentration, decreased Mo diffusion seems to be compensated for in part by increased sorption at the outer G13 surface, as can be concluded from the 0.2-mM sorption experiment (Figure 13.2b). This is proved by an XPS study of a less crystalline goethite.[17]

TABLE 13.3
Molybdenum content of G13 after different incubation
times and Mo/Fe ratio at surface of G13 as well as Mo
sorption density (calculated). Values in parentheses ( )
give standard deviation

Time (h)		Mo Content (mg g^{-1})	Mo/Fe	Mo Sorption Density (nm^{-2})
12	G13−	8.4 (0.2)	0.22	8.0
	G13+	6.2 (1.2)	0.15	5.5
24	G13−	10.2 (0.3)	0.24	8.7
	G13+	8.2 (1.1)	0.20	7.3
48	G13−	12.4 (0.7)	0.27	9.0
	G13+	9.5 (0.5)	0.22	8.0

MOLYBDENUM/DOC COSORPTION

As mentioned above, C coatings of iron oxide might affect Mo sorption not only because of the pore-clogging effect, but also due to Mo complexation by organic molecules, which are sorbed at the surface of iron oxides. Different studies imply that Mo may be complexed by soil organic matter. For example, Karimian and Cox[33] observed that Mo sorption to soils is correlated positively with the content of organic carbon.

Bibak and Borggaard[34] suggest that Mo forms complexes with humic acids, while Xie and McKenzie[35] conclude that cation bridging is responsible for Mo bonding to organic substances. From complex chemistry research, it is known that Mo is complexed by organic molecules, providing two hydroxyl groups that are in ortho position to each other.[36] Molybdenum is octahedrally coordinated in such complexes, that is, it has the coordination number 6. For example, gallic acid that originates in soils from lignin reduce shows high Mo complexation capacity. Seen against this, complexation of Mo by DOC and Mo–DOC cosorption seem to be possible. However, as far as we know, there are no studies dealing with Mo–DOC interactions. Using dialysis and p-jump experiments, we studied Mo sorption by soil organic molecules.

DIALYSIS EXPERIMENT

To study Mo–DOC interactions we used dialysis membranes. A dialysis tube that is permeable for free Mo but not for Mo–DOC complexes is filled with DOC and Mo-containing solution. The tube is then inserted into a polyethylene bottle filled with deionized water. According to the diffusion principle, an equilibrium between the solution inside and outside the dialysis membrane will be reached according to the free Mo concentration. Based on the differences between total Mo concentration inside and outside, complexed Mo can be calculated.

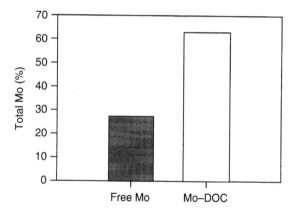

FIGURE 13.3 Portions of complexed and free Mo in presence of DOC.

For our experiment, we filled the dialysis tube with 50 mL Mo + DOC solution (Mo: 0.2 mM as Na_2MoO_4, DOC: 60 mg C L^{-1}, pH 4) and inserted it into a PE bottle containing 150 mL deionized water. Molybdenum concentration inside and outside the tube was determined after 2, 4, and 7 days of equilibration by furnace AAS (Varian SpectrAA 800 Z). To account for possible Mo polymerization effects, we also studied Mo dialysis of DOC-free solutions. The treatments were conducted with one paralleling the others.

According to Jones,[37] Mo polymerization, which might result in different Mo concentrations inside and outside the tube, should be taken into account above 1 mM Mo at pH 3.0 and an ionic strength of 1 M. Molybdenum polymerization decreases with increasing pH and decreasing ionic strength. Thus, we provide that in our experiments polymerization reactions can be neglected. This is assured by the result of the − DOC treatment: dialysis of the Mo solution without DOC results in equal Mo concentrations outside and inside the dialysis tube, even after 2 d of equilibration.

In the presence of DOC, Mo total concentration was higher inside (0.11 mM ± 0.01) than outside (0.03 mM ± 0.004) the tube. The change in concentrations could now be observed along with time. Accordingly, equilibrium obviously has already been reached after 2 d. The result implies that DOC has a high affinity for Mo (Figure 13.3), as shown for other heavy metals (e.g., McBride et al.[8]).

PRESSURE-JUMP EXPERIMENTS

Principle of p-Jump Technique

With the p-jump technique, very fast kinetic reactions with half-lives in the range of milliseconds can be analyzed. This method has been suggested, for example, by Hachiya et al.[38] to examine chemisorption processes at the surface of minerals that are in general very fast.[26]

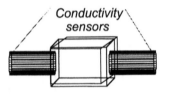

Sample Cell:
5 g goethite dm^{-3}
! C coatings
100 mg Mo dm^{-3}
pH 4.0

Conductivity
sensors

Reference Cell:
C-free goethite
NaNO$_3$
C-coated goethite
supernatant solution

Autoclave:
Brass membrane at top
applied pressure: 13 MPa
Temperature: 5–30°C

FIGURE 13.4 P-jump experimental setup.

During p-jump experiments, pressure is applied to a sorbent/sorbate suspension. As the equilibrium constant is pressure dependent, this results in perturbation of the chemical equilibrium. After a short time, a new equilibrium according to the high pressure will have been reached. Now the pressure is released again, and ambient pressure will be adjusted within microseconds. At that time, chemical relaxation is measured via conductivity measurement.[39]

Figure 13.4 shows a very simplified model of a p-jump apparatus. For pressure application, water is pumped into the autoclave. The cell that contains the study suspension is connected to the autoclave via an elastic PE membrane, which enables pressure transformation from the autoclave to the sample cell. To account for the pressure dependence of conductivity, a second cell is enclosed (reference cell) that contains a nonrelaxing solution with the same conductivity as the solution of the sample cell. A brass membrane is placed at the top of the autoclave that bursts at a pressure of 13.5 MPa.

Experimental Conditions

Pressure-jump experiments were conducted with Mo goethite suspensions. The same goethite (G13) described above was used in uncoated and coated form. The suspensions were equilibrated for 24 h prior to p-jump analyses. Figure 13.4 shows the experimental conditions.

Modeling p-Jump Results

Because the distance from equilibrium is very small, reaction kinetics can be described by a first-order equation irrespective of reaction order. It has been shown for different oxyanions that sorption/desorption kinetics studied by p-jump can be modeled by the combination of two first-order terms (e.g., Grossl et al.[40]). Thus, we fit Eq. 13.3 to our results:

$$C = C_1 \times e^{-t/r1} + C_2 \times e^{-t/r2} + C_3 \qquad (13.3)$$

where C is the relative conductivity, r_1 and r_2 are the relaxation times, t is the time, and C_1, C_2, and C_3 are constants.

Effects of Carbon Coating on Mo Chemisorption

The results of all experiments could be fitted very well by Eq. 13.3. Each measurement was done with 10 replicates. The variation coefficients were below 10%. Slow and fast relaxation times differ clearly. The fast relaxation time was about tenfold smaller than one-tenth the size of the slow one (Figures 13.5 and 13.6). According to Strehlow and Knoche[41] in this case, short relaxation times can be calculated irrespective of slow processes.

Based on single p-jump experiments, it is not possible to identify rate-limiting processes. Thus, to prove whether relaxation times represent chemisorption processes, we measured the temperature dependence of the relaxation time. As shown in Figures 13.5 and 13.6, rapid relaxation of pure goethite suspension is strongly temperature dependent, and no effects of temperature on slow relaxation can be observed. The energy of activation of the fast relaxation derived from temperature dependence of rate constants (Arrhenius equation, Atkins,[42] p. 716 ff.) is 76 kJ mol^{-1}, that is, in the range of chemical reactions. Thus, we suggest that the rate-determining step for fast relaxation is chemisorption of Mo to the iron oxide, while the low-energy slow relaxation is due to mixing and transport phenomena within the suspension.

No influence of carbon coating on fast relaxation can be observed (Figure 13.5). This implies that chemisorption to iron oxides is not affected by sorbed organic carbon. However, slow relaxation as well as the temperature dependence of the slow relaxation time change as a consequence of carbon coating (Figure 13.6). The energy of activation of the slow relaxation of coated samples is 63 kJ mol^{-1}. We hypothesize that complexation of Mo to C coatings, being a slow process compared to chemisorption to iron oxides, overlaps the mixing process observed for Mo/goethite suspensions

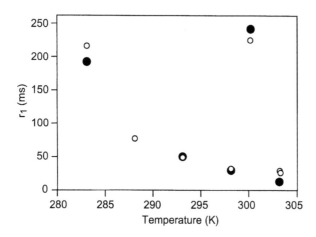

FIGURE 13.5   Temperature dependence of fast relaxation time.

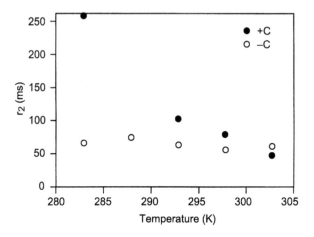

FIGURE 13.6   Temperature dependence of slow relaxation time.

without C. However, for more mechanistic interpretation, further experiments are needed to differentiate between DOC–goethite- and DOC–Mo-controlled relaxation. Nevertheless, our results seem to confirm the hypothesis of Mo–DOC cosorption.

CARBON COATINGS AS A DIFFUSION BARRIER FOR ANIONS

Our results suggest that diffusion of Mo into the pores of iron oxides results in Mo fixation and limits Mo mobilization. Carbon coatings on iron oxides may clog the pores of iron oxides and slow down Mo penetration. This is expected to be true for other anions, such as PO_4 or AsO_4.

In addition, interactions between sorbed organic molecules and anions might "catch" anions and keep them from diffusing to the inner surfaces of iron oxides. Thus, when studying anion sorption to iron oxides, sorbed carbon should be taken into account, as it may explain the discrepancy between results from model experiments with pure goethites and field observations.

ACKNOWLEDGMENTS

We wish to thank H.D. Narres and A. Pohlmeier, Forschungszentrum Jülich, for helping us with the p-jump experiments. We extend our thanks to F. Simon, Institute of Polymer Research, who conducted the XPS analyses.

REFERENCES

1. Schwertmann, U. and Cornell, R.M., *Iron Oxides in the Laboratory,* VCH, Weinheim, 1991.

2. Reisenauer, H.M., Tabikh, A.A., and Stout, P.R., Molybdenum reactions with soils and the hydrous oxides of iron, aluminum and titanium, *Soil Sci. Am. Proc.,* 26, 23, 1962.

3. Goldberg, S. and Forster, H.S., Factors affecting molybdenum adsorption by soils and minerals, *Soil Sci.,* 163, 109, 1998.

4. Xie, R.J., McKenzie, A.F., and Lou, Z.J., Causal modeling pH and phosphate effects on molybdate sorption in three temperate soils, *Soil Sci.,* 155, 385, 1994.

5. Plass, W., Molybdänmangel bei Sulfat- und zeitweisem Nitrat-Überangebot—Ein hypothetischer Beitrag zum Waldsterben in Westdeutschland, *Geoökodynamik,* 4, 19. 1983 (in German).

6. Lang, F. and Kaupenjohann, M., Molybdenum fractions and mobilization kinetics in acid soils, *Z. Pflanzenernähr, Bodenkunde,* 162, 309–314, 1999.

7. Lang, F. and Kaupenjohann, M., Molybdenum at forest sites: contents and mobility, *Canadian J. Forest Res.,* 30, 1034–1040, 2000.

8. Smith, B.H. and Leeper, G.W., The fate of applied molybdate in acidic soils, *J. Soil Sci.,* 20, 246, 1969.

9. McBride, M., Sauvé, S., and Hendershot, W., Solubility control of Cu, Zn, Cd and Pb in contaminated soils, *Euro. J. Soil Sci.,* 48, 337, 1997.

10. Kaiser, K. and Zech, W., Dissolved organic matter sorption as influenced by organic and sesquioxide coatings and sorbed sulfate, *Soil Sci. Soc. Am. J.,* 62, 129, 1997.

11. Lang, F., *Molybdän auf sauren Waldstandorten,* Bayreuther bodenkundl, Berichte, Bd. 45, 1995.

12. Barrow, N., Leahy, P.J., Southey, I.N., and Purser, D.B., Initial and residual effectiveness of molybdate fertilizer in two areas of southwestern Australia, *Aust. J. Agric. Res.,* 36, 579, 1985.

13. Barrow, N.J., Testing a mechanistic model, I. The effects of time and temperature on the reaction of fluoride and molybdate with a soil, *J. Soil Sci.,* 37, 267, 1986.

14. Onken, B.M. and Adriano, D.C., Arsenic mobility in soil with time under saturated and subsaturated conditions, *Soil Sci. Soc. Am. J.,* 61, 746, 1997.

15. Strauss, C.R., Brümmer, G.W., and Barrow, N.J., Effects of crystallinity of goethite: Rates of sorption and desorption of phosphate, *Euro. J. Soil Sci.,* 48, 101, 1997.

16. Madrid, L. and Arambarri, P., Adsorption of phosphate of two iron oxides in relation to their porosity, *J. Soil Sci.,* 36, 523, 1985.

17. Mayer, L.M., Relationships between mineral surfaces and organic carbon concentrations in soils and sediments, *Chem. Geol.,* 114, 347, 1994.

18. Lang, F. and Kaupenjohann, M., Influence of carbon coatings in the Mo fixation by iron oxides, *Euro. J. Soil Sci.,* 1999 (submitted).

19. Barrett, E., Joyner, L., and Halenda, P., Determination of pore volume and area distribution in porous substances, I. Computations from nitrogen isotherms, *J. Am. Chem. Soc.,* 73, 373, 1951.

20. Pennell, K.D., Boyd, S.A., and Abriola, L.M., Surface area of soil organic matter reexamined, *Soil Sci. Soc. Am. J.,* 59, 1012, 1995.

21. Lang, F. and Kaupenjohann, M., Molybdenum in forest soils: Infinite sink extractability and its relevance for Mo uptake, *Soil Sci. Am. J.,* 2000 (submitted).

22. Zhou, M., Rhue, R.D., and Harris, W.G., Phosphorus sorption characteristics of Bh and Bt horizons from sandy coastal plain soils, *Soil Sci. Soc. Am. J.,* 61, 1364, 1997.

23. Van der Zee, S.E.A.T.M., Fokking, L.G.J., and Van Rimsdijk, W.H., A new technique for assessment of reversibly adsorbed phosphate, *Soil Sci. Soc. Am. J.,* 51, 599, 1987.

24. He, Z.L., Yang, X., Yuan, K.N., and Zhu, Z.X., Desorption and plant availability of phosphate sorbed by some important minerals, *Plant and Soil,* 162, 89, 1994.

25. Ruan, H.D. and Gilkes, R.J., Kinetics of phosphate sorption and desorption by synthetic aluminous goethite before and after thermal tranformation to hematite, *Clay Minerals,* 31, 63, 1996.
26. Hodges, S.C. and Johnson, G.C., Kinetics of sulfate adsorption and desorption by Cecil soil using miscible displacement, *Soil Sci. Soc. Am. J.,* 51, 323, 1987.
27. Sparks, D.L., *Kinetics of Soil Chemical Processes,* Academic Press, San Diego, pp. 210, 1989.
28. Barrow, N.J., A brief discussion on the effect of temperature on the reaction of inorganic anions with soil, *J. Soil Sci.,* 43, 37, 1992.
29. Ma, Y.B. and Uren, N.C., The effects of temperature, time and cycles of drying and rewetting on the extractability of zinc added to a calcerous soil, *Geoderma,* 75, 89, 1997.
30. Fischer, L., Zur Mühlen, E., Brümmer, G.W., and Niehus, H., Atomic force microscopy investigation of the surface topography of a multidomain porous goethite, *Euro. J. Soil Sci.,* 47, 329, 1996.
31. Briggs, D. and Shea, M.P., *Practical Surface Analysis by AES and XPS,* John Wiley, New York, 1990.
32. Schwertmann, U. and Cornell, R.M., *Iron Oxides,* VCH, Weinheim, 1996.
33. Karimian, N. and Cox, F.R., Adsorption and extractability of molybdenum in relation to some chemical properties of soils, *Soil Sci. Soc. Am. J.,* 42, 757, 1978.
34. Lang, F. and Kaupenjohann, M., Characterization of Mo sorption to iron oxides: Apparent versus mechanistical equilibrium constant, 2000 (in prep.).
35. Bibak, A. and Borggaard, O.K., Molybdenum adsorption by aluminum and iron oxides and humic acid, *Soil Sci.,* 158, 323, 1994.
36. Xie, R.J., and McKenzie, A.F., Molybdate sorption desorption in soils treated with phosphate, *Geoderma,* 48, 321, 1991.
37. Gmelin, *Gmelin Handbook,* Mo Suppl., Vol B 3b, Springer Verlag, New York, 1995, 163.
38. Jones, L.H.P., The solubility of molybdenum in simplified systems and aqueous soil solution, *J. Soil Sci.,* 8, 313, 1957.
39. Hachiya, K., Ashida, M., Sasaki, M., Kann, H., Inoue, T., and Yasunaga, T., Study of the kinetics of adsorption-desorption of Pb on Al_2O_3 surface by means of relaxation techniques, *J. Phys. Chem.,* 83, 14, 1979.
40. Bernasconi, C.F., *Relaxation Kinetics,* Academic Press, New York, 1976.
41. Grossl, P.R., Eick, M., Sparks, D.L., Goldberg, S., and Ainsworth, C.C., Arsenate and chromate retention mechanisms on goethite, 2. Kinetic evaluation using a p-jump relaxation technique, *Environ. Sci. Technol.,* 31, 321, 1997.
42. Strehlow, H. and Knoche, W., *Fundamentals of Chemical Relaxation,* VCH, Weinheim, 1977.
43. Atkins, P.W., *Physical Chemistry*, Oxford University Press, Oxford, 1990.

14 Speciation and Sorption of Lead (II) in Soils

Alexander Ponizovsky and Eugene Mironenko

ABSTRACT

Lead contamination of soils is caused by the supply of various compounds whose properties essentially influence the mobility and availability of the element. Some mechanisms of lead binding in soils can be evaluated basing on the data on the content and forms of this trace metal in soils and chemical properties of its compounds. Lead 'species' in soils, estimated by the procedures presented in the literature, have no unique chemical interpretation. Each method gives only some 'operational' values that can be compared only with that obtained with the same procedure for the soils of similar properties. The 'exchangeable' fraction cannot be taken as the amount of Pb(II) that can be exchanged, 'carbonate' fraction – as $PbCO_3$, and so on. Basing on the data obtained with the existing methods of fractionation up to 10-70% of the total amount of lead in acid soils can exist in exchangeable form, bound either with inorganic or organic compounds. In neutral soils, the greatest part of lead may be 'bound to Fe-Mn oxides' and 'organic' fraction. In slightly alkaline soils, lead is retained as 'carbonate', 'organic', and 'residual' fractions. At pH <6, ion exchange can be the main process of lead retention by soils. The equivalent exchange between Ca^{2+} and Pb^{2+} occurred in soils containing halloysite, montmorillonite, vermiculite, or Al-humus complexes. Some 'extra' sorption was observed in soils abandoned with allophane, imogolite, iron oxides, or vermiculite-chlorite intergrades. Lead can be sorbed in the amounts that exceed the cation exchange capacity. This phenomenon hardly could be interpreted either in terms of inclusion of a single charged hydroxy species such as $PbOH^+$, or by taking into account the precipitation of lead(II) hydroxide.

INTRODUCTION

Contamination with heavy metals (HM) has become the ultimately troubling problem all over the world, especially in the developed countries. Lead is one of the most toxic heavy metals. Entering into the environment and subsequently into the food chain, this metal causes heavy poisoning of both animals and humans. Therefore, it is very important to monitor the paths and forms of lead migration, to forecast transport of its compounds in the environment, and to develop approaches for preventing the dangerous impact of this pollutant.

To do this it is necessary to estimate not only the total amount of lead, but also its chemical forms that may reside in soil or be bound to it. The data on lead speciation in soils are numerous in the literature, but are hardly mutually compatible. Lead binding in soils is referred to as ion exchange, formation of surface complexes, precipitation of slightly soluble salts, and so forth.

The object of this study is to estimate the mechanisms of lead binding in soils based on the data on the content and forms of this trace metal in soils and the chemical properties of its compounds.

To estimate the properties of lead compounds in soils, it is worthwhile to evaluate the paths of entry and the distribution of this metal in soils.

PATHS AND MAGNITUDES OF LEAD ENTRY TO SOILS

World production of metallic lead and its compounds, according to one review,[1] was estimated at about 2 million tons (recalculated to pure metal) in the 1970s. Lead was used in storage batteries (35%), the production of tetraethyl lead (10%), paints (5%), cable casing (10%), and construction (5%). The rest was used to manufacture ammunition, in polygraphy, in radiation protection, and so forth.

Soils are subject to lead contamination from the waste of polymetallic ore mining and smelting, lead admixtures in fertilizers, use of tetraethyl lead as an addition to motor fuel, and the manufacture and processing of storage batteries.[2,3] In some areas, soil contamination by lead is caused by horticulture applications of insecticides containing lead arsenate $PbHAsO_4$.[4] A specific contamination problem arises at shooting ranges,[5] hunting areas, and battlefields where metallic lead is spread in soil as more or less fragmented pieces.

The combustion of ethylated motor fuel containing tetraethyl lead (TEL), ethylene dibromide, ethyl bromide, and other additives, with a content of lead up to $0.78\ g\,L^{-1}$,[6] led to the emission of lead (II) bromide and oxide that contaminated soils and plants near busy roads.[7,8] In the U.S. alone, from 1926 to 1985, use of TEL caused the emission of more than 7 million tons of lead into the environment.[6] In the 1970s in the U.S., cars combusted annually about 250,000 tons of TEL. Since 1970, the production of TEL has decreased, but in 1984 the atmospheric emission of lead in the U.S. was assessed at 39,000 tons, 90% of which was released due to combustion of ethylated fuel, 5% in smelting, and 4% in incineration.[9,10]

In recent years, changes in technology have redistributed lead production among different branches, but the figures quoted are worthy of attention as they reflect a situation that has existed for a rather long time.

These data, together with those on lead emissions in other countries and presented in numerous publications (e.g., *Report on Lead Contamination . . . in the Russian Federation . . .*[11]), reveal a very dangerous picture of global contamination with this heavy metal.

LEAD CONTENT IN SOILS

Lead content in various soils is said to be between 2 to 300 mg kg^{-1}.[3,12-15] In noncontaminated soils, the mean values are taken to be from 15 to 17 mg kg^{-1} for light soils, and from 17 to 22 mg kg^{-1} for heavy textured soils.[16] These concentrations are higher than those in rocks, where lead content varies from 0.1 mg kg^{-1} (ultra-basic rocks) to 8 mg kg^{-1} (basic rocks) and 20 mg kg^{-1} (acid rocks). Sedimentary rocks usually contain about 20 mg kg^{-1} of lead.[17]

The variation of lead content in different noncontaminated soils is not very high. In the main types of soils of the European part of the former Soviet Union, in territories remote from coal and ore mines and smelters, lead content varied from 15 to 47 mg kg^{-1}.[18] The highest values were observed in Dernovo-podzolic, gray forest soils, and leached Chernozem (from 33 to 35 mg kg^{-1} in the upper layers), and the lowest were estimated for Podzolic soil of light texture on the fluvioglacial deposits (23 mg kg^{-1} in the upper layer) and in red earth soil (Krasnozem) from Georgia (10 mg kg^{-1}). The highest lead content was estimated in the humic accumulative horizons, especially in forest litter, the rooting zone, and the topsoil horizon. The second maximum in the Podzolic soil was usually observed in the illuvial horizon. In Chernozems, distribution down the profile was more homogeneous, although in some cases Pb(II) content increased at the alkaline geochemical barrier at the upper boundary of the calcareous horizon. In alluvial Dernovic soils (pH < 6 or pH > 6, 0.2–2% organic matter), lead concentration was 2 to 19 mg kg^{-1}; in meadow alluvial and bog soils (up to 9–30% organic matter), it was 5 to 27 mg kg^{-1}.[19]

The highest lead concentration is usually observed in the surface horizon. In regions of intensive industrial and agricultural development, this distribution is usually attributed to environmental contamination. For example, lead content in the surface horizon 0–20 cm in soils of the intensively used regions of Switzerland varied from 13.5 to 40.5 mg kg^{-1} (median 23.8 mg kg^{-1}), but in the subsoil it was 7.8 to 18.3 mg kg^{-1} (median value 12.1 mg kg^{-1}).[20] Vogel and his colleagues suggest that aerosol emissions from traffic, industry, and incineration plants are mostly responsible for increased lead contents in the surface horizons. In cambic arenosol on glacifluvial deposits under the forest in the forest floor in southern Norway (O horizon, pH 3.7), Berthelsen and his coworkers[21] observed a lead content of 163 mg kg^{-1}. It was essentially higher than that in the lower horizon E (9 to 14 mg kg^{-1}) and in horizon Bs1 (pH 4.8) (10 to 18 mg kg^{-1}). The authors attribute this to airborne deposition of heavy metals as well as significant accumulation of lead by plants. In some cases, researchers have observed other patterns of lead distribution in contaminated soils. In the acid brown soils around former Pb–Zn–Cu mines in the U.K. (pH < 5, between 3 and 11% organic matter and 24 to 43% clay), Pb content distribution in the 5-cm layers of the upper 25 cm of the profile was homogeneous.[22] Merrington and Alloway[22] explain this by the strong retention of the metal in the soil, which prevents its leaching.

In noncontaminated soils, the highest lead content is also observed in the litter and in the surface horizon. For example, in the mountain soil profiles in the Sudety Mountains in Poland, the maximal lead content was estimated in the litter.[23] A similar phenomenon was observed in the soils of the European part of Russia,[18] Israel,[24] and other regions.[25] The authors explained it by the extensive transfer of the element by plants from the soil-forming rock. This distribution can hardly be caused by some global atmospheric precipitation of lead. Only in soils southeast of Lake

Baikal (Siberia) was lead content rather homogeneous down the soil profile, from 16 to 60 mg kg^{-1} (mean value 34 mg kg^{-1}) and close to the Pb(II) concentration in the underlying rocks.[26]

Based on these data, one may conclude that lead is slightly mobile in soils and may be transferred mainly by plants. Even in noncontaminated areas it usually accumulates in the upper horizon, and atmospheric precipitates do not leach it down. However, bulk concentration of the element and its profile distribution do not enable us to guess how it is bound in the soil and what can influence its mobility.

LEAD SPECIES IN SOILS

Various approaches have been suggested to estimate the species of lead in soils (Table 14.1). Most researchers use the procedures recommended by Tessier et al.[27] and Sposito et al.[28] Some authors apply the method of Zeien and Brummer[29] or modify the one suggested by McLaren and Crawford[30] for fractionation of copper in soils.[31] In several publications, exchangeable lead is estimated with the technique commonly used to determine exchangeable alkaline and alkaline-earth cations, i.e., by extraction with 1-N NH$_4$Ac at pH 7.0[32] or pH 4.8.[33,34]

As can be seen from Table 14.1, the approaches used by different authors are very different. The ability of each treatment to extract only one chemical species of soil lead is doubtful. Moreover, the kind of extracting solution as well as the conditions and the number of treatments can essentially influence the results. For example, one can hardly expect that a single extraction with 1-M NH$_4$Ac can replace all of the exchangeable lead (II), since this treatment extracts from the soil no more than 50% even of alkali-earth and alkaline-exchangeable cations,[35] which are sorbed essentially less strongly than Pb(II). On the other hand, Gorbatov and Zyrin[36] suggest that lead (II) extracted with NH$_4$Ac can be exchangeable, as the acetic ion forms relatively stable complexes with Pb(II), with stability constants K_1 = 145, K_2 = 810, and K_3 = 2950, respectively.[1] According to Gorbatov and Zyrin,[36] 0.5-M or 1-M

TABLE 14.1
Solubility (S) and Solubility Products (SP) of Some Lead (II) Compounds at Temperature t, °C

Compound	S (mol L^{-1})	SP	t (°C)	Reference
Pb(OH)$_2$	5.5×10^{-5}	1×10^{-15}	25	Handbook . . .[80]
PbCl$_2$	2.34×10^{-2}	2.12×10^{-5}	25	Handbook . . .[80]
PbCO$_3$	$0.74^a \times 10^{-5}$	2.74×10^{-11}	18	Handbook . . .[80]
		3.3×10^{-14}		Abel[1]
PbHPO$_4$	1.6×10^{-5}	1.2×10^{-10}	25	Handbook . . .[80]
PbSO$_4$	1.8×10^{-4}	1.6×10^{-8}	25	Handbook . . .[80]
PbS	2.9×10^{-17}	8×10^{-28}	25	Abel[1]

$Ca(NO_3)_2$ is more applicable for estimating exchangeable lead (II) content in soils, though soils are much less selective for Ca^{2+} than for Pb(II), and even after seven treatments with this solution not all the exchangeable Pb(II) is replaced.

The extraction with 1-M $CaCl_2$ may be of similar efficiency. However, most researchers apply only one treatment with 1-M $CaCl_2$, really estimating only equilibrium amounts of exchangeable cations in the soil suspension. A single treatment with 0.5-M $Ca(NO_3)_2$ replaces, for example, no more than half of the available exchangeable lead (II) from Brown forest soil and Serozem.[36] A single extraction by 1-M $NaNO_3$ replaces only 15% of the lead (II) that can be replaced with 0.5-M $Ca(NO_3)_2$. Treatment with 1-M KNO_3 is hardly more efficient for such replacement.

Treatment with 4-M HNO_3 is hardly specific to estimate only lead sulfide and may also dissolve some other lead compounds.

Data on lead speciation in some soils are presented in Tables 14.2 and 14.3 to evaluate the influence of soil properties, especially pH, CEC, and organic carbon content, on the distribution of the metal between various forms. A discrepancy can be observed between the presented data and common knowledge of the chemical properties of lead substances. For example, in Thermic Mollic Albaqualfs (Kansas) with pH 5.2, lead was estimated in the fraction obtained by treatment with 0.05-M Na_2EDTA (Table 14.3). This fraction should be called carbonate according to the method of Sposito et al.,[28] despite the fact that solid carbonates cannot exist at this pH value. The same should be said about the 14% of carbonate lead fraction estimated with the procedure of Tessier et al.[27] in Podzol soil (pH 4.1) (Table 14.2).

It can be concluded that lead fractions determined by extraction procedures are only operationally defined, their chemical meaning is very limited, and different data may be matched only if they are obtained with the same method. Estimations of lead fractions by methods other than those suggested by Tessier et al.[27] and Sposito et al.[28] are less numerous. Some are mentioned below to evaluate speciation of the element in some other soils.

In light calcareous, slightly alkaline (pH 7.35 to 7.80) soils of Caracas (Venezuela) with predominant micas and kaolinite, fractionation with a slightly modified McLaren and Crawford[30] procedure revealed less than 0.7% lead in the exchangeable fraction (treatment with 1 N $CaCl_2$), 20 to 50% in the carbonate fraction (2.5% acetic acid extract), 20 to 40% as the element bound with organic matter (revealed with $Na_4P_2O_7$), and 22 to 37% in the residual fraction (HF extract).[31]

In slightly acidic or neutral contaminated soils near Legnitsa and Glogov, Poland, fractionation of lead with the Zeien and Brummer[29] method revealed 1 to 5% exchangeable lead (of the total amount) in the surface horizon (pH_{KCl} 5.6 to 7.4). Two to 12% of Pb(II) was determined to be in the exchangeable fraction in the subsurface horizon (pH_{KCl} from 5.1 to 5.8), but the majority of the element in both horizons was in the organic fraction and bound with Mn oxides.[37]

Acidic sandy podzolic soils of North Belgium (pH_{H2O} 3.4 to 4.8, 50 to 82 g kg^{-1} organic matter) contained from 15 to 20% of exchangeable lead fraction, estimated by treatment with 1 N NH_4Ac.[32] The same procedure applied to acid (pH_{H2O} from 2.8 to 3.9) organic soils sampled in Finland, from west of Helsinki near the roads, revealing from 40 to 70% of exchangeable lead fraction.[38]

TABLE 14.2

Methods Suggested to Estimate Chemical Speciation of Lead in Soils

		Treatment				
N	Fraction	Tessier et al.[27]	Sposito et al.[28]	McLaren & Crawford[30] modified by Garcia-Miragaya[31]	Zeien & Brummer[29]	Zyrin et al.,[32] Andersson[33]
1	Mobile				1 M NH$_4$NO$_3$,	
2	Exchangeable	1.0 M CaCl$_2$ or 1.0 M MgCl$_2$, 2.0 pH 7 (1 h)	0.5 N KNO$_3$ (16 h)	1 N CaCl$_2$	10 M NH$_4$Ac, pH 6	1 N NH$_4$Ac, pH 4.8
3	Sorbed		H$_2$O (2 h, 3 treatments)			
4	Bound to carbonate	1.0 M NaOAc, pH 5 (5 h)	0.05 M Na$_2$-EDTA (6 h)	2.5%	CH$_3$COOH	
5	Reducible (bound to Fe–Mn oxides)	0.04 M NH$_2$OH • HCl in 25% CH$_3$COOH (5 h at 96°C)				
6	Bound to MnOx				M NH$_2$OH • HCl +1 M NH$_4$Ac, pH 6	
7	Bound to amorphous FeOx			0.1 M NH$_4$ Oxal + 0.175 M NH$_4$Oxal, pH 3.25	0.2 M NH$_4$Oxal, pH 3.25	
8	Bound to crystalline FeOx			—	0.2 M NH$_4$Oxal + ascorbic acid, pH 3.25	
9	Oxidizable (bound to organic matter)	0.02 M HNO$_3$ + 30% H$_2$O$_2$ (pH 2) (2 h at 85°C), then 3.2 M NH$_4$OAc in 20% HNO$_3$	0.5 N NaOH (16 h)	0.1 N Na$_4$P$_2$O$_7$	0.025 M NH$_4$-EDTA, pH 4.6	
10	Residual	Conc. HF + conc. HClO$_4$, heating and evaporation to dryness		HF	65% HClO$_4$	
11	Sulfide		4 M HNO$_3$ (16 h, 80°C)			

TABLE 14.3
Influence of soil properties on distribution of available lead between various forms estimated according to procedure of Tessier et al.[27]

Soil	pH[a]	CEC (cmol$_c$ kg^{-1})	C (g kg^{-1})	Exchangeable	Carbonates	Fe-Mn oxides	Organic	Residual
						Fractions of Lead (%)		
Light, sandy soil, Upper Silesia, Poland[81]	6.5 (?)	7.1	13.8	0.5	3.5	15	19.2	61.6
	4.1 (1 M KCl)	10.4	12.6	10	14	52	8	8
Podzol (Haplortods)[81]	6.5 (1 M KCl)	23.5	31.6	0	15	48	19	6
	3.7 (1 M KCl)	7.9	8.9	1	11	62	7	12
Brown (Haplumbrepts)[81]	6.9 (1 M KCl)	24.2	21.1	0	2	59	15	18
Mulberry field[5] soil, horizon A, Annaka, Japan[1]	5.5 (H$_2$O)	14.6	21	6.5	6.5	40.3	33	13

[a]Method used to measure pH is in parentheses.

It can be seen that, in acid soils, a pronounced part of lead is present in the exchangeable fraction. Only in the mountain soils of Sudety (pH$_{H2O}$ 3.2 to 5.3) was lead estimated mainly in the organic and residual fractions as determined with the Zeien and Brummer[29] method.[23]

It is difficult to compare the data presented in Tables 14.2 and 14.3, because even pH values were obtained with different methods. Nevertheless, in spite of all the discrepancies between the methods, the data allow us to derive some general conclusions:

- In slightly alkaline soils, lead is retained mainly as the carbonate, organic, and residual fraction.
- In neutral soils, most lead may be bound to Fe–Mn oxides and as the organic fraction.
- In slightly acidic and acid soils, from 10 to 70% lead can exist in exchangeable form.

The philosophy and operational aspects of trace metal extraction and fractionation in soils were initially designed to be useful if these procedures provided reproducible statistical correlations between the extracted amounts of trace metal and plant uptake or growth response.[28] The available procedures do not provide exact identification of chemical substances that are responsible for lead retention in soils, although the relation between plant availability and lead forms needs extended study and more detailed analysis.

LEAD CONTENT AND SPECIES IN SOIL SOLUTIONS AND LYSIMETER WATERS

Solid lead compounds in soils interact with the soil solution, so the concentration of dissolved Pb(II) should reflect the composition of these compounds. On the other hand, the dissolved lead is supplied to the plant roots and transfers down the soil profile.

Pb(II) concentration in soil solutions and water extracts usually varies from 0.1 to 10 μg L^{-1}.[16,19] A concentration of the metal in water extract 1:2 up to 35,000 μg L^{-1} was observed only in the soils contaminated with lead (II) arsenate.[13]

In noncontaminated soils, Pb(II) content in soil solution is usually low. In soil solutions of the litter at the Hubbard Brook Experimental Forest (U.S.), the content of dissolved lead did not exceed 6 nmol L^{-1} as inorganic ions and complexes,[39] and a similar concentration was bound with colloidal particles. In Bh and Bs horizons of this soil, concentration of the ionic Pb(II) was about 0.4 nmol L^{-1}, and that of bound with colloidal particles was 0.1 to 0.2 nmol L^{-1}. Wang and Benoit[39] suggest that lead leaches out of forest litter, mainly due to the migration of solid and colloid particles.

In lysimetric waters (LW), Pb(II) concentrations are usually of the same magnitude as those in soil solutions. Berggren[40] revealed in LW of brown forest soil in Sweden 2×10^{-2} to 5×10^{-2} μmol L^{-1} (4 to 10 mg L^{-1}) Pb, 75 to 87% of which was bound with organic matter. Increased metal concentration from 50 to 60 mg L^{-1} with essential seasonal variation was estimated in the LW in Podzolic acid brown earth soil, under spruce forest near Goettingen (Germany).[41]

Much higher Pb(II) contents were observed in noncontaminated gray forest soil and leached Chernozem, where concentration of Pb(II) ions in LW varied throughout the year from 0.02 to 0.1 mg L^{-1}, and that of lead bound with the colloid particles was 1.0 to 1.6 mg L^{-1}.[18]

MECHANISMS OF LEAD (II) BINDING IN SOILS

Lead (II) retention in soils is attributed to precipitation of Pb(II) salts and ion exchange with silicates, metal oxides–hydroxides, and organic matter.[14,42] To evaluate the significance of each process, it is worthwhile to compare the properties of soil lead with those of different lead compounds.

NATURAL LEAD COMPOUNDS SUPPLIED TO SOILS BY INDUSTRIAL AND AGRICULTURAL ACTIVITY

The most important lead ore is galena PbS, which is widely distributed throughout the world. Other ore minerals are anglesite $PbSO_4$, cerussite $PbCO_3$, pyromorphite $PbCl_2 \bullet 3Pb_3(PO_4)_3$, and mimetesite $PbCl_2 \bullet 3Pb_3(AsO_4)_3$.[1]

Metallic lead is relatively stable under natural conditions because air and water vapors form a thin protective cover of oxycarbonate on the metal surface. Water and dissolved oxygen produce lead hydroxides that cause toxicity of the water that contacts the lead-covered surfaces. However, if the water contains small amounts of carbonates or silicates, corrosion of the lead cover becomes negligible. The solubility of metallic lead in pure water without air at 24°C is about 311×10^{-6} g L^{-1},[1,3] so the mobility and toxicity of the fragments of this metal at shooting ranges and hunting areas located on nonacid soils should be low.

Lead (II) oxide, PbO, appears, for instance, when the metal is heated in air. Other known oxides are PbO_2 and Pb_3O_4. The former is used in storage batteries; the latter is an ingredient in some paints. The only known lead oxide–hydroxide is $3PbO \cdot H_2O$.

Lead (II) salts together with alkalis produce base salts. For instance, when sodium hydroxide is added to a solution of lead nitrate, two salts are formed, namely $Pb(NO_3)_2 \cdot Pb(OH)_2$ and $Pb(NO_3)_2 \cdot 5Pb(OH)_2$. Even at pH 12, which significantly exceeds the equivalence point, the hydroxide $Pb(OH)_2$ is not formed.[43] Nevertheless, many authors postulate formation of this species, in particular in soils, and present the solubility of some compound with the operational formula of $Pb(OH)_2$.[1]

When alkali is added to a solution of lead acetate with air passing through, the base carbonate $3Pb(CO_3)_2 \cdot 2Pb(OH)_2$ precipitates. Lead (II) sulfate or chloride boiled together with sodium carbonate produces $2Pb(CO_3)_2 \cdot Pb(OH)_2$. This salt is used as a pigment for white paints.

There are several hydrated complexes of lead (II), namely $[Pb(OH)_3]^-$, $[Pb_4(OH)_4]^{4+}$, $[Pb_2(OH)]^{3+}$, $[Pb_3(OH)_4]^{2+}$, $[Pb(OH)]^+$, $[Pb_6(OH)_8]^{4+}$, and $[Pb(OH)_2]^0$. Oxyhalides Pb_2OCl_2, $Pb_3O_2Cl_2$, and Pb_3OCl_4 appear as natural minerals. Lead (II) sulfate precipitates from solutions of lead salts interacting with sulfate. Lead (II) nitrate is well soluble in water; however, unlike most nitrates, it generates the relatively stable complex $[Pb(NO_3)]^+$ (K = 15.1).

Lead (II) hydrophosphate $PbHPO_2$ appears as the natural mineral monetite. It is a slightly soluble salt. Several apatites are known with the common composition $3[Pb_3(PO_4)_2] \cdot PbX_2$, where X is F, Cl, Br, or OH. Lead chlorapatite appears in nature as chloropyromorphite.

Carbonate $PbCO_3$ is found as cerussite. It is formed when soluble lead salts interact with carbonates of alkaline metals in the cold (base salts are formed on heating). This salt may also be obtained by shaking suspensions of less soluble lead salts in the cold, in the presence of sodium carbonate. The solubility of the precipitate increases appreciably in the presence of dissolved CO_2.

One of the least soluble lead species, sulfide PbS, is found as galena. Subjected to the action of atmospheric oxygen, it oxidates. Thus, being exposed on the air, galena covers with the crust of anglesite, $PbSO_4$, which gradually transforms into cerussite, $PbCO_3$.[44] The solubility of some lead (II) compounds is presented in Table 14.4. Comparison of Pb(II) concentrations in soil solutions and LW with the solubility of lead (II) salts (Table 14.4), reveals that concentration of the element in the liquid phase of soils may be controlled by the solubility of only lead carbonate, sulfide, and pyromorphite.

TABLE 14.4

Influence of soil properties on distribution of available lead between various forms estimated according to procedure of Sposito et al.[28]

Soil	CEC pH[a]	C (cmol$_c$ kg^{-1})	Exchange-able + (g kg^{-1})	Sorbed	Organic	Carbonate	Sulfide
			Fractions of Lead (%)				
Thermic Mollic Albaqualfs, Kansas[82]	5.2 (H$_2$O)	n.d.	19	12	0	34	53
	8.1 (H$_2$O)	n.d.	8	11	4.5	35	49
Aquic and Typic Hapludults, Delaware[83]	5.3	3.54	17.6	5.4	7.7	52	35
	5.0 (H$_2$O)	4.1	15.2	6.8	3.6	57	33
Thermic Typic Haploxeralf (Greenfield), California, amended with sludge[84]	7.1 (sat. paste)	8.7	8	2.1	5.3	69	23.6

n.d.—No data

[a]Method used to measure pH is in parentheses.

Santillan and Jurinak[45] and Gorbatov[46] revealed constant levels of lead concentration in the solution phase in soil suspensions of different soil:water ratios that are specific to solutions saturated with respect to some particular compound. Similar data are reported by Jopony and Young.[47] However, this phenomenon may be caused by some other mechanism of lead retention.

Lead carbonate, like all the other carbonates, can be stable only at elevated pH. Stability of Pb(II) sulfide, as mentioned above, is limited by anaerobic conditions. One of the least soluble minerals, chloropyromorphite, $Pb_5(PO_4)_3Cl$, with a solubility of about 0.18 to 20 mg L^{-1}, can precipitate only in the presence of phosphate. Under the other conditions, lead retention should be caused not by precipitation, but by some other processes.

Berthelsen et al.[21] report very low mobility of Pb(II), even in soil horizons with pH_{H2O} 3.4 to 3.7 that does not increase under the influence of acid deposits. These data also support the conclusion that lead in acid soils is not bound in hydroxide and carbonate minerals.

LEAD (II) SORPTION BY IRON AND MANGANESE OXIDES–HYDROXIDES, CLAY MINERALS, AND SOIL ORGANIC MATTER

Extensive studies of Pb(II) interaction with iron oxides were carried out by McKenzie.[48] An increase in pH of 2 pH units, from 3.5 to 5.5, resulted in an increase in adsorption from a low level up to 100% of the amount added. Sorption on Mn oxides (e.g., birnessite) was significantly less influenced by pH increase, and Pb(II) binding did not increase for samples with greater specific surface. Twenty to 93% of lead sorbed by freshly precipitated Mn oxides could not be extracted with 2.5% acetic acid, but after aging for 28 weeks this nonextractable fraction increased to 37 to 100% of sorbed Pb(II). Similar values for fresh and aged goethite were 10 to 44% and 19 to 62% Pb, respectively. Thus, one may conclude that interaction of lead (II), even with pure Mn and Fe oxides, results in formation of both strongly and weakly retained fractions.

The mechanism of lead (II) sorption can by estimated from the balance between sorbed and displaced ions. To preserve constant charge of the surface, Pb^{2+} retention should be accompanied by displacement of some other ions. The ratios of moles of H^+ released per mole metal ion sorbed for manganese hydroxides was found to be close to 1.0 for birnessite and 1.3 for cryptomelane. The ratios for iron oxides (hematite and goetite) ranged from 1.3 to 2.48 The author suggests that in acid, conditions (pH 4) Pb(II) are sorbed according to the equation

$$SOH^0 + M^{2+} = [SO^- - M^{2+}] + H^+ \qquad (14.1)$$

where SOH^0 is an uncharged surface site and the term in square brackets is the adsorbed form of the cation M^{2+}. An increase in the adsorption of the monohydroxo complex can be expected with increasing pH:

$$SOH^0 + M^{2+} + H_2O = [SO^- - MOH^+] + 2H^+ \qquad (14.2)$$

until eventually surface-induced interfacial precipitation of metal hydroxide or surface-induced hydrolysis of the adsorbed ions occurs as

$$S + M^{2+} + 2H_2O = [S - M(OH)_2] + 2H^+. \qquad (14.3)$$

For iron oxides, a sharp increase in adsorption over a narrow range of pH was attributed to the attaining of a pH value at which the adsorption of MOH^+ becomes possible (Eq. 14.2). Nevertheless, to adopt the hypothesis of surface precipitation of Pb hydroxide (Eq. 14.3) one needs to know the properties of this compound and how far its solubility is close to that of the pure Pb(II) hydroxide presented in Table 14.4.

TABLE 14.5
Selected Soil Properties

Soil	Texture	pH (H₂O)	CEC (mmol₍ kg⁻¹)	Organic C (g kg⁻¹)	Specific surface (m²g⁻¹)
Dernovo-Podzolic	Silt loam	6.50	12.2	20	44.6
Gray Forest	Silt loam	6.86	15.2	18	58.7
Leached Chernozem	Fine loam	6.10	29.9	27	109

The interaction of Pb(II) with soils should be much more complex than that with oxides. Various soils were observed to sorb Pb^{2+} from 0.014-M $PbCl_2$ at pH 4.8 to 7 in the amounts that exceed the initial contents of exchangeable Ca^{2+}.[49] High amounts of sorbed $(Ca^{2+} + Pb^{2+})$ (S_{Ca+Pb}) may be attributed to the inclusion of a single charged hydroxy species such as $PbOH^+$. The authors calculated that the observed 20 to 40% increase in $S^{Ca + Pb}$ requires the hydrolysis of more than half the Pb^{2+} present. However, the high S_{Ca+Pb} values were revealed at low rather than high equilibrium pH values. Abd-Elfattah and Wada[49] state that it is unlikely that hydrolysis occurs at pH values as low as 4.9 to 5.5 to the extent mentioned above, so the idea that the presence of single charged hydroxy species accounts for the increase of S_{Ca+Pb} was dismissed. The adsorption of "extra" Pb(II) that replaced H^+, which has not been replaced by Ca^{2+} in the initial soils at pH 7, was considered as an alternative. The equivalent exchange between Ca^{2+} and Pb^{2+} occurred in soils containing halloysite, montmorillonite, vermiculite, or Al-humus complexes, whereas the extra sorption was found in soils in which allophane, imogolite, iron oxides, or vermiculite–chlorite intergrades play a role as cation-exchange materials.

Extra retention by soils of another heavy metal, copper(II), was observed by Ponizovsky and his coworkers.[50] The experiments revealed that the total amount of Cu(II) retained is close to the sum of Ca^{2+} and H^+ displaced from the soil.

Abd-Elfattah and Wada[49] calculated that all the studied solutions were undersaturated with respect to lead hydroxide. The formal selectivity coefficient of the $Ca^{2+} - Pb^{2+}$ ion exchange decreased with the increase of the sorbed Pb(II) fraction. Only from 2 to 13% of exchange sites exhibited the highest selectivity for lead (II) in all the studied samples. The selectivity of samples with respect to Pb^{2+} decreased in the sequence

Fe oxides, halloysite, imogolite > humus, kaolinite > montmorillonite.

According to Abd-Elfattah and Wada,[49] soils containing humus, as compared with other soils, showed no particular indication that humus is a very strong adsorbent for lead (II), but very selective adsorption of Pb(II) was observed on iron oxides.

Sorption of cations in general has been discussed in terms of nonspecific and specific adsorption.[51] Nonspecific adsorption refers to the binding of cations by coulombic interaction in the diffuse electric double layer. Specific sorption refers to the sorption of cations in the inner layer forming coordination bonds to surface metal atoms via O atoms or OH groups.[52] Abd-Elfattah and Wada[49] concluded that exchange sites responsible for extra adsorption did not show a particularly high selectivity for Pb^{2+} cations.

Some authors discuss whether any other reactions could be responsible for Pb(II) retention in soils. Hildebrand and Blum[53] attribute lead (II) retention by clay minerals at pH > 7 to interaction with -SiOH and -AlOH groups. Zyrin and his colleagues[54] suggest that Pb^{2+} ions penetrate the interlayer space of montmorillonite and alter its structure. X-ray diffraction analysis revealed structural alterations of montmorillonite and illite (hydromica) after saturation with Pb(II) of bentonite and clay fraction of Chernozem, where the content of smectites is high. In kaolinite, saturation with Pb(II) did not result in an increase in interlayer spaces. In contrast, Garcia-Miragaya[31] insists that diffusion of Pb(II) ions into the octahedral positions of a crystal lattice of layer silicates is doubtful because their dimensions are too large. The author attributes the existence of a residual fraction of lead to the occlusion of the Pb^{2+} ions by soil silicates.

Another type of interaction could be the formation of some compounds (complexes) with organic matter, sorbed on the surface of solid particles. Bourg[55] suggests that in soils the impact of the clay fraction on lead (II) sorption is insignificant compared with that of organic matter.

Zyrin and his colleagues[54] report that humic acid (HA), separated from the soil, can strongly bind Pb(II), and only 62.5% of lead retained by HA can be taken as an exchangeable fraction. The residual part could be displaced only with 1 M HNO_3. The authors suggest that this acid replaces Pb(II), forming complexes with HA. Pb(II) cations can be bound with HA through two carboxylic or OH- groups.[56] Nondissociated, slightly acidic OH groups of sugars, phenols, etc. could also participate in the formation of such complexes.

According to Harter,[57] differences between soil samples in lead adsorption could not be explained on the basis of CEC differences. For any soil at a given pH, the horizon with the highest CEC usually exhibited a higher sorption, but the relationship could not be quantified.

INFLUENCE OF SOLUTION COMPOSITION AND COMPLEX FORMATION IN SOLUTION PHASE ON LEAD (II) SORPTION BY SOILS AND RELEASE OF SORBED METAL

Lead (II) sorption by soils depends on the anionic composition of the solution.[58] Greater sorption was observed from chloride solutions as compared with nitrate solutions.[59] An increase in Cl^- concentration resulted in an increase in lead retention; the authors suggested that it was caused by sorption of $PbCl^+$ ions.

In soil solutions, lead (II) can occur both as free Pb^{2+} ions and in complexes with inorganic (OH-, Cl-, CO_3^{2-}, PO_4^{3-}, etc.) or organic ligands, including humic acids.[60] Sludge waters applied for irrigation can contain artificial complexons. That is why in soil solutions the metal was revealed in cation, anion, and neutral complexes.[42,61] Neutral Pb(II) complexes were observed, e.g., in soil solutions from acid soils.[62]

Soluble complexes of Pb(II) with humic substances from soils or sewage sludges were detected when lead concentration was low compared with humic substances. The complexes were more stable than those formed by humic substances with Cu(II) and Zn(II).[63] Some insoluble compounds precipitate with increased lead (II)-to-humic-acids ratios.[19,64] Sludge waters can contain different substances forming complexes with Pb(II), so they may essentially modify the mobility of the element in contaminated soils.

Complex formation in the solution should influence lead (II) sorption by soils. This phenomenon has been observed in the study of model systems with artificial complexes. Thus, EDTA prevents sorption of Pb(II) by quartz[65] and bentonite.[66] Metal complexes with EDTA may not sorb on negatively charged surfaces.[67]

Extraction of lead with chelating agents has been extensively studied as a promising method for remediation of soils contaminated with heavy metals. Therefore, leaching with EDTA was suggested for removal of Pb from paddy soils polluted by mine waste waters.[68,69] Leaching of heavy metals bound in oxides and silicates is limited by the slow rate of dissolution of these substances.[70] In batch experiments, Kedziorek et al.[71] were able to remove with 0.01-M EDTA about 70% of total lead from the soil of an agricultural area near Lille, France, polluted due to smelting operations. In a column experiment, percolation of six pore volumes of 1 mM EDTA leached from the soil only 6% of the lead amount that was extracted in batch conditions, whereas the same volume of 10 mM EDTA removed 40% of lead. Total soil lead content in the column in this experiment was 0.66 mM and the pore volume was 50 mL, so six pore volumes contained 0.3 mmol EDTA when using a 1-mM solution, and 3 mmol EDTA when its concentration was 10 mM.

Nitriltriacetic acid (NTA) forms with Pb(II) a charged complex, PbNTA$^-$, that can be resorbed due to the formation of coordination bonds similar to those observed for free cations. This property of PbNTA$^-$ was called metal-like behavior.[72] The addition of a neutral electrolyte (NaClO$_4$) decreased the ability of the NTA solution to leach out Pb(II) from contaminated soil.[70] There was no reasonable explanation for this phenomenon.

Anionic biologically produced surfactants (biosurfactants) with high complexing ability, monorhamnolipid, are produced by *Pseudomonas aeruginosa*, with a critical micelle-forming concentration of 0.1 mM and pKa 5.6.[73] The operational pK value was found to be 6.5. This substance was suggested for removal of lead from soils. Pure water removed only 1% of sorbed lead (II) from the soil, but addition of 50 and 80 mM of the biosurfactant increased this amount to 28 and 42% of sorbed lead, respectively.

RATE OF LEAD SORPTION/RELEASE AND TRANSFORMATION OF LEAD COMPOUNDS IN SOILS

According to Zyrin and his colleagues,[54] the interaction of solute containing lead salts with soil in batch experiments is relatively fast and the equilibrium is achieved in several hours to several days. In the experiments of Petruzzelli et al.,[74] 24 h were sufficient to achieve equilibrium lead (II) sorption isotherms in soil at 20°C and pH 3.8. After this period, no appreciable changes in metal concentration in suspension were observed. In experiments with 1:10 soil suspensions (pH 7.35–7.80, 0.76% to 1.2% CaCO$_3$), lead concentration in the liquid phase remained practically constant after 100 h of shaking.[31] Merrington and Alloway[22] equilibrated the 1:25 suspensions at 20°C for 3 h. Garcia-Delgado et al.[75] and Wang and Benoit[39] did so for 48 h, noticing after 3 h that there were no changes in the composition of phases. Kinetics of lead sorption fitted to the pseudo first-order equation with a half-decay period of 0.8 to 2 min.[74] Displacement of Pb(II) from bentonite with glycine[77] was also fast, and equilibrium metal concentration was achieved after 2 h.

In contrast, transformation of slightly soluble lead compounds in soils is rather slow. Tsaplina[77] spiked the surface of Dernovo-podzolic soil with lead oxide in the amount 500 mg metal kg^{-1} soil per 10-cm soil layer. Under meadow vegetation after four years, the soil contained 53% of lead in the initial oxide form. Under the forest eight years after spiking, the soil still contained 28% of lead as oxide. According to

Obukhov and Tsaplina,[78] after three years of a model experiment, the vertical migration of lead (II) in Dernovo-podzolic soil did not exceed 10 to 15 cm.

When sandy neutral (pH 6.5) soil (Upper Silesia, Poland) was spiked with 100 μg $PbCO_3$ g^{-1} soil, the salt dissolved slowly, and after three years lead (II) was found in Fe–Mn oxide and organic forms, but the amount of the exchangable fraction did not increase.[79]

CONCLUSIONS

1. Lead contamination of soils can be caused by the supply of various compounds whose properties may essentially influence the mobility and availability of the element.

2. Lead species in soils, estimated by the procedures presented in the literature, have no unique chemical interpretation. Each method gives only some operational values that can be compared only with those obtained with the same method for soils of similar properties. The exchangeable fraction cannot be taken as the amount of Pb(II) that can be exchanged, the carbonate fraction as $PbCO_3$, and so on. The fraction bound with Fe–Mn oxides could actually be lead oxide or carbonate, and the residual fraction could be either Pb(II) sulfide or Pb(II) penetrated into the interlayer space of clay minerals or occluded by silicates.

3. Based on the data obtained with existing methods of fractionation, in acid soils the essential part of lead, from 10 to 70% of the total amount, can exist in exchangeable form, bound with either inorganic or organic compounds. In neutral soils, the greatest part of lead may be bound to Fe–Mn oxides and the organic fraction. In slightly alkaline soils, lead is retained as carbonate, organic, and residual fractions.

4. At pH < 6, ion exchange can be the main process of lead retention by soils. The equivalent exchange between Ca^{2+} and Pb^{2+} occurred in soils containing halloysite, montmorillonite, vermiculite, or Al–humus complexes, whereas some extra sorption was observed in soils in which allophane, imogolite, iron oxides, or vermiculite–chlorite intergrades played a role as cation-exchange materials. Lead can be sorbed in amounts that exceed the cation exchange capacity. This phenomenon could hardly be interpreted either in terms of inclusion of a single charged hydroxy species such as $PbOH^+$ or by taking into account the precipitation of lead (II) hydroxide.

5. Interaction of Pb(II) solutions with soil suspensions at pH < 6 is rather fast, and equilibrium is attained in a few hours or days. However, dissolution and transformation of slightly soluble lead compounds, e.g., lead carbonate or oxide, in soils appear to be slow processes.

ACKNOWLEDGMENT

The authors are grateful to the Russian Foundation on Basic Research for financial support of this study (Grant no. 98-04-48533).

REFERENCES

1. Abel, E.V., Lead, in *The Chemistry of Germanium, Tin and Lead,* Pergamon International Library of Science, Technology, Engineering and Social Studies, Pergamon Press, 1975, 105.
2. Glazovskaya, M.A., Ed., *Geochemistry of Heavy Metals in Natural and Technogenic Landscapes,* Moscow, Moscow University Pub., 1983 (in Russian).
3. Polyansky, N.G., *Lead,* Moscow, Nauka, 1986 (in Russian).
4. Kalbasi, M., Peryea, F.J., Lindsay, W.L., and Drake S.R., Measurement of divalent lead activity in lead arsenate contaminated soils, *Soil Sci. Soc. Amer. J.,* 59, 1274, 1995.
5. Rooney, C.R., McLaren, R.G., and Cresswell, R.J., Forms and phytoavailability of lead in soil contaminated with lead shot, in *4th Intl. Conf. on Biogeochemistry of Trace Elements,* 23–26 June, 1997. Berkeley, CA, 289, 1997.
6. Nriagu, J.O., The rise and fall of leaded gasoline, *Sci. Total Environment,* 92, 13, 1990.
7. Ganley, J.J. and Springer, G.S. Physical and chemical characteristics of particulates in spark ignition engine exhausts, *Environ. Sci. Technol.,* 8, 340, 1974.
8. Hababi, K., Characterization of particulate matter in vehicles' exhausts, *Environ. Sci. Technol.,* 7, 223, 1973.
9. Travis, C.C. and Hester, S.T., Global chemical pollution, *Environ. Sci. Technol.,* 25, 815, 1991.
10. Nriagu, J.O. and Pacyna, J.M., Quantitative assessment of worldwide contamination of air, water, and soils by trace metals, *Nature,* 333, 134, 1988.
11. *Report on Lead Contamination of the Environment in the Russian Federation and its Impact on the Public Health,* Moscow, REFIA, 1997 (in Russian).
12. Prikhodko, N.N., Vanadium, chrome, nickel, and lead in the soils of Pri-Tizsa lowland and Zakarpatye, *Agrokhimiya,* 4, 95, 1977 (in Russian).
13. Bowen, H.J.M., *Environmental Chemistry of the Elements,* Academic Press, New York, 1979.
14. Kabata-Pendias, A. and Pendias, H., *Trace Elements in Soils and Plants,* CRC Press, Boca Raton, 1992, 365.
15. Xingfu Xian and Shokohifard, G.I., Effect of pH on chemical forms and plant availability of cadmium, zinc, and lead in polluted soils, *Water Air Soil Poll.,* 45, 267, 1989.
16. Kabata-Pendias, A., Biogeochemistry of lead, in *Olow w srodowisku, Problemy ekologiczne i metodyczne,* Warsaw, 1998, 9.
17. *Gornaya Encyclopedia (Handbook of Rocks),* Sovyetskaya Encyclopedia, Moscow, 1989 (in Russian).
18. Zolotaryeva, B.N., Skripnitchenko, I.I., Geletyuk, N.I., Sigayeva, E.V., and Piunova, V.V., Content and distribution of heavy metals (lead, cadmium and mercury) in the soils of the European part of the USSR, in *Genezis, plodorodiye i melioratsiya pochv, Pushchino,* 77, 1980 (in Russian).
19. Stepanova, E.A., Korobova, E.M., Orlov, D.S., and Petrovskaya, I.V., Lead in the soils of the remote zone of Chernobyl power plant accident, *Pochvovedeniye,* 10, 61, 1990 (in Russian).
20. Vogel, H., Desaules, A., and Hani, H., Heavy metals contents in the soils of Switzerland, *Intl. J. Environ. Anal. Chem.,* 46, 3, 1992.
21. Berthelsen, B.O., Ardal, L.A., Steinnes, E., Abrahamsen, G., and Stuanes, A.O., Mobility of heavy metals in pine forest soils as influenced by experimental acidification, *Water Air Soil Poll.,* 73, 29, 1994.
22. Merrington, G. and Alloway, B.J., The flux of Cd, Cu, Pb, and Zn in mining polluted soils, *Water Air Soil Poll.,* 73, 333, 1994.

23. Kabala, C., Szerszen, L., and Karczewska, A., Total concentration and forms of lead in forest soils from Sudety Mountains, in *Olow w srodowisku, Problemy ekologiczne i metodyczne,* Warsaw, 1998, 25.

24. Foner, H.A., Anthropogenic and natural lead in soils in Israel, *Israel J. Earth Sci.* 42, 29, 1993.

25. Krosshavn, M., Steinnes, E., and Varskog, P., Binding of Cd, Cu, Pb and Zn in soil organic matter with different vegetational background, *Water Air Soil Poll.,* 71, 185, 1993.

26. Kashin, V.K. and Ivanov, G.M., Lead in the soils of the southwest Trans-Baikal region, *Pochvovedeniye,* 12, 1502, 1998 (in Russian).

27. Tessier, A., Campbell, P.G.C., and Bisson, M., Sequential extraction procedure for the speciation of particular trace metals, *Anal. Chem.,* 51, 844, 1979.

28. Sposito, G., Lund, L.J., and Chang, A.C., Trace metal chemistry in arid-zone field soils amended with sewage sludge, I. Fractionation of Ni, Cu, Zn, Cd and Pb in solid phases, *Soil Sci. Soc. Am. J.,* 46, 260, 1982.

29. Zeien, H. and Brummer, G., Chemische Extractionen zur Bestimmung von Schwermetallbindungsformen in Boden. Mitteilgn. *Dtsch. Bodenkundl. Geselsch.,* 59, 505, 1989.

30. McLaren, R.C. and Crawford, D.V., Studies of soil copper, 1. The fractionation of copper in soils, *J. Soil Sci.,* 24, 172, 1973.

31. Garcia-Miragaya, J., Levels, chemical fractionation, and solubility of soil lead in roadside soils of Caracas, Venezuela, *Soil Sci.,* 138, 147, 1984.

32. Scokart, P.O., Meeus-Verdinne, K., and De Borger, R., Mobility of heavy metals in polluted soils near zinc smelters, *Water Air Soil Poll.,* 20, 451, 1983.

33. Zyrin, N.G., Obukhov, A.I., and Motuzova, G.V., Forms of trace elements in soils and methods to study them, in *Proc. X International Congress of Soil Science,* Moscow, 2, 1974 (in Russian).

34. Andersson, A., On the determination of ecologically significant fractions of some heavy metals in soils, *Swed. J. Agr. Res.,* 6, 1, 1976.

35. Ponizovsky, A.A. and Polubesova, T.A., Seasonal variations of soil solution composition and soil surface properties in gray forest soil, *Pochvovedeniye,* 12, 36, 1990 (in Russian).

36. Gorbatov, V.S. and Zyrin, N.G., Comparison of extracting solutions applied to displace exchangeable cations of heavy metals from soils, *Vestnik Moskovskogo Universiteta,* Ser. 17, Soil Science, 2, 22, 1987 (in Russian).

37. Karczewska, A., Forms of lead in polluted soils as determined by single and sequential extraction, in *Olow w srodowisku, Problemy ekologiczne i metodyczne,* Warsaw, 69, 1998.

38. Vuorinen, A., Lead distribution and "forms" in the environment near the roads of southern Finland, *Vestnik Moskovskogo Universiteta,* Ser. 17, Soil Science, 1, 33, 1986 (in Russian).

39. Wang, E.X. and Benoit, G., Fate and transport of contaminant lead in Spodosols: A simple box model analysis, *Water Air Soil Poll.,* 95, 381, 1997.

40. Berggren, D., Speciation of aluminum, cadmium, copper, and lead in humic soil solutions—A comparison of the ion exchange column procedure and equilibrium dialysis, *Intern. J. Environ. Anal. Chem.,* 35, 1, 1989.

41. Lamersdorf, N.P., The behaviour of lead and cadmium in the intensive rooting zone of acid spruce forest soils, *Toxicolog. Environ. Chem.,* 18, 239, 1989.

42. Tills, A.R. and Alloway, B.J., The speciation of lead in soil solution of very polluted soils, *Environ. Tech. Letters,* 4, 529, 1983.

43. *Gmelins Handbook der anorganischen Chemie,* 8. auflage, B. 47, Teil C 4, 1961.

44. *Khimicheskaya Entsiklopedia (Handbook of Chemistry),* Entsiklopedia Publishers, Moscow, 1965.

45. Santillan, M.J. and Jurinak, J.J., The chemistry of lead and cadmium in soil: Solid phase formation, *Soil Sci. Soc. Amer. J.,* 39, 851, 1975.

46. Gorbatov, V.S., Stability and transformation of heavy metal (Zn, Pb, Cd) oxides in soils, *Pochvovedeniye*, 1, 35, 1988 (in Russian).

47. Jopony, M. and Young, S.D., The solid–solution equilibria of lead and cadmium in polluted soils, *European J. Soil Sci.*, 45, 59, 1994.

48. McKenzie, R.M., The adsorption of lead and other heavy metals on oxides of manganese and iron, *Aust. J. Soil Res.*, 18, 61, 1980.

49. Abd-Elfattah, A. and Wada, K., Adsorption of lead, copper, zinc, cobalt, and cadmium by soils that differ in cation-exchange materials, *J. Soil Sci.*, 32, 271, 1981.

50. Ponizovsky, A.A., Mironenko, E.V., and Studenikina, T.A., Copper retention as affected by the complex formation with tartaric and fulvic acids, in *Fate and Transport of Heavy Metals in the Vadose Zone*, Selim, H.M. and Iskandar, I.K., Eds., Ann Arbor Press, Chelsea, MI, 1999, 109.

51. Scheidegger, A.M. and Sparks, D.L., A critical assessment of sorption–desorption mechanisms at the soil mineral/water interface, *Soil Sci.*, 161, 813, 1996.

52. Bowden, J.W, Posner, A.M., and Quirk, J.P., Ionic adsorption on variable charge mineral surfaces, Theoretical-charge development and titration curves, *Austral. J. Soil Res.*, 15, 121, 1977.

53. Hildebrand, E.E. and Blum, W.E., Lead fixation by clay minerals, *Naturwissenschaften*, 61, 169, 1974.

54. Zyrin, N.G., Serdyukova, A.V., and Sokolova, T.A., Lead sorption and state of the sorbed element in soils and soil constituents, *Pochvovedeniye*, 39, 1986 (in Russian).

55. Bourg, A.C.M., Metals in Aquatic and Terrestrial Systems: Sorption, Speciation, and Mobilization, in *Chemistry and Biology of Solid Waste*, Salomons, W. and Forstner, U., Eds., Springer Verlag, Berlin, 1988, 3.

56. Soldatini, G.F., Riffaldi, R., and Levi-Minzi, R., Pb adsorption by soils, I. Adsorption as measured by the Langmuir and Freundlich isotherms, *Water Air Soil Poll.*, 6, 11, 1976.

57. Harter, R.D., Effect of soil pH on adsorption of lead, copper, zinc, and nickel, *Soil Sci. Soc. Am. J.*, 47, 47, 1983.

58. Reed, B.E. and Cline, S.R., Retention and release of lead by a very fine sandy loam, I. Isotherm modelling, *Sep. Sci. Technol.*, 29, 12, 1529, 1994.

59. Barrow, N.J., *Reactions with Variable Charge Soils*, Martinus Nijhoff, Dordrecht, The Netherlands, 1987.

60. Schnitzer, M. and Skinner, S.J.M., Organo-metallic interactions in soils, 7. Stability constants of Pb^{++}-, Ni^{++}-, Mn^{++}-, Co^{++}-, and Mg^{++}- fulvic acid complexes, *Soil Sci.*, 103, 247, 1967.

61. Gregson, S.K. and Alloway, B.J., Gel permeation chromatography studies on the speciation of lead in soil solution of heavily polluted soils, *J. Soil Sci.*, 35, 56, 1984.

62. Alloway, B.J., Tills, A.R., and Morgan H., The speciation and availability of cadmium and lead in polluted soils, in *Trace Substances Environ. Health—XVIII. Proc. University of Missouri, 18th Ann. Conf.*, 4–7 June, 1984, Columbia, Missouri, 1984, 187.

63. Tyler, G., Balsberg, A.M., Bengtsson, G., Baath, E., and Trank, L., Heavy-metal ecology of terrestrial plants, microorganisms, and invertebrates, *Water Air Soil Poll.*, 47, 189, 1989.

64. Steinbrich, A. and Turski, R., Wiazanie Cu, Zn: Pb przez kwasy huminowe wyizolowane z gleb i osadow sciekowych, *Rocz glebozn*, 37, 2–3, 1986 (in Polish).

65. Vuceta, J. and Morgan, J.J., Chemical modelling of trace metals in fresh waters: Role of complexation and adsorption, *Environ. Sci. Tech.*, 12, 1302, 1978.

66. Guy, R.D. and Chakrabarti, D.L., Studies of metal–organic interactions in model systems pertaining to natural waters, *Can. J. Chem.*, 54, 2600, 1976.

67. Barrow, N.J., Bowden, J.W., Posner, A.M., and Quirk, J.P., Describing the adsorption of copper, zinc, and lead on a variable charge mineral surface, *Aust. J. Soil Res.* 19, 309, 1981.

68. Kobayashi, J., Morii, F., and Muramato, S., Removal of cadmium from polluted soil with the chelating agent EDTA, in *Trace Substances in Environment and Health,* XIII, University of Missouri, 179, 1974.

69. Alloway, B.J. and Morgan, H., The behaviour and availability of Cd, Ni, and Pb in polluted soils, in *International Conference on Contaminated Soils,* 11–15 November 1985, Utrecht, Assink, J.W. and van der Brink, W.J., Eds., Martinus Nijhoff, Dordrecht, The Netherlands, 1986, 101.

70. Elliott, H.A. and Brown, G.A., Comparative evaluation of NTA and EDTA for extractive decontamination of Pb-polluted soils, *Water Air Soil Poll.,* 45, 361, 1989.

71. Kedziorek, M.A.M., Dupuy, A., Bourg, A.C.M., and Compere, F., Leaching of Cd and Pb from a polluted soil during percolation of EDTA: Laboratory column experiments modeled a nonequilibrium solubilization step, *Environ. Sci. Technol.,* 32, 1609, 1988.

72. Benjamin, M.M. and Leckie, J.O., Multiple-site adsorption of Cd, Cu, Zn, and Pb on amorphous iron oxihydroxide, *J. Colloid Interface Sci.,* 79, 209, 1981.

73. Herman, D.C., Artiola, J.F., and Miller, R.M., Removal of cadmium, lead, and zinc from soil by rhamnolipid biosurfactant, *Environ. Sci. Technol.,* 29, 2280, 1995.

74. Petruzzelli, G., Lubrano, L., Petronio, B.M., Gennaro, M.C., Vanni, A., and Liberatori, A., Soil sorption of heavy metals as influenced by sewage sludge addition, *J. Environ. Sci. Health,* A29, 31, 1994.

75. Garcia-Delgado, R.A., Garcia-Herruzo, J.M., Rodriguez-Maroto, J.M., and Vereda, C., Influence of soil carbonates on lead fixation, *J. Environ. Sci. Health,* A31, 2099, 1996.

76. Fischer, K., Rainer, C., Bieniek, D., and Kettrup, A. Desorption of heavy metals from typical soil components (clay, peat) with glycine, *Intl. J. Anal. Chem.,* 46, 53, 1992.

77. Tsaplina, M.A., Transformation and transport of lead, cadmium, and zinc oxides in Dernovo-podzolic soil, *Pochvovedeniye,* 45, 1994 (in Russian).

78. Obukhov, A.I., and Tsaplina, M.A., Migration and transformation of Pb compounds in Dernovo-podzolic soils, in *Migratsiya zagryzn. veshchestv v pochvach i sopredel'nych sredach, Proc. 5th Conf., Obninsk, 1987,* Leningrad, 194, 1989 (in Russian).

79. Chlopecka, A., Bacon, J.R., Wilson, M.J., and Kay, J., Forms of cadmium, lead, and zinc in contaminated soils from southwest Poland, *J. Environ. Qual.,* 25, 69, 1996.

80. *Handbook of Chemistry (Spravochnik khimika),* 2nd Edn., Khimiya, Moscow–Leningrad, 3, 1965, 233 (in Russian).

81. Chlopecka, A., Forms of Cd, Cu, Pb, and Zn in soil and their uptake by cereal crops when applied jointly as carbonates, *Water Air Soil Poll.,* 87, 297, 1996.

82. Abdel-Saheb, I., Schwab, A.P., Banks, M.K., and Hetrick, B.A., Chemical characterization of heavy-metal contaminated soil in southeast Kansas, *Water Air Soil Poll.,* 78, 73, 1994.

83. Sims, J.T. and Kline, J.S., Chemical fractionation and plant uptake of heavy metals in soils amended with co-composted sewage sludge, *J. Environ. Quality,* 20, 387, 1991.

84. Sposito, G., Levesque, C.S., LeClair, J.P., and Chang, A.C., Trace metal chemistry in arid-zone field soils amended with sewage sludge, III. Effect of time on the extraction of trace metals, *Soil Sci. Soc. Am. J.,* 47, 898, 1983.

ndex

Milton Keynes UK
Ingram Content Group UK Ltd.
UKHW040448071024
449327UK00020B/1080